INTRODUCTION TO
MATHEMATICS
WITH MAPLE

P ADAMS

K SMITH

R VÝBORNÝ

The University of Queensland, Australia

INTRODUCTION TO
MATHEMATICS
WITH MAPLE

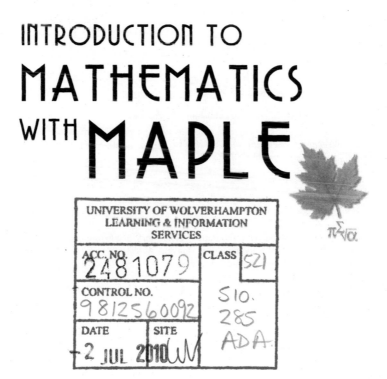

World Scientific

NEW JERSEY • LONDON • SINGAPORE • BEIJING • SHANGHAI • HONG KONG • TAIPEI • CHENNAI

Published by

World Scientific Publishing Co. Pte. Ltd.

5 Toh Tuck Link, Singapore 596224

USA office: Suite 202, 1060 Main Street, River Edge, NJ 07661

UK office: 57 Shelton Street, Covent Garden, London WC2H 9HE

British Library Cataloguing-in-Publication Data
A catalogue record for this book is available from the British Library.

INTRODUCTION TO MATHEMATICS WITH MAPLE

ISBN 981-238-931-8
ISBN 981-256-009-2 (pbk)

Printed by FuIsland Offset Printing (S) Pte Ltd, Singapore

To
Narelle, Helen and Aťa
for their understanding and support
while we were writing this book.

Preface

I attempted mathematics, and even went during the summer of 1828 with a private tutor to Barmouth, but I got on very slowly. The work was repugnant to me, chiefly from my not being able to see any meaning in the early steps in algebra. This impatience was very foolish, and in after years I have deeply regretted that I did not proceed far enough at least to understand something of the great leading principles of mathematics, for men thus endowed seem to have an extra sense.

Charles Darwin, Autobiography (1876)

Charles Darwin wasn't the last biologist to regret not knowing more about mathematics. Perhaps if he had started his university education at the beginning of the 21st century, instead of in 1828, and had been able to profit from using a computer algebra package, he would have found the material less forbidding.

This book is about pure mathematics. Our aim is to equip the readers with understanding and sufficiently deep knowledge to enable them to use it in solving problems. We also hope that this book will help readers develop an appreciation of the intrinsic beauty of the subject!

We have said that this book is about mathematics. However, we make extensive use of the computer algebra package *Maple* in our discussion. Many books teach pure mathematics without any reference to computers, whereas other books concentrate too heavily on computing, without explaining substantial mathematical theory. We aim for a better balance: we present material which requires deep thinking and understanding, but we also fully encourage our readers to use *Maple* to remove some of the laborious computations, and to experiment. To this end we include a large number of *Maple* examples.

Most of the mathematical material in this book is explained in a fairly traditional manner (of course, apart from the use of *Maple*!). However, we depart from the traditional presentation of integral by presenting the Kurzweil—Henstock theory in Chapter 15.

Outline of the book

There are fifteen chapters. Each starts with a short abstract describing the content and aim of the chapter.

In Chapter 1, "Introduction", we explain the scope and guiding philosophy of the present book, and we make clear its logical structure and the role which *Maple* plays in the book. It is our aim to equip the readers with sufficiently deep knowledge of the material presented so they can use it in solving problems, and appreciate its inner beauty.

In Chapter 2, "Sets", we review set theoretic terminology and notation and provide the essential parts of set theory needed for use elsewhere in the book. The development here is not strictly axiomatic—that would require, by itself, a book nearly as large as this one—but gives only the most important parts of the theory. Later we discuss mathematical reasoning, and the importance of rigorous proofs in mathematics.

In Chapter 3, "Functions", we introduce relations, functions and various notations connected with functions, and study some basic concepts intimately related to functions.

In Chapter 4, "Real Numbers", we introduce real numbers on an axiomatic basis, solve inequalities, introduce the absolute value and discuss the least upper bound axiom. In the concluding section we outline an alternative development of the real number system, starting from Peano's axioms for natural numbers.

In Chapter 5, "Mathematical Induction", we study proof by induction and prove some important inequalities, particularly the arithmetic-geometric mean inequality. In order to employ induction for defining new objects we prove the so-called recursion theorems. Basic properties of powers with rational exponents are also established in this chapter.

In Chapter 6, "Polynomials", we introduce polynomials. Polynomial functions have always been important, if for nothing else than because, in the past, they were the only functions which could be readily evaluated. In this chapter we define polynomials as algebraic entities rather than functions and establish the long division algorithm in an abstract setting. We also look briefly at zeros of polynomials and prove the Taylor Theorem for polynomials in a generality which cannot be obtained by using methods of calculus.

In Chapter 7, "Complex Numbers", we introduce complex numbers, that is, numbers of the form $a + bi$ where the number i satisfies $i^2 = -1$. Mathematicians were led to complex numbers in their efforts of solving algebraic equations, that is, of the form $a_n x^n + a_{n-1} x^{n-1} + \cdots + a_0 = 0$, with the a_k real numbers and n a positive integer (this problem is perhaps more widely known as finding the zeros of a polynomial). Our introduction follows the same idea although in a modern mathematical setting. Complex numbers now play important roles in physics, hydrodynamics, electromagnetic theory and electrical engineering, as well as pure mathematics.

In Chapter 8, "Solving Equations", we discuss the existence and uniqueness of solutions to various equations and show how to use *Maple* to find solutions. We deal mainly with polynomial equations in one unknown, but include some basic facts about systems of linear equations.

In Chapter 9, "Sets Revisited", we introduce the concept of equivalence for sets and study countable sets. We also briefly discuss the axiom of choice.

In Chapter 10, "Limits of Sequences", we introduce the idea of the limit of a sequence and prove basic theorems on limits. The concept of a limit is central to subsequent chapters of this book. The later sections are devoted to the general principle of convergence and more advanced concepts of limits superior and limits inferior of a sequence.

In Chapter 11, "Series", we introduce infinite series and prove some basic convergence theorems. We also introduce power series—a very powerful tool in analysis.

In Chapter 12, "Limits and Continuity of Functions", we define limits of functions in terms of limits of sequences. With a function f continuous on an interval we associate the intuitive idea of the graph f being drawn without lifting the pencil from the drawing paper. The mathematical treatment of continuity starts with the definition of a function continuous at a point; this definition is given here in terms of a limit of a function at a point. We develop the theory of limits of functions, study continuous functions, and particularly functions continuous on closed bounded intervals. At the end of the chapter we touch upon the concept of limit superior and inferior of a function.

In Chapter 13, "Derivatives", we start with the informal description of a derivative as a rate of change. This concept is extremely important in science and applications. In this chapter we introduce derivatives as limits, establish their properties and use them in studying deeper properties of functions and their graphs. We also extend the Taylor Theorem from polynomials to power series and explore it for applications.

In Chapter 14, "Elementary Functions", we lay the proper foundations for the exponential and logarithmic functions, and for trigonometric functions and their inverses. We calculate derivatives of these functions and use these for establishing important properties of these functions.

In Chapter 15, "Integrals", we present the theory of integration introduced by the contemporary Czech mathematician J. Kurzweil. Sometimes it is referred to as Kurzweil—Henstock theory. Our presentation generally follows Lee and Výborný (2000, Chapter 2).

The Appendix contains some examples of *Maple* programs. Finally, the book concludes with a list of References, an Index of *Maple* commands used in the book, and a general Index.

Notes on notation

Throughout the book there are a number of ways in which the reader's attention is drawn to particular points. Theorems, lemmas[1] and corollaries are placed inside rectangular boxes with double lines, as in

> **Theorem 0.1 (For illustrative purposes only!)**   *This theorem is referred to only in the* Preface, *and can safely be ignored when reading the rest of the book.*

Note that these are set in slanting font, instead of the upright font used in the bulk of the book.

Definitions are set in the normal font, and are placed within rectangular boxes, outlined by a single line and with rounded corners, as in

> **Definition 0.1 (What is mathematics?)**   There are almost as many definitions of what mathematics is as there are professional mathematicians living at the time.

---

[1]A lemma is sometimes known as an auxiliary theorem. It does not have the same level of significance as a theorem, and is usually proved separately to simplify the proof of the related theorem(s).

The number before the decimal point in all of the above is the number of the chapter: numbers following the decimal point label the different theorems, definitions, examples, etc., and are numbered consecutively (and separately) within each chapter. Corollaries are labeled by the number of the chapter, followed by the number of the theorem to which the corollary belongs then followed by the number of the individual corollary for that theorem, as in

---

Corollary 0.1.1 (Also for illustrative purposes only!) *Since* Theorem 0.1 *is referred to only in the* Preface, *any of its corollaries can also be ignored when reading the rest of the book.*

Corollary 0.1.2 (Second corollary for Theorem 0.1) *This is just as helpful as the first corollary for the theorem!*

Corollaries are set in the same font as theorems and lemmas.

Most of the theorems, lemmas and corollaries in this book are provided with proofs. All proofs commence with the word "***Proof.***" flush with the left margin. Since the words of a proof are set in the same font as the rest of the book, a special symbol is used to mark the end of a proof and the resumption of the main text. Instead of saying that a proof is complete, or words to that effect, we shall place the symbol □ at the end of the proof, and flush with the right margin, as follows:

Proof. This is not really a proof. Its main purpose is to illustrate the occurrence of a small hollow square, flush with the right margin, to indicate the end of a proof. □

Up to about the middle of the 20th century it was customary to use the letters 'q.e.d.' instead of □, q.e.d. being an abbreviation for *quod erat demonstrandum*, which, translated from Latin, means *which was to be proved.*

To assist the reader, there are a number of Remarks scattered throughout the book. These relate to the immediately preceding text. There are also Examples of various kinds, used to illustrate a concept by providing a (usually simple) case which can show the main distinguishing points of a concept. Remarks and Examples are set in sans serif font, like this, to help distinguish them.

Remark 0.1   The first book published on calculus was Sir Isaac Newton's *Philosophiae Naturalis Principia Mathematica* (Latin for *The Mathematical Principles of Natural Philosophy*), commonly referred to simply as *Principia*. As might be expected from the title, this was in Latin. We shall avoid the use of languages other than English in this book.

Since they use a different font, and have additional spacing above and below, it is obvious where the end of a Remark or an Example occurs, and no special symbol is needed to mark the return to the main text.

Scissors in the margin

Obviously, we must build on some previous knowledge of our readers. Chapter 1 and Chapter 2 summarise such prerequisites. Chapter 1 contains also a brief introduction to *Maple*. Starting with Chapter 3 we have tried to make sure that all proofs are in a strict logical order. On a few occasions we relax the logical requirements in order to illustrate some point or to help the reader place the material in a wider context. All such instances are clearly marked in the margin (see the outer margin of this page), the beginning by scissors pointing into the book, the end by scissors opening outwards. The idea here is to indicate readers can skip over these sections if they desire strict logical purity. For instance, we might use trigonometric functions before they are properly introduced,[2] but then this example will be scissored.

Exercises

There are exercises to help readers to master the material presented. We hope readers will attempt as many as possible. Mathematics is learned by doing, rather than just reading. Some of the exercises are challenging, and these are marked in the margin by the symbol (!). We do not expect that readers will make an effort to solve all these challenging problems, but should attempt at least some. Exercises containing fairly important additional information, not included in the main body of text, are marked by (i). We recommend that these should be read even if no attempt is made to solve them.

---

[2]Rather late in Chapter 14

Acknowledgments

The authors wish to gratefully acknowledge the support of Waterloo Maple Inc, who provided us with copies of *Maple* version 7 software. We thank the Mathematics Department at The University of Queensland for its support and resourcing. We also thank those students we have taught over many years, who, by their questions, have helped us improve our teaching.

Our greatest debt is to our wives. As a small token of our love and appreciation we dedicate this book to them.

Peter Adams
Ken Smith
Rudolf Výborný

The University of Queensland, February 2004.

Contents

Chapter 1

Introduction

In this chapter we explain the scope and guiding philosophy of the present book, we try to make clear its logical
structure and the role which *Maple* plays in the book. We
also wish to orientate the readers on the logical structure
which forms the basis of this book.

1.1 Our aims

Mathematics can be compared to a cathedral. We wish to visit a small
part of this cathedral of human ideas of quantities and space. We wish
to learn how mathematics can be built. Mathematics spans a very wide
spectrum, from the simple arithmetic operations a pupil learns in primary
school to the sophisticated and difficult research which only a specialist
can understand after years of long and hard postgraduate study. We place
ourselves somewhere higher up in the lower half of this spectrum. This can
also be roughly described as where University mathematics starts. In natural sciences the criterion of validity of a theory is experiment and practice.
Mathematics is very different. Experiment and practice are insufficient for
establishing mathematical truth. Mathematics is deductive, the only means
of ascertaining the validity of a statement is logic. However, the chain of
logical arguments cannot be extended indefinitely: inevitably there comes a
point where we have to accept some basic propositions without proofs. The
ancient Greeks called these foundation stones *axioms*, accepted their validity without questioning and developed all their mathematics therefrom. In
modern mathematics we also use axioms but we have a different viewpoint.
The axiomatic method is discussed later in section 2.5. Ideally, teaching
of mathematics would start with the axioms, however, this is hopelessly

impractical at any level of instruction.

We begin our serious work in Chapter 3. New concepts are introduced by rigorous definitions and theorems are proved. We believe that in our exposition, set theoretical language and notation is not only convenient but desirable. This poses a problem in that set theory should be established axiomatically and we are in no position to do so here. So we patched it up; Chapter 2 contains all that the reader needs, but not at a rigorous level. Hence scissors in the margin[1] appear only after Chapter 2. We hope that at some later stage the reader will fill in this gap but we also give a warning not to do so now: axiomatic set theory is more difficult than anything we do here.

We shall aid our computations by a computer program called *Maple*. This program is supposed to be user friendly, the name can be thought of as an acronym for MAthematics and PLEasure, or MAthematical Programming LanguagE. The truth is that *Maple* was developed at the University of Waterloo in Canada, and the creators of *Maple* wanted to give it a name with a distinctive Canadian flavour. Ever since computers were invented they have been very powerful instruments for solving problems which required large scale numerical calculations. However, in the the last decades there were programs invented that can manipulate a variety of symbolic expressions and operate on them. These programs have various names; we shall call them computer algebra systems. For instance, such a program can provide you with a partial quotient and remainder of division of two polynomials. *Maple*, if properly asked, will tell you that $x^6 + 1$ divided by $x^2 + x + 1$ is $x^4 - x^3 + x - 1$ with remainder 2. It can also give you an approximation of

$$\frac{1 - x^2}{1 + x^2}$$

by a polynomial of fourth degree as $1 - 2x^2 + 2x^4$. *Maple* is one of the computer algebra systems; others include MACSYMA, MuMath, MATLAB, and Mathematica. *Maple* is now widely used in engineering, education and research; clearly its knowledge is useful. This book is not about *Maple*, however when we arrive at some point where *Maple* can be usefully employed we show the reader how to do it. We hope that after readers finish reading this book it will be easy for them to adjust to another computer algebra system, if they need to. Some very basic facts about how to use *Maple* are explained in the rest of this Chapter. *Maple* 'knows' more math-

---

[1]The notation of using scissors in the margin is explained in the preface.

ematics than we can ever hope of managing, but this is not always an advantage. It is like having a servant who is better educated than oneself. It can give us an answer in terms of more advanced mathematics than we know. When we encounter this we shall show how to overcome it. Clearly, we cannot use *Maple* when building our theory; this would be circuitous. Hence readers should understand that scissors are applied automatically whenever *Maple* appears. *Maple* has proved itself as a very powerful tool; it helped solved research problems which were intractable by paper and pencil manipulation. However, computer algebra systems are only human inventions and can fail. As an example we present a graph of the function f, where $f(x) = x + \sin 7x$, as produced by *Maple* (see Figure 1.1).

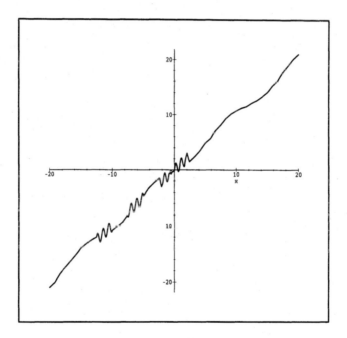

Fig. 1.1   Graph of $x + \sin 7x$

The graph, because of its irregularity, is clearly wrong. Here it is easy to spot the error, but on another occasion it can be more insidious. It is therefore necessary to be careful and critical when looking at output from a computer algebra system. Results are not necessarily correct simply because they come from the computer. The use of computer in mathematics does not free anybody from the need of thinking.

1.2 Introducing *Maple*

1.2.1 *What is* Maple*?*

Maple is a powerful mathematical computer program, designed to perform a wide variety of mathematical calculations and operations. It can do simple calculations, matrix operations, graphing, and even symbolic manipulations, such as finding the derivative or integral of a function. It can also solve a variety of equations such as finding zeros of a polynomial or to solve linear as well as some nonlinear systems of equations.

Maple is a mathematical computer program, designed to perform a variety of mathematical calculations and operations on symbolic as well as numeric entities.

1.2.2 *Starting* Maple

We presume that the readers can handle basic tasks on a computer, be it their own PC or a large computer at an educational institution with a Unix operating system. In particular, we presume that the readers know how to handle files and directories and do some simple editing of a file. The differences between *Maple* use on a PC or on a Unix or Linux machine are minimal and we shall comment only on the essential ones. Before you can start *Maple* you must be sure that *Maple* is installed on your machine. At a large educational institution this will be almost automatic, for your home PC you can obtain *Maple* from a scientific software supplier or from Waterloo Maple at

> `http://www.maplesoft.com/sales/student_intro.html`.

Once the PC has has finished booting, you can begin your *Maple* session. To start *Maple*, use the mouse to point at the *Maple for Windows* icon, and double click the left-hand mouse button. Alternatively, on a Unix or Linux machine you give the command `xmaple` after you login.[2]

---

[2]There is also the command `maple`, which will open *Maple* on a Unix machine without the graphical interface. We do not recommend to use it because of the limited display. However it can be useful, for example, if you are connected to a Unix machine through

When *Maple* starts, a window appears with a number of menus at the top (such as **File, Edit, Format,** and so on). Near the top left-hand side of the screen is the *Maple* prompt, which is a > character, followed by a flashing cursor, which is a | character.

To run *Maple*, double click the left-hand mouse button on the *Maple for Windows* icon. On a Unix or Linux machine give the command xmaple.

1.2.3 *Worksheets in* Maple

An important concept in *Maple* is that of a *worksheet*. During a *Maple* session, the computer screen will display your commands (and comments), together with *Maple* output, graphs and other pictures. All of this material is collectively called a *worksheet*.

Usually, when you use *Maple*, you will be wanting to enter a new set of commands. By default, *Maple* opens an empty worksheet as soon as you run the program. Any commands you enter will be executed, thus forming your worksheet.

When you have finished using *Maple*, you will be given the option of saving your worksheet to a file. Sometimes you will not need to do this, and will not save a copy of the worksheet. However, if you have not finished your work, or if you wish to use the results at a later date, you may decide to save the worksheet to a floppy or hard disk on your computer. Save files from *Maple* in the usual way; note that the standard filename extension for a *Maple* worksheet file is *.mw* or *.mws*.

If a worksheet has already been saved during a previous *Maple* session then you can load a copy of that worksheet into *Maple*. That is, rather than starting a new worksheet, you can reload an existing worksheet, and modify it as required.

To load an existing worksheet, you first need to know the name of the worksheet, and where it has been saved (for example, on a floppy disk, or on the PC network). Next, use the **File** menu, near the top of the screen, and select the **Open** menu option. Again this is done in the standard way. Once the existing worksheet has been loaded, you can modify and execute commands, add comments, and perform your choice of *Maple* operations. Of course, when you have finished you may choose to save the modified

---

a modem. Most often in such a case the graphics are not available anyhow. If you use Linux then xmaple can be applied only if Xwindows is being used.

worksheet so you can use it again.

> A *Maple* worksheet is the name given to the text and
> graphics from a *Maple* session.
> *Maple* opens a new, empty worksheet by default, but you
> can load an existing worksheet using the File menu.
> You will be asked if you wish to save your current
> worksheet before you exit *Maple*.

1.2.4　*Entering commands into* Maple

Whether you are creating a new worksheet, or reloading an existing one,
you will probably want to enter some new commands into your worksheet.

Whenever *Maple* displays the flashing | cursor, it is waiting for you to
type a command. You can use menu items to get *Maple* to do the things
you want, but you will also often need to enter your commands as text. This
is done by typing your desired command, and (almost always!) ending the
command with a semicolon (;) or a colon (:). Press the `Enter` (or `Return`)
key to execute the command.

If you end the command with a semicolon, then *Maple* performs the
command, and displays the result. If you end the command with a colon,
then *Maple* performs the command, but no result is displayed. The colon
is useful for intermediate calculations, but most of the time you will use a
semicolon. Sometimes you might forget the semicolon or colon, and press
the `enter` key. Nothing will happen, until you enter a semicolon (on the
next line).

> Most *Maple* commands end with a semicolon or colon,
> followed by the enter key.
> Don't forget!
> But if you do forget, type the semicolon or colon
> on the next line.

Using arithmetic in *Maple* is discussed in detail in Section 1.4, but the
following examples should be fairly clear. For example, you might wish to
evaluate the sum $1+3+5$. Your *Maple* session will appear as the following:

```
>   1+3+5;
```

9

The command entered into *Maple* is `1+3+5;` (note the semicolon!), and the answer from *Maple* is **9**. *Maple* displays your command near the left-hand side of the screen, and the result is displayed in the centre.

When using *Maple* commands, please note the following points:

- *Maple* is case sensitive. Most of the in-built *Maple* functions are in lowercase letters. Be careful to choose the correct case for your commands!
- *Maple* usually ignores spaces in your input commands.
- Be careful to spell the commands correctly, and use the required punctuation.
- Be particularly careful to use the correct brackets, in the correct places.
- If you make any mistakes while typing your command, use the `backspace` key to correct the error. You can also highlight something and remove it by using **Cut** from the **Edit** menu.
- Sometimes, a command you type may take a long time to execute. You can interrupt execution (and return to the *Maple* prompt) by typing `Control-C`, sometimes written `Ctrl-C`, (that is, you press the `control` key, at the same time as pressing the C key).

Note that you can enter multiple commands on the same line, provided you use a semicolon or colon at the end of each command. For example, to evaluate $1 + 3 + 5$, 2^{10} and 64×123, enter the following commands. Note the use of $*$ for multiplication and $\char`\^$ for exponentiation.

```
>   1+3+5; 2^10; 64*123;
```

9

1024

7872

If you forgot the semicolon and hit the **enter** key, you can complete the command on the next line by typing semicolon and the **enter** key there.

> 16*4

> ;

                              64

1.2.5 *Stopping* Maple

When you have finished, you can quit *Maple* in a number of ways. The easiest is to select the **Exit** command in the **File** menu.

> **To exit *Maple*, select Exit from the File menu!**

1.2.6 *Using previous results*

Often, you will want to use the results of immediately previous calculations in subsequent calculations. The percentage character % refers to the previous result, two %% refers to the second-last result, and %%% refers to the third last result. In older versions of *Maple* the double quote character " is used instead of %. To find out which of the % or " to use on your machine type ?ditto at the *Maple* prompt. To evaluate 2^4, then $2^4 \times 5$, then $2^4 \times 5 + 2^4$:

> 2^4;

                              16

> %*5;

                              80

> %+%%;

                              96

> **Type %, %% and %%% to use the results of
> the three most recent calculations.**

Sometimes, you will need to use many previous results, not just the last three. This can be done via the **history** function. You can use the **help** command to learn more about the **history** command. Using **help** is discussed in detail in Section 1.3, but typing **?history** will give you the required information.

1.2.7 *Summary*

In summary:

- double click on the *Maple for Windows* icon or give the command **xmaple**
- type your commands, ended with a semicolon or colon
- make sure the syntax, spelling and punctuation are correct
- exit using **Exit** from the **File** menu

1.3 Help and error messages with *Maple*

1.3.1 *Help*

Maple contains a great deal of on-line help. This can be accessed via the **help** command, usually entered as the question mark **?** character.

> **Use the ? command for helpful information.
> Use ?name for help on the command name.**

You can use the **?** command in a number of ways. The following table summarises the more common uses (some of which will only become important after you have used *Maple* more extensively).

Command	Help Topic
?	General help
?name	Help on the topic **name**, or list all topics which begin with **name**
?index	An index of available help topics
?library	Standard library functions
?datatypes	Basic data types
?expressions	*Maple* expressions
?statements	*Maple* statements

Other useful help information can be accessed via the commands info, usage, related. These commands all give more information about a specified function. A very convenient way of using help is to open the **Help** menu by clicking on **Help** located in the upper right corner of the screen. If you know exactly the name you are looking for choose **Topic Search** otherwise use **Full Text Search** from the **Help** menu.

1.3.2   *Error messages*

When *Maple* encounters an error, it usually displays an error message. These error messages are intended to be helpful: please read them, and think about what is being said.

The following examples show some error messages from *Maple*. In each case, the message is reasonably simple to understand. In the first case, we have tried to divide by 0, which is (of course) impossible. In the second case, the expression which has been entered is not valid: we have missed out a number or a variable.

```
>   1/0;
```

`Error, numeric exception: division by zero.`

```
>   3*4+;
```

`Error, ';' unexpected.`

Note that in this case, the *Maple* input cursor will be positioned in the command you entered, at the place where *Maple* thinks the error occurred. In the next two examples we failed to type * to indicate multiplication. The results are surprising.

```
>   (3+2)4;
```

Error, unexpected number

```
>   4(3+2);
```

$$4$$

In the first instance *Maple* gave the correct error message. However, in the second example *Maple* gave a wrong answer and did not issue an error message. The moral of this is simple, always try to enter your command precisely and carefully. A wrong input can return a wrong result without any warning.

> **Pay attention to the error messages!**
> *Maple* **tries to make them meaningful, often with a pointer to the location of the problem.**

1.4   Arithmetic in *Maple*

1.4.1   *Basic mathematical operators*

As many of the previous examples have shown, *Maple* can be used for arithmetical calculations, and supports all of the usual mathematical operations. Furthermore, *Maple* follows the usual rules regarding order of calculations. For example, multiplication is evaluated before addition, and so on. The following table summarises many of the common arithmetical operations.

Mathematical Operation	Description	Mathematical example	*Maple* usage	Result
$+$	Addition	$1 + 3 + 5$	1+3+5	9
-	Subtraction	$9 - 5 - 3$	9-5-3	1
$\times$	Multiplication	2×5	2*5	10
/	Division	$8/2$	8/2	4
^	Exponentiation	2^5	2^5	32
!	Factorial	$5!$	5!	120
$\sqrt{}$	Square root	$\sqrt{25}$	sqrt(25)	5

1.4.2   *Special mathematical constants*

Maple also knows a number of useful mathematical constants, which can be used in your calculations. Two of the most important ones are shown in the following table. Note the capital letter in `Pi`.

Mathematical name	Description	*Maple* usage	Approximate value
π	Area of the unit circle	`Pi`	3.141592
e	Exponential constant	`exp(1)`	2.7182818

> **Maple supports all of the common mathematical operations, and special mathematical constants.**

1.4.3   *Performing calculations*

When performing calculations, you must be sure to:

- include suitable brackets. For example, when using `sqrt`, whatever you want to take the square root of must be enclosed in round brackets.
- match up your opening and closing brackets. For example, `sqrt(4)+5` is not the same as `sqrt(4+5)`.
- always use an asterisk character `*` for multiplication. To evaluate $2x\sqrt{4x}$, you must enter `2*x*sqrt(4*x)`.
- ensure that you have entered your expressions with appropriate precedence. For example, `4*3+5` is different to `4*(3+5)`. When in doubt, use brackets.
- enter divisions correctly, using a `/` character. To evaluate $\dfrac{2 \times 5 + 6}{3 \times 2}$, you would enter `(2*5+6)/(3*2)`.

> **Always put the multiplication sign in!**
> **Always use brackets for function arguments!**
> **Be careful with precedence rules in calculations!**
> **Ensure that your brackets match up,**
> **and are correctly located!**

Consider the following *Maple* session. Note that if a `#` character appears anywhere in a line, then the rest of the line is regarded as a comment by

Maple, and is ignored. You should use comments to make your calculations more readable.

```
>   # Forgetting brackets around function arguments
>   # gives an error
>   # Try to find the square root of 4
>   sqrt 4;

Error, unexpected number.

>   # The argument of the square root function
>   # needs to be placed in parentheses
>   # Evaluate the square root of 4
>   sqrt(4);


                        2

>   # Evaluate (the square root of 4)+5;
>   sqrt(4)+5;


                        7

>   # Evaluate the square root of (4+5)
>   sqrt(4+5);


                        3

>   # Evaluate (2*5+6) divided by (3*2)
>   (2*5+6)/(3*2);


                        8
                        -
                        3
```

1.4.4 *Exact versus floating point numbers*

The *Maple* session given in the previous section shows that *Maple* evaluates $(2 * 6 + 6)/(3 * 2)$ to $\frac{8}{3}$, rather than to 2.66667. This is important. *Maple* will (within the limitations of the current version of the software) give you the exact result rather than an approximation. The number of digits may be large but *Maple* will produce all the digits.

```
>    (9^9)^9;
```

19662705047555 29136180759085269121162831034509442147669273154155 3\
7966391196809

Note the backslash at the end of the line telling you that the number continues on the next line.

If you want an approximation you must ask for it. One way to do this is to type the numbers in your command with decimal points. Any number which contains a decimal point is called in *Maple* a floating point number, FP number for short. The name refers to the way in which these numbers are stored and handled by *Maple*.[3] For sake of brevity we shall refer to numbers without decimal points as exact numbers. For example, integers and fractions are *exact* numbers. So are $\sqrt{3}$ and π. If only exact numbers enter a computation, *Maple* will produce an exact result. It is important to realize that for *Maple*, $\frac{1}{10}$ and .1 are distinct entities, not because of their appearance, but because of the way they are treated by *Maple*. If you enter .1 into *Maple* you give tacitly some instructions which are absent when entering $\frac{1}{10}$. Here is our first example with FP numbers.

```
>    1+sqrt(3);
```

$$1 + \sqrt{3}$$

```
>    1.0+sqrt(3.0);
```

$$2.732050808$$

It is sufficient to write

```
>    1+sqrt(3.0);
```

$$2.732050808$$

---

[3]Neither here nor anywhere else do we attempt to explain the internal workings of *Maple*.

However

> `1.0+sqrt(3);`

$$1.0 + \sqrt{3}$$

---

produces a correct result but not the one you wanted.

You can control the precision of the decimal output via the `Digits` command (note: capital D!), which allows you to view and set the number of significant figures:

> `Digits;`

$$10$$

> `sqrt(132.);`

$$11.48912529$$

> `Digits:=20; sqrt(132.);`

$$Digits := 20$$

$$11.480125293076057320$$

> **Use `Digits` to customise the number of significant figures given in the answer to a calculation.**

The next two examples show why exact arithmetic is preferable. In the first example we set digits to three. This is not essential; a similar thing can happen with any number of digits.

> `Digits:=3:1/(22-7*sqrt(2)-7*sqrt(3));`

$$\frac{1}{22 - 7\sqrt{2} - 7\sqrt{3}}$$

```
>   1/(22-7*sqrt(2.0)-7*sqrt(3.0));
```

Error, division by zero

We did not ask for division by zero, however, within the precision of three decimals the denominator is approximately equal to zero and this causes *Maple* to issue an error message. If the above number appeared somewhere in a long calculation the process would stop. This would not happen with exact arithmetic. Increasing the number of digits helps, but the result using *Maple's* default precision of ten digits is correct only to seven digits.

```
>   Digits:=10:1/(22-7*sqrt(2.0)-7*sqrt(3.0));
```

$$-41.92768397$$

```
>   Digits:=55:1/(22-7*sqrt(2.0)-7*sqrt(3.0));
```

$$-41.9276846833037394942018274452836316691110746186476 7765$$

In the next example one would expect that after taking the root and exponentiation the original number would be returned: however, calculation with floating point numbers is not exact!

```
>   27.0^(1/8);
```

$$1.509803648$$

```
>   (%)^8;
```

$$26.99999993$$

The imperfection of the last two examples is *not* some fault in *Maple's* design, it is instead inherent in the nature of approximate calculations. We would encounter the same phenomena if we performed the calculations with paper and pencil.

If all of the entries in a calculation are exact, you can still obtain a decimal point approximation, using the evalf() function. Evalf stands for EVALuate with Floating point arithmetic. The first argument to the `evalf()` function must be the calculation you want to be evaluated. Optionally, a second argument tells *Maple* how many significant digits are required.

```
>   evalf(sqrt(132));
```

$$11.48912529$$

```
>   # Evaluate the square root with 30 significant digits.
>   evalf(sqrt(132),30);
```

$$11.489125293076057319701229364$$

```
>   sqrt(132);
```

$$2\sqrt{33}$$

```
>   # Recall that % gives the result of
>   # the most recently entered computation.
>   evalf(%);
```

$$11.48912529$$

You can use the `convert()` function to convert a FP number to an exact fraction. More generally, you can use `convert()` to convert numbers or expressions to many different formats (such as binary, hexadecimal, and so on). Type `?convert;` for more information on this useful command.

```
>   # Give Pi to 20 significant digits,
>   # and write the result as an exact fraction.
>   convert(evalf(Pi,20),fraction);
```

$$\frac{103993}{33102}$$

> Use `evalf()` to produce a floating point result.
> Use `convert()` to convert from one format to another.

1.5   Algebra in *Maple*

In the previous examples, *Maple* has largely been used as a calculator: an
expression is entered, and *Maple* calculates an answer. However, *Maple* is
much more powerful than this. You can use any letters or combinations of
letters (except for system constants such as Pi) as variables, and you can
then manipulate algebraic expressions in many ways. For example, you can
find sums and products, expand and factorise expressions, and so on.

1.5.1   *Assigning variables and giving names*

The **assignment** operator := is used to assign values to variables. Then
whenever you use that variable, *Maple* will replace it with the expression or
value which has been assigned to the variable. This allows interim results
to be calculated, and then used in subsequent calculations.

```
>   # Assign the values 5^2 and (3^3)/4
>   # to the variables a1 and a2
>   a1:=5^2; a2:=3^3/4;
```

$$a1 := 25$$

$$a2 := \frac{27}{4}$$

```
>   # Look at the values of a1 and a2
>   a1; a2;
```

$$25$$

$$\frac{27}{4}$$

```
>   # Use the values of a1 and a2
>   a1*a2;
```

$$\frac{675}{4}$$

Here are two more examples with assigning algebraic expressions rather than numbers.

```
>   u:=2*x+4;
```

$$u := 2x + 4$$

```
>   3*u/2;
```

$$3x + 6$$

```
>   v:=x^2-1;
```

$$v := x^2 - 1$$

```
>   z:=x-1;
```

$$z := x - 1$$

```
>   v/z;
```

$$\frac{x^2 - 1}{x - 1}$$

When a value is assigned to a variable, this value remains assigned until it is explicitly cleared. You must remember this: if you have assigned a value to a variable, each time you use that variable in a subsequent expression, the assigned value will be substituted into the expression. To clear the values of variables, you can use the **restart** command, which completely clears all variables (having the same result as if you exit *Maple*, then started it again). To clear the values of individual variables, you assign the variable its own name, with the right-hand side enclosed in single quotes. It is a good habit to begin a second worksheet with the **restart** command, otherwise the variables from the previous worksheet would be still assigned.

```
>   # Assign a value to the variable a2
>   a2:=27/4;
```

$$a2 := \frac{27}{4}$$

```
>   # Clear the variable a2,
>   # forgetting its current value.
>   a2:='a2';
```

$$a2 := a2$$

The assignment operator can be used to name almost any object which *Maple* recognizes. This can save a lot of typing by recalling the object by its assigned name either alone or as an argument of a command.

```
>   p:=Pi^3+5!;
```

$$p := \pi^3 + 120$$

```
>   eq:=x^2+5=p;
```

$$eq := x^2 + 5 = \pi^3 + 120$$

```
>   poly:=x^5+x+1;
```

$$poly := x^5 + x + 1$$

```
>   factor(poly);
```

$$(x^2 + x + 1)(x^3 - x^2 + 1)$$

> Use := to assign values to variables.
> Use restart to clear the values of all variables.
> Clear an individual variable by assigning the variable's
> name to it, for example, x1:='x1';.
> Use the assignment operator := to name mathematical
> objects.

1.5.2   *Useful inbuilt functions*

Maple provides you with a large number of very useful inbuilt functions. These can be used to manipulate expressions which you have defined. There is a command `simplify`. However, simplification can mean one thing on one occasion and something altogether different on another. Which is simpler,

$x^4 - 1$ or $(x+1)(x-1)(x^2+1)$? *Maple* simplifies only when there cannot be any doubt: x is undoubtedly simpler than $x + 0$. Therefore the decision how to manipulate is often left with the user: *Maple* provides a range of options: expand, factor, normal, collect, sort and combine.

1.5.2.1 *expand()*

The expand() function is used to distribute products over sums:

```
>   # Define A=x^2-1 and B=x^2+1
>   A:=x^2-1; B:=x^2+1;
```

$$A := x^2 - 1$$

$$B := x^2 + 1$$

```
>   # Evaluate A*B
>   A*B;
```

$$(x^2 - 1)(x^2 + 1)$$

```
>   # Expand the previous result
>   expand(%);
```

$$x^4 - 1$$

```
>   # Expand (x^2-1)(x^2+1)-(x-1)^2
>   expand((x^2-1)*(x^2+1)-(x-1)^2);
```

$$x^4 - 2 - x^2 + 2x$$

```
>   # Expand (x+y+z)^2-(x-y-z)^2
>   expand((x+y+z)^2-(x-y-z)^2);
```

$$4xy + 4xz$$

```
>   # Expand (1+2x)^6
>   expand((1+2*x)^6);
```

$$1 + 12\,x + 60\,x^2 + 160\,x^3 + 240\,x^4 + 192\,x^5 + 64\,x^6$$

1.5.2.2 *factor()*

The **factor()** function is used to factor a polynomial with integer, rational or algebraic number coefficients. **factor()** is often used to reduce complicated expressions to a more manageable size.

```
>    # Assign x^2-1 to f, in other words name x^2-1 as f
>    f:=x^2-1;
```

$$f : x^2 - 1$$

```
>    # Factorise f
>    factor(f);
```

$$(x - 1)(x + 1)$$

```
>    # Factorise x^4-x^2+2x-2
>    factor(x^4-x^2+2*x-2);
```

$$(x - 1)(x^3 + x^2 + 2)$$

1.5.2.3 *simplify()*

The **simplify()** function is used to write a complicated expression in a simpler form. Often *Maple* leaves the expression unchanged and leaves the decision how to proceed with the user.

```
>    # Assign (sin x)^2 + (cos x)^2 to T
>    T:=sin(x)^2+cos(x)^2;
```

$$T := \sin(x)^2 + \cos(x)^2$$

```
>    simplify(T);
```

1

Note that the *Maple* command `sin(x)^2` is evaluated as `(sin(x))^2` and not as `sin((x)^2)`. Here *Maple* follows the standard rules of precedence in evaluating functions before applying other mathematical operations. If you are in any doubt, insert additional parentheses to avoid possible errors.

```
>   # Assign (x+1)^3 to C
>   C:=(x+1)^3;
```

$$C := (x+1)^3$$

```
>   # Assign (x-1)^3 to C1
>   C1:=(x-1)^3;
```

$$C1 := (x-1)^3$$

```
>   C-C1;
```

$$(x+1)^3 - (x-1)^3$$

```
>   # Simplify the previous result
>   simplify(%);
```

$$6x^2 + 2$$

1.5.2.4 *normal()*

The `normal()` function is used to write expressions in normalised form; that is, a numerator over a common denominator:

```
>   # Normalise 1/x + 1/(x-1)
>   normal(1/x + 1/(x-1));
```

$$\frac{2x-1}{x(x-1)}$$

```
>   # Normalise 1/(x-1) + x/(x-1)^2
>   normal(1/(x-1) + x/(x-1)^2);
```

$$\frac{2\,x - 1}{(\,x - 1\,)^2}$$

1.5.2.5   *collect()*

The collect() function is used to collect coefficients of like powers of a specified variable; this variable must be indicated, otherwise *Maple* does not know what to collect. In the following example, the final expression has been collected into coefficients of the variable x:

```
>   # Define f as (x+(x-y)^2)^2
>   f:=(x+(x-y)^2)^2;
```

$$f := \left(x + (\,x - y\,)^2\right)^2$$

```
>   expand(f);
```

$$x^2 + 2\,x^3 - 4\,x^2\,y + 2\,x\,y^2 + x^4 - 4\,x^3\,y + 6\,x^2\,y^2 - 4\,x\,y^3 + y^4$$

```
>   collect(%,x);
```

$$x^4 + (\,2 - 4\,y\,)\,x^3 + (\,-4\,y + 1 + 6\,y^2\,)\,x^2 + (\,2\,y^2 - 4\,y^3\,)\,x + y^4$$

1.5.2.6   *sort()*

The sort() function is used to sort the terms of an expression into descending order of power of a specified variable.

```
>   # Assign x^5-3x^3+x^7 to f
>   f:=x^5-3*x^3+x^7;
```

$$f := x^5 - 3\,x^3 + x^7$$

```
>   sort(f,x);
```

$$x^7 + x^5 - 3\,x^3$$

```
>   # Assign x~3y+x^4y^4+x^2+x^6y^2 to g
>   g:=x^3*y+x^4*y^4+x^2+x^6*y^2;
```

$$g := x^3\,y + x^4\,y^4 + x^2 + x^6\,y^2$$

```
>   sort(g,y);
```

$$x^4\,y^4 + x^6\,y^2 + x^3\,y + x^2$$

1.5.2.7 *combine()*

The `combine` command reduces a given expression into a single term, if possible, or transforms it into a more compact form. Sometimes, similarly as with simplify, it leaves the original expression unchanged, leaving the user with the option to try another command. The following examples illustrate the use of `combine`.

```
>   combine(sqrt(12)*sqrt(3));
```
$$6$$

```
>   combine(sqrt(3+sqrt(5))*sqrt(3-sqrt(5)));
```
$$(\frac{1}{2}\,\sqrt{10} + \frac{1}{2}\,\sqrt{2})\,(\frac{1}{2}\,\sqrt{10} - \frac{1}{2}\,\sqrt{2})$$

```
>   expand(%);
```
$$2$$

```
>   combine((cos(x))^2-(sin(x))^2);
```
$$\cos(2\,x)$$

The `combine` command reduces integer powers of sin and cos into linear expressions in terms of sin and cos of multiple arguments. The result is often more extensive than the original expression.

```
>   combine(cos(x)^5);
```

$$\frac{1}{16}\cos(5\,x) + \frac{5}{16}\cos(3\,x) + \frac{5}{8}\cos(x)$$

```
>   combine(cos(x)^2+sin(x)^3);
```

$$\frac{1}{2}\cos(2\,x) + \frac{1}{2} - \frac{1}{4}\sin(3\,x) + \frac{3}{4}\sin(x)$$

1.5.2.8 *subs()*

The subs command makes a substitution, it can be used to substitute a value or an expression. It does not simplify the result, it leaves it to the user to choose an appropriate command for further processing.

```
>   subs(x=1,sqrt(x^2+11*x+24));
```

$$\sqrt{36}$$

```
>   simplify(%);
```

$$6$$

```
>   subs(x=t-3,x^2+6*x+1);
```

$$(t-3)^2 + 6\,t - 17$$

```
>   expand(%);
```

$$t^2 - 8$$

---

combine(), expand(), factor(), simplify(), normal(), collect(), sort() and subs() **are all useful for manipulating mathematical expressions.**

1.6   Examples of the use of *Maple*

Example 1.1 (simplify; rationalize; factor) *Maple* can simplify some quite complicated expressions

---

```
>   simplify(sqrt(2*sqrt(221)+30));
```

$$\sqrt{13} + \sqrt{17}$$

On some similar occasions *Maple* does not simplify, the reason being that there is another command more appropriate: for example, `rationalize` is one such command.

```
> simplify(1/(2+sqrt(3)));
```
$$\frac{1}{2+\sqrt{3}}$$
```
> rationalize(%);
```
$$2-\sqrt{3}$$

Another example of the same kind is:

```
> simplify(2*a^2+5*a*b+2*b^2);
```
$$2\,a^2 + 5\,a\,b + 2\,b^2$$
```
> factor(%);
```
$$(2\,a + b)\,(a + 2\,b)$$

Example 1.2 (`solve`)   *Maple* can solve equations for you. The basic command is `solve`. We shall learn more about solving equations and the command `solve` in Chapter 8, but here is a simple example.

```
> eq:=1/(x+1)+1/(x-1)=1;
```
$$eq := \frac{1}{x+1} + \frac{1}{x-1} = 1$$
```
> solve(eq);
```
$$1 - \sqrt{2},\, 1 + \sqrt{2}$$

We trust that the readers do not need *Maple* to solve such simple equations, the next example is more sophisticated. We use *Maple* to solve a Diophantine equation, this means that the solution must be in integers.

Example 1.3 (Diophantine equation)   An example of a Diophantine equation is

$$1234567x - 7654321y = 1357924680.$$

The command for solving a Diophantine equation is `isolve`.

```
> isolve(1234567*x-7654321*y=1357924680);
```

$$\{x = 4484627 + 7654321\,\_N1,\, y = 723149 + 1234567\,\_N1\}$$

The entry $\_N1$ indicates an arbitrary integer, there are infinitely many so-lutions. The amount of work with pencil and paper to find this solution would be considerable. *Maple*'s answer is instantaneous.[4] It is easy to see that the Diophantine equation $3\,x - 2\,y = n$, where n is a given integer, has a solution (among infinitely many others) $x = n$, $y = n$. *Maple*'s answer is not very helpful:

```
> isolve(3*x-2*y=n);
```

$$\{x = \_N1,\, n = 3\,\_N1 - 2\,\_N2,\, y = \_N2\}$$

This means that if we choose x arbitrarily and y also then $n = 3\,x - 2\,y$, which is correct but not helpful. What can we do with $1234567\,x + 7654321\,y = n$? As we might expect, blind use of isolve will not help.

```
> isolve(1234567*x-7654321*y=n);
```
$$\{x = \_N1,\, y = \_N2,\, n = 1234567\,\_N1 - 7654321\,\_N2\}$$

However we can solve the equation

```
> isolve(1234567*u-7654321*v=1);
```

$$\{v = 661375 + 1234567\,\_N1,\, u = 4100528 + 7654321\,\_N1\}$$

---

and then clearly $x = un$ and $y = vn$

---

[4] *Maple* will even tell you how long it took to solve an equation or to execute another command. See time in the Help file.

We have seen that *Maple* can save a lot of tedious work. However, even in a situation when *Maple* is unable to solve the problem directly, it can be used with advantage.

At the beginning of this chapter we said that *Maple* can divide polynomials: we mentioned dividing $x^6 + 1$ by $x^2 + x + 1$. There are two commands, `rem` to give the remainder and `quo` to give the quotient. The following *Maple* session shows how to use these commands. Note that three entries in brackets are required in both commands.

```
>   rem(x^6+1,x^2+x+1,x);
```

$$2$$

```
>   quo(x^6+1,x^2+x+1,x);
```

$$x^4 - x^3 + x - 1$$

The third argument in both commands is important as the next example shows.

```
>   rem(u^2+v^2,u-2*v,u);
```

$$5\,v^2$$

```
>   rem(u^2+v^2,u-2*v,v);
```

$$\frac{5}{4}\,u^2$$

Maple can also find the remainder and quotient for division of integers. The commands are `irem` and `iquo`.

```
>   irem(987654,13);
```

$$5$$

```
>   iquo(987654,13);
```

$$75973$$

So far all the commands we have used were available directly on the command line. These commands reside in the main part of *Maple* called the kernel. More specialized or not very frequently used commands are grouped in packages. This helps with efficiency and speed of *Maple*. The list of packages can be obtained by first clicking on **Help** (located in the upper right corner of the screen) and choosing **Topic Search** from the menu which opens. In the dialog window which then appears fill in **packages** and click OK. Another window opens and there you can click on **index,package**. There are literally dozens of packages. In this section we shall use only two, the number theory package and the finance package. To obtain the list of commands available in a package issue a command `with(package)`. For instance, if we issue a command `with(geometry)` *Maple* will produce a long list of various commands relating to geometry, and make these commands available. The first command in the list is *Apollonius*. Using **Help** will tell us how to use this command to solve the problem of finding a circle which is tangent to three given circles, a problem formulated and solved by Apollonius from Perge in third century BC. In a similar fashion we can employ any command from the above list after the command `with(geometry)` has been issued. If we know the name of the command and the package we can use the command on the command line. For instance, we can find all positive divisors of a number using the number theory package `numtheory` and the command `numtheory[divisors]`, or the number of positive divisors using the command `numtheory[tau]` (note the use of [and]).

> `numtheory[divisors](144);`

$$\{1, 2, 3, 4, 6, 8, 9, 12, 16, 18, 24, 36, 48, 72, 144\}$$

> `numtheory[tau](144);`

$$15$$

Alternatively, we can activate the command

> `with(numtheory,divisors);`

$$[divisors]$$

and then use the short form of the command thus

> `divisors(144);`

$$\{1, 2, 3, 4, 6, 8, 9, 12, 16, 18, 24, 36, 48, 72, 144\}$$

Mathematics for finance is contained in the `finance` package.

Example 1.4 (Mortgage)   A 35 year-old man is confident that he can save \$12,000 a year. He wishes to buy a house and stop payments at 55, when he plans to retire. The current interest rate is 6%. In order to find out how much he can borrow he uses the command `annuity` from the finance package. From a mathematical point of view, mortgage and annuity are the same. A company advances the money for the mortgage and the consumer pays back installments; in an annuity, in contrast, the individual makes a down payment and receives regular income from a finance institution.

```
>   with(finance,annuity);
```
$$[annuity]$$
```
>   annuity(12000,0.06,20);
```
$$137639.0546$$

This amount seems insufficient for a house of the desired standard and quality. The man decides to extend the life of the mortgage to see what happens.

```
>   annuity(12000,0.06,30);
```
$$165177.9738$$

This is a sufficient amount. However, paying the mortgage for ten more years is not an attractive idea. To see by how much the installments must be increased for a mortgage of \$165,000, use the following commands.

```
>   solve(annuity(x,.06,20)=165000);
```
$$14385.45190$$

Clearly, as long as it is affordable, it is far better to increase the cash payments rather than to increase the life of the mortgage.

The Help file contains details of how to adapt the annuity for a more realistic situation of monthly or fortnightly payments.

Exercises

ⓘ **Exercise 1.6.1**   *The command* time(X) *will tell you how much time (in seconds) was needed by Maple to execute* X. *Assign* a, b, c *to be* $2^{100}7^{50}$, $3^{100}5^{50}$, *and* 10^{100}, *respectively. Solve the Diophantine equation* $ax + by = c$. *Determine how much time was needed.* [Hint: On the authors' machine using *Maple* 7 it was .039 sec].

Exercise 1.6.2   *Use Maple to factorize* $x^{11} + x^4 + 1$.

Exercise 1.6.3   *Find the quotient and remainder for*

(1) 987654321, 123456789;
(2) 100000, 17.

Exercise 1.6.4   *Find the quotient and the remainder for the polynomials*

(1) $x^{10} + 5x + 7$, $x^4 + 1$;
(2) $x^{20} + 1$, $x^9 + 2x + 7$.

Exercise 1.6.5   *What are the monthly installments for a mortgage of* $200,000 *over a period of 40 years at interest rate of 6.5%?*

Chapter 2

Sets

In this chapter we review set theoretic terminology and notation. Later we discuss mathematical reasoning.

2.1 Sets

The word set as used in mathematics means a collection of objects. It is customary to denote sets by capital letters like M, M_1, etc., below. At this stage the concept of a set is best illuminated by examples, so we list a few sets and name them for ease of reference.

Table 2.1   Some examples of sets

M_1, the set of all rational numbers greater than 1;
M_2, the set of all positive even integers;
M_3, the set of all buildings on the St Lucia campus of the University of Queensland;
M_4, the set of all readers of this book;
M_5, the set of all persons who praise this book;
M_6, the set of all persons who condemn this book;
M_7, the set of all even numbers between 1 and 5.

In these examples we have used the word 'all' to make it doubly clear that, for instance, *every* rational number greater than 1 belongs to M_1. But usually, in mathematics, if someone mentions the set of rational numbers greater than 1 he or she means the set of all such numbers, and this is automatically understood. Thus with a similar understanding we would say that M_2 is the set of positive even integers.

If an object x belongs to the set M, we say that x *is an element of* M or that x *lies in* M, and we denote this by writing $x \in M$. If x is not an element of M we write $x \notin M$. A set M is defined by specifying which elements lie in M.

One way of describing a set is to list its elements, separated by commas, and enclose them in braces (curly brackets); for example, $M_7 = \{2, 4\}$, $M_2 = \{2, 4, 6, 8, \dots\}$.

It is worth pointing out that the elements in a set can be other sets. Thus the set $A = \{2, \{x, y\}\}$ has two elements; one is the number 2, and the other is a set with two elements, x and y. These could be some mathematical entities, or they could be sets, whose elements are other sets, whose elements .... Furthermore, the order in which the elements are listed inside $\{\dots\}$ is irrelevant; thus $M_7 = \{4, 2\}$ is the same set as M_7 defined in Table 2.1. Also, any repetition of elements is ignored, so that we could equally well write $M_7 = \{4, 4, 2, 2, 2, 4, 2, 4\}$. The essential part of the definition of M_7 is that *both* 2 and 4 must lie inside $\{\dots\}$, and nothing else is allowed there.

A *finite set* has only a finite number of elements. Thus M_3 and M_4 are finite, whereas M_1 and M_2 are infinite (that is, they are not finite). Two sets M and N are considered to be *equal* if they consist of the same elements. If all elements of a set A belong to a set B then we say that A is a *part* of B, or that A is a *subset* of B. This is denoted by $A \subset B$. Obviously $A \subset A$ always. The relation $A \subset B$ is often referred to as an *inclusion*. Thus in the above list it is obvious that $M_2 \subset M_1$. If all people were honest and rational it would be also true that $M_5 \subset M_4$ and $M_6 \subset M_4$; but in the world as it is this is not certain. If $A \subset B$ and A is not equal to B we say that A is a *proper subset* of B. This relation between A and B is denoted by $A \subsetneq B$.

It is also convenient to introduce the *empty set*. This is a set which has no elements, and we regard it as a finite set. The symbol $\emptyset$ is used to denote the empty set. We could also write $\emptyset = \{\}$. At the first encounter the empty set may look a little strange but the introduction of the empty set has advantages; for example, in the above list M_4, M_5 and M_6 need not contain any elements at all.[1] The introduction of the empty set allows us to talk freely about M_4, M_5 and M_6 without worrying whether or not they

---

[1] But since you have read this footnote it seems that M_4 contains at least one element.

contain an element. We also agree that $\emptyset \subset M$ for any set M.

If $M \subset N$ and $N \subset M$ then clearly M and N consist of the same elements and hence $M = N$. In particular cases later, where we prove that $M = N$ we often do so by showing that $M \subset N$ and $N \subset M$.

We now introduce some further notation. If P is a certain property then the set of all elements having the property P is denoted by $\{x;\ x$ has property $P\}$; for example,

$$M_2 = \{x;\ x \text{ is a positive even integer}\},$$

$$M_7 = \{x;\ x \in M_2, 1 < x < 5\}.$$

2.1.1 *Union, intersection and difference of sets*

Given two sets M and N we can define a new set, denoted by $M \cup N$, consisting of all elements belonging to either M or N (or both). The set $M \cup N$ is called the *union* of M and N (see Figure 2.1), and clearly $M \cup N = N \cup M$.

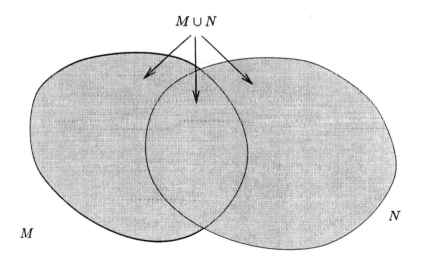

Fig. 2.1   Union of two sets

We define another new set by taking elements common to both M and N; these form a set called the *intersection* of M and N which is denoted by $M \cap N$ (see Figure 2.2). Clearly $M \cap N = N \cap M$. Using the notation introduced above we have $M \cup N = \{x;\ x \in M \text{ or } x \in N\}$ and $M \cap N = \{x;\ x \in M \text{ and } x \in N\}$. If $M \cap N = \emptyset$ then we say that M and N are *disjoint*.

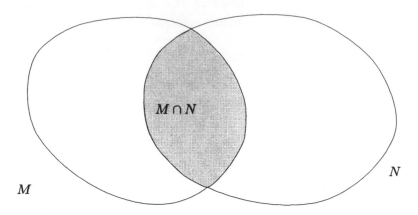

Fig. 2.2   Intersection of two sets

Yet another set which can be formed from two sets is the difference. If M and N are sets then the *difference* $M \setminus N$ is, by definition, the set of elements lying in M but not in N, that is, $M \setminus N = \{x;\ x \in M,\ x \notin N\}$. This is illustrated in Figure 2.3. It is clear that $M \setminus N \neq N \setminus M$, unless $M = N$, in which case $M \setminus N = \emptyset$.

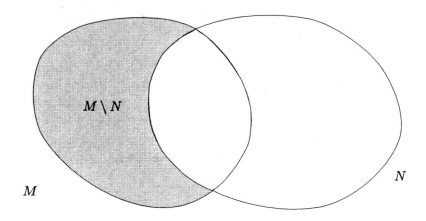

Fig. 2.3   Difference of two sets

Given three sets L, M and N we can form two new sets $A = (L \cup M) \cup N$ and $B = L \cup (M \cup N)$. For example, for B we form first the union $M \cup N = Q$ and then use Q to form $L \cup Q$. It is easy to see that each of the new sets A, B consists of those elements belonging to at least one of the sets L, M, N. Hence $(L \cup M) \cup N = L \cup (M \cup N)$ and we write simply $L \cup M \cup N$. The same convention is used for intersections; for example, the intersection of four sets A, B, C, D may be denoted simply by $A \cap B \cap C \cap D$.

The corresponding ideas for 5 sets, or 25, or any finite number, presents no new difficulties but some compact notation is needed. Thus $A_1 \cap A_2 \cap A_3 \cap A_4$ is denoted by $\bigcap_{i=1}^{4} A_i$, and in the same way, $\bigcup_{i=1}^{7} B_i$ is the union of the seven sets B_1, B_2, ..., B_7. The notation is extended to refer to an infinite number of sets so that $\bigcup_{i=1}^{\infty} C_i$ is the set of those elements contained in at least one of the sets C_1, C_2, C_3, ... and $\bigcap_{k=1}^{\infty} D_k$ is the set of those elements contained in all the sets D_k, $k = 1, 2, 3, \ldots$. For example, if C_i is the set of positive proper fractions with denominator i then the union $\bigcup_{i=1}^{\infty} C_i$ is the set of rational numbers between 0 and 1. And if D_k is the set of integers less than or equal to k then the intersection $\bigcap_{k=1}^{\infty} D_k$ consists of the set of negative integers, together with 0 and 1, that is, $\bigcap_{k=1}^{\infty} D_k = \{1, 0, -1, -2, -3, \ldots\}$.

Exercises

Exercise 2.1.1   *Use the definitions*
$$A \cup B = \{x;\ x \in A \text{ or } x \in B\},$$
$$A \cap B = \{x;\ x \in A \text{ and } x \in B\},$$
$$A \setminus B = \{x;\ x \in A,\ x \notin B\},$$

to prove $(A \cap B) \cup (A \setminus B) = A$.

(i) **Exercise 2.1.2**   *The set of subsets of A is called the power set of A. For example, the power set of $\{1, 2\}$ is $\{\{\}, \{1\}, \{2\}, \{1, 2\}\}$. Use Maple to create the power set of $\{1, 2, 3, a, b, M\}$.* [Hint: Use **Help** to find the command `combinat[powerset]`.]

2.1.2   *Sets in* Maple

Notation for sets in *Maple* is the same as in set theory. The empty set
is denoted by {} (the common mathematical symbol for the empty set, $\emptyset$,
cannot be entered from the keyboard). Instead of the symbols $\cup$, $\cap$, \ we
write `union`, `intersect`, `minus`, respectively. Hence

```
>   {1,2,a};
```

$$\{1, 2, a\}$$

```
>   A:={19, 25, 31, 37, 43, 49, 55};
```

$$A := \{19, 25, 31, 37, 43, 49, 55\}$$

```
>   B:={23,29,31,37,41,43,47,53,59,61,67,71};
```

$$B := \{23, 29, 31, 37, 41, 43, 47, 53, 59, 61, 67, 71\}$$

```
>   A union B;
```

$$\{19, 23, 25, 29, 31, 37, 41, 43, 47, 49, 53, 55, 59, 61, 67, 71\}$$

```
>   A intersect B;
```

$$\{31, 37, 43\}$$

```
>   A minus B;
```

$$\{19, 25, 49, 55\}$$

```
>   B minus A;
```

$$\{23, 29, 41, 47, 53, 59, 61, 67, 71\}$$

```
>   ({a,b} union {x}); ({1,2,x} minus {1});
```

$$\{b,\ x,\ a\}$$

$$\{2,\ x\}$$

In ordinary life there is no distinction between 1 and 1.0. We know that *Maple* makes a distinction, consequently

---

```
>   ({1} intersect {1.0}); ({1} union {1.0});
```

$$\{\}$$

$$\{1,\ 1.0\}$$

2.1.2.1 *Expression sequences, lists*

An expression is a basic entity in *Maple*; roughly speaking, it is a string of characters which can be entered into *Maple*. An expression sequence is an ordered *n*-tuple of expressions separated by commas. For instance john, 1, Kai!Hyâ, set, x+1/x, 1, 1 is an expression sequence. An expression sequence enclosed in square brackets constitutes a list. Order and repetition of expressions in lists or expression sequences do matter: [1, 2], [2, 1] and [1, 2, 1] are all distinct. Sets, lists and expression sequences are all similar: the difference lies in the way *Maple* handles them. You can make an intersection of two sets but not of two lists. You can combine two lists into a new list in which the first element of the second list follows the last element of the first list. The following are some examples of using sets, lists and expression sequences in *Maple*.

```
>   # First we name a few sets, lists and
>   # expression sequences.
>   A:={a,b,c}; B:={john,3!,x!,Kathy};
>   C:=[x,y,z,z,z];Z:=1,a,1,b;
```

$$A := \{a,\ b,\ c\}$$

$$B := \{6,\, x!,\, john,\, Kathy\}$$

$$C := [x,\, y,\, z,\, z,\, z]$$

$$Z := 1,\, a,\, 1,\, b$$

```
>   # the command convert changes a set into a list
>   # or vice-versa
>   convert(A,list);convert(C,set);
```

$$[a,\, b,\, c]$$

$$\{x,\, y,\, z\}$$

```
>   # enclosing an expression sequence in appropriate
>   # symbols converts the sequence to a list or set,
>   # respectively.
>   [Z];{Z};
```

$$[1,\, a,\, 1,\, b]$$

$$\{1,\, a,\, b\}$$

```
>   # the nops command counts the number of elements
>   # in a list or a set,
>   # but cannot be applied to an expression sequence.
>   nops(B);nops(C);
```

$$4$$

$$5$$

```
>   # the n-th element can be extracted from a
>   # list or an expression sequence by
>   # attaching [n] to the list
>   # or expression sequence, respectively.
>   # Note that a set does not have an n-th element.
>   # Maple will return a result, giving what it thinks
>   # is the n-th element, and this need not agree
>   # with what you might guess is the n-th element.
>   A[2];B[1];Z[3];
```

$$b$$

$$6$$

$$1$$

```
>   # The command op(2,A) has the same effect as A[2].
>   # However, the op command is more powerful
>   # and more flexible.
>   # It can extract all elements, or  some chosen
>   # elements from a list or a set.
>   # Some examples on the use of op follow.
>   op(2..4,C); op(B); [op(A), op(3..4,B)];
```

$$y, z, z$$

$$6, x!, john, Kathy$$

$$[a, b, c, john, Kathy]$$

```
>   # The command seq can be used to generate
>   # an expression sequence according to some formula.
>   seq(i^3,i=2..5);seq(n^2+n+7,n=0..3);
```

$$8, 27, 64, 125$$

7, 9, 13, 19

```
>  {seq(n^2+n+7,n=1..30)}intersect{seq(n^2+n+11,n=1..30)};
```

$$\{13\}$$

Exercises

Exercise 2.1.3   *Combine the lists $L = [1, 3, 5]$ and $M = [1, 2, 3, 4]$ into a single list using the* op *command.*

Exercise 2.1.4   *Find the number of distinct elements in a list by first converting it to a set.*

Exercise 2.1.5   *Find the number of elements in the power set of A, where A has 17 elements.* [Hint: See Exercise 2.1.2. Use **nops**!]

2.1.3   *Families of sets*

Instead of saying that A is a set of sets we prefer to say that A is a *system* or a *family of sets*. Later we shall encounter situations where to each element l of a set L there is associated a set C_l. In such a situation we say that the *family* of sets C_l is *indexed* by the set L. It is natural to denote this indexed family by $\{C_l; l \in L\}$. If $\{C_l; l \in L\}$ is an indexed family then by $\bigcup_{l \in L} C_l$ we understand the union of the family C_l, that is, the set of all elements belonging to at least one set C_l.

Similarly, the intersection $\bigcap_{l \in L} A_l$ is by definition the set of all elements belonging to every set A_l. A family of sets can be indexed by any set, not necessarily by a subset of positive integers. If, for instance, A_l is the set of points on the straight line between the points $(l, 0)$ and $(l, 1)$ in the plane and L is the set of real numbers between 0 and 1 then $\bigcup_{l \in L} A_l$ is the interior of the square with vertices $(0, 0)$, $(1, 0)$, $(1, 1)$ and $(0, 1)$.

2.1.4 *Cartesian product of sets*

If A and B are sets then the set of ordered pairs of elements, (a, b),[2] the first from A and the second from B, is called the *cartesian product* of A and B and denoted by $A \times B$. Expressing this in symbols, we have $A \times B = \{(a, b); a \in A, b \in B\}$. Two ordered pairs (a, b) and (c, d) are defined to be equal if and only if $a = c$ and $b = d$. It is natural to define $A \times B \times C$ as the set of ordered triplets (a, b, c) with $a \in A, b \in B$ and $c \in C$. One would perhaps expect that $A \times B \times C = (A \times B) \times C = A \times (B \times C)$. However this is not strictly correct since $(A \times B) \times C$ is a set of pairs of things in which the first thing is a pair of elements and the second thing is a single element; $((a, b), c)$ is its typical member, whereas a typical member of $A \times B \times C$ is a triplet (a, b, c). In this book we shall not make this fine distinction between $(A \times B) \times C$ or $A \times (B \times C)$ and $A \times B \times C$. In the same way the idea of a cartesian product of four or any finite number of sets should be clear. For $A \times A$ we write A^2, and for $A \times A \times \cdots \times A$ with n factors we write A^n.

Warning: Realise that $A \times B$ generally is not equal to $B \times A$.

2.1.5 *Some common sets*

We assume that the reader is, at least on an intuitive basis, familiar with real numbers. The theory of real numbers (including, for example, the definition of integers, rationals, etc.) is systematically reviewed in Chapter 4. For ease of reference we name here some subsets of reals as follows. The set of all real numbers will be denoted consistently by $\mathbb{R}$.

$\mathbb{N}$ is the set of natural numbers, $\mathbb{N} = \{1, 2, 3, \ldots\}$;

$\mathbb{N}_0$ is the set of non-negative integers, $\mathbb{N}_0 = \{0, 1, 2, 3, \ldots\}$;

$\mathbb{Z}$ is the set of integers, $\mathbb{Z} = \{0, 1, -1, 2, -2, 3, -3, \ldots\}$;

$\mathbb{Q}$ is the set of rationals, $\mathbb{Q} = \left\{\dfrac{p}{q}; p \in \mathbb{Z}, q \in \mathbb{N}\right\}$;[3]

$\mathbb{P}$ is the set of positive reals, $\mathbb{P} = \{x; x \in \mathbb{R}, x > 0\}$.

---

[2]Using set notation, the ordered pair (a, b) can be written $(a, b) = \{a, \{a, b\}\}$. In fact this equation can be used to *define* an ordered pair.

[3]A more precise formulation of the definition of $\mathbb{Q}$, using terminology introduced so far, would read $\mathbb{Q} = \{(p, q); p \in \mathbb{Z}, q \in \mathbb{N}\}$

Exercises

Exercise 2.1.6   *Give examples of sets M, N, P, Q such that $M \subset N$ and neither $P \subset Q$ nor $Q \subset P$.*

Exercise 2.1.7   *For the sets listed in Table 2.1 show that $M_1 \cup M_2 = M_1$, $M_3 \cap M_4 = \emptyset$.*

Exercise 2.1.8   *In this exercise A, B, C are arbitrary sets. Discover which of the following relations are correct. Prove the correct ones and give examples (strictly, counter-examples) to show that the others are false.*

(1) $A \cap (B \cup C) = (A \cap B) \cup (A \cap C)$;
(2) $(A \cup B) \cap C = (A \cap C) \cup (B \cup C)$;
(3) $A \cup (B \cap C) = (A \cup B) \cap (A \cup C)$;
(4) $A \cup A = A$, $A \cap A = A$;
(5) $(A \cap C) \cup B = (A \cup B) \cap (C \cup B)$.

Exercise 2.1.9   *Let $A_n = \{x; \ x \in \mathbb{N}, \ x$ a multiple of $n\}$, $B_l = \{x; \ x \in \mathbb{R}, \ x < l\}$, $L = \{x : \ x \in \mathbb{R}, \ x > 0\}$. Find $\bigcup\limits_{n=2}^{\infty} A_n$, $\bigcap\limits_{n=2}^{\infty} A_n$, $\bigcup\limits_{l \in L} B_l$, $\bigcap\limits_{l \in L} B_l$.*

Exercise 2.1.10   *What can you say about the sets A, B if $A \cup B = A \cap B$?*

(i) **Exercise 2.1.11**   *Prove that*

(1) $(A \setminus B) \cap C = (A \cap C) \setminus (B \cap C)$;
(2) $(A \setminus B) \cup (A \setminus C) = A \setminus (B \cap C)$;
(3) $(A \setminus B) \cap (A \setminus C) = A \setminus (B \cup C)$.

(i) **Exercise 2.1.12**   *Parts (2) and (3) of Exercise 2.1.11 can be generalized to families of sets. Prove that, if L is an index set, A_l a family indexed by L, and X is any set, then*

$$\bigcup_{l \in L} (X \setminus A_l) = X \setminus \bigcap_{l \in L} A_l;$$

$$\bigcap_{l \in L} (X \setminus A_l) = X \setminus \bigcup_{l \in L} A_l;$$

(These relations are known as de Morgan's Rules.)

(i) **Exercise 2.1.13**   *Let A_l and B_l be two families of sets indexed by a set L. Prove that*

$$\bigcup_{l \in L} A_l \setminus \bigcup_{l \in L} B_l \subset \bigcup_{l \in L} (A_l \setminus B_l).$$

Also show by an example that the inclusion cannot, in general, be replaced by equality.

2.2   Correct and incorrect reasoning

The ability to reason correctly is essential in mathematics. Care is often needed, as the following fallacious argument shows. Let b be an arbitrary real number, and let c be the same number, that is, $c = b$. Multiply this equation by c and subtract b^2, giving

$$c^2 - b^2 = cb - b^2 \, ;$$

that is,

$$(c + b)(c - b) = b(c - b), \tag{2.1}$$

and hence, dividing by $c - b$ we have

$$c + b = b, \tag{2.2}$$

and therefore $c = 0$.

We appear to have proved that an arbitrary real number c must be zero. As this is absurd we must look for an explanation. The error occurred when passing from Equation (2.1) to Equation (2.2), and involved an incorrect application of the following theorem.

If $yx = zx$ and $x \neq 0$ then $y = z$.

We applied this for $x = c - b$ and overlooked the fact that $c - b = 0$.

Consider the previous theorem again. It is of the form: 'If ... then ...'. The part of the sentence starting with 'If' is called the *hypothesis* (or assumption(s) or condition(s)) and the phrase starting with 'then' is called the *conclusion* (or assertion(s)). The moral of the above example is that when applying a theorem one must make sure that the hypothesis is satisfied.

Sometimes a mathematical theorem is not expressed in the form 'If ... then ...', but even with such a variation of form the hypothesis can be identified. For example the theorem

> **T:** *The product of two consecutive integers is divisible by two.*

can be rephrased to exhibit the hypothesis and the conclusion more clearly as

> **T:** *If n is an integer then the number $n(n + 1)$ is divisible by 2.*

If the conclusion and the hypothesis are interchanged a new theorem is obtained, which is called the *converse* of the original theorem. The theorems **P** and **C** below are converses of each other.

> **P:** *If a right-angled triangle has hypotenuse c and other sides a, b then $a^2 + b^2 = c^2$.*
>
> **C:** *If the sides of a triangle a, b, c satisfy $a^2 + b^2 = c^2$ then the triangle is right-angled and c is its hypotenuse.*

The converse is different in content from the original theorem. In the above example it happens that both theorems are true. The next example is different.

> **E:** *If an integer is divisible by six then it is an even number.*
>
> **S:** *If an integer is even then it is divisible by six.*

Clearly **S** says something quite different from **E**; moreover **S** is false while **E** is true.

Summary: A piece of mathematical knowledge is usually stated in the form of a theorem, which has a hypothesis and a conclusion. If these are interchanged a new theorem is obtained. It is wrong to assume that the converse holds simply because the original theorem did. If one suspects that the converse is true and wishes to use it, then one must prove it.

Actually, the converses of many important theorems are true. In such cases instead of stating the two separate theorems we use the phrase 'if and only if', and thus combine both theorems into one statement. For example, **P** and **C** together read

> **PC:** *The triangle with sides a, b, c is right-angled with hypotenuse c if and only if $a^2 + b^2 = c^2$.*

The abbreviation 'iff' is sometimes used for 'if and only if'.

Exercises

Exercise 2.2.1   *State the following theorems in the form 'If ... then ...'.*

(1) *The diagonals of a rhombus are perpendicular.*
(2) *Every algebraic equation of degree one or higher has a solution.*
(3) *Every two positive integers have a greatest common divisor.*
(4) *Grandfather's knee pains whenever it rains.*

Exercise 2.2.2   *Give an example of a valid theorem whose converse is false.*

Exercise 2.2.3   *Give an example of a true theorem with a true converse. Use the phrase 'if and only if' to combine both theorems into one statement.*

2.3   Propositions and their combinations

What has been said in the last section is really only a vague introduction to certain ideas. We now consider them more carefully.

The typical structure of a theorem has already been mentioned: 'If something, then some other things'. But what are the 'things' in question? The first point is that they cannot be any phrase or sentence (in English) like 'Hooray!', 'How are you?', 'Go home!', 'The colour seven is tropical.' They must be capable of being true or false (but not both). We shall call such sentences *propositions*.

If A and B are propositions, then each may be true or false, and there are four possible cases in relation to the truth of both (see Table 2.2).

Table 2.2   Truth and falsity of propositions

I. A is true, and so is B;
II. A is true, but B is not;
III. A is not true, but B is;
IV. A is not true, and neither is B.

Consider the following example.

A: John, the engineering student, achieved top grades.

B: John, the engineering student, has passed the year.

In this example case II cannot occur. If A and B are propositions, and case II does not occur, we say that A *implies* B, or that B *follows from* A, and we denote this by $A \Rightarrow B$. Such a relation between two propositions is called an *implication*. We may express the implication $A \Rightarrow B$ in other words by saying *If A, then B*; here A is the hypothesis and B is the conclusion. The reader will remember from Section 2.2 that when $A \Rightarrow B$ then B need not imply A. In our example, if John has all top grades he passes (so $A \Rightarrow B$), but if he passes he need not have all top grades, so $B \Rightarrow A$ is not true.

By forming an implication $A \Rightarrow B$ we combine two propositions A and B into a new one; *if A then B*. This new proposition is true if case II in Table 2.2 does not occur.

After having made precise the meaning of the implication $A \Rightarrow B$, it is time to reflect and realize that in ordinary life many people do not understand implication the way mathematicians do. For example, our interpretation of $A \Rightarrow B$ means that for a false proposition A the implication $A \Rightarrow B$ is true no matter what proposition B is. This is clear, since if A is a false proposition and B is any proposition, the case II does not occur (neither does I). Consequently, *if 3 is less than 1 then all numbers are equal* is a true implication, and so is *if 3 is less than 1 then not all numbers are equal*.

Implication is an example of combining two propositions into a new one. There are many ways of forming new propositions. The *negation* of a proposition C is a proposition which is true if C is false and false if C is true. The negation of C is usually denoted by *not C* or $\neg C$. For example, if D is the proposition *the number six is even* then $\neg D$ is *the number six is not even*. The success of many proofs hinges upon a correct formulation of the proposition $\neg C$ corresponding to a given C. For instance, if C means *every blonde girl has a handsome suitor* then $\neg C$ means *it is not true that every blonde girl has a handsome suitor* or, in other words, *there is at least one blonde girl who either has no suitor or her suitor is not handsome*.

Given two propositions A and B we can form a new proposition A *and* B. For example, if A is *16 is even* and B is *16 is positive* then A *and* B means *16 is even and positive*. We shall agree that the proposition A *and* B is true if *both* propositions A and B are true. We denote the proposition A *and* B by $A \wedge B$.

The last combination of two propositions A and B we consider is A *or*

B. It is taken to mean that at least one of the propositions A and B holds, that is to say case IV in Table 2.2 does not occur. Instead of A *or* B we may sometimes say *either A or B*. We denote the proposition A *or* B by $A \vee B$.

Again, in ordinary life, some people may understand the proposition A *or* B differently.[4] (They would say 'either A or B or both' to cover the meaning of our 'A or B'.) If A is *three is greater than zero* and B is *three is equal to zero* then $A \vee B$ is *three is greater than or equal to zero* and is a true proposition. Clearly, in this example, case II occurs, case IV does not occur, and A *or* B is true (by our convention).

If $A \Rightarrow B$ and $B \Rightarrow A$ we say that A and B are *equivalent* and denote this by writing $A \Leftrightarrow B$. Using the terminology from Section 2.2 we say that A holds if and only if B holds. The relation $A \Leftrightarrow B$ is called *equivalence*.

The truth of the implication $A \Rightarrow B$ is often proved by showing that $\neg B \Rightarrow \neg A$ since both implications mean that case II of Table 2.2 does not occur. Similarly, the equivalence $A \Leftrightarrow B$ is often proved by proving both $A \Rightarrow B$ and $\neg A \Rightarrow \neg B$.

In this book we shall use the signs $\Rightarrow$ or $\Leftrightarrow$ for concise and convenient recordings of many theorems and mathematical statements. For instance we can state the theorem from Section 2.2 like this:
For real numbers x, y, z the implication

$$(xy = xz \text{ and } x \neq 0) \Rightarrow y = z \tag{2.3}$$

holds (that is, the implication is true). Another example of this kind is

$$x > 1 \Rightarrow x > 0. \tag{2.4}$$

This certainly looks like a true implication.

Unfortunately, there is a snag in (2.4). Strictly speaking $x > 1$ is not a proposition (in terms of our earlier definition); this is because it is not possible to decide whether or not $x > 1$ is true unless we know what x actually is. In order to circumvent this difficulty we specify the precise meaning of (2.4). We shall say that (2.4) (or a similar 'implication', for example, (2.3)) holds (is true) if it becomes a true implication after substitution of an arbitrary real number into (2.4) (or numbers x, y, z into (2.3)). Sometimes in statements like (2.3) or (2.4) we shall substitute only natural

---

[4] As in the question 'Would you like tea or coffee?' The implication here is '...tea or coffee but not both'. In mathematical logic this is known as the *exclusive or*, but will not be used in this book.

numbers or elements of a certain set, but it will be clear from the context what substitutions are envisaged.

Now we are in a better position to appreciate the convention that

$$\text{false } A \Rightarrow \text{ any } B.$$

If we substitute either $x = 1/2$ or $x = -1$ into (2.4) then the hypothesis is false and the conclusion is true in the first case and false in the second. The implication, however, stays always true.

Exercises

Exercise 2.3.1   *Let A be 2 is odd and B be 3 is odd. Decide whether or not the following implications hold.*

(1) $A \Rightarrow B$;
(2) $\neg A \Rightarrow \neg B$;
(3) $A \Rightarrow \neg B$;
(4) $B \Rightarrow A$;

Exercise 2.3.2   *In implication (2.3) in the text substitute*

(1) $x = 1$, $y = z = 2$;
(2) $x = 2$, $y = 2$, $z = 3$;
(3) $x = 0$, $y = 3$, $z = 9$;

and verify that the resulting implication is true in each case.

(i) **Exercise 2.3.3**   *Prove that $(A \Leftrightarrow B)$ if and only if $(B \Leftrightarrow A)$.*

Exercise 2.3.4   *Prove that for positive numbers a and b, $(a \leq b) \Leftrightarrow (a^2 \leq b^2)$ by showing $(a \leq b) \Rightarrow (a^2 \leq b^2)$ and $\neg(a \leq b) \Rightarrow \neg(a^2 \leq b^2)$.* (Inequalities are systematically treated in Chapter 4, Section 4.2.)

Exercise 2.3.5   *Explain why the word 'implication' appears in quotation marks in one place on page 49.*

Exercise 2.3.6   *Verify that the following implications are correct.*

(1) $A \Rightarrow (A \vee B)$;
(2) $\neg(A \wedge B) \Leftrightarrow (\neg A) \vee (\neg B)$;
(3) $[(A \Leftrightarrow B) \wedge (B \Leftrightarrow C)] \Rightarrow (A \Leftrightarrow C)$;

Exercise 2.3.7   *If M, N are sets prove*

(1) $(M \setminus N) \cup N = N \Leftrightarrow M \subset N$;

(2) $M \cup N \subset N \Leftrightarrow M \subset N$;

(3) $(M \setminus N) \cup (N \setminus M) = \emptyset \Leftrightarrow M = N$.

Exercise 2.3.8   *If A and B are propositions state the negation of $A \vee B$ and $A \Rightarrow B$. (Example: the negation of $A \wedge B$ is $\neg A \vee \neg B$.)*

Exercise 2.3.9   *Decide which of the following propositions are true and which are false. Also state, in words, the negations of these propositions.*

(1) *There is an integer x such that for all integers y the equation $xy = y$ holds.*

(2) *For all integers y there is an integer x such that the equation $xy = y$ holds.*

(3) *For all integers y there is an integer x such that the equation $xy = y^2$ holds.*

(4) *There is an integer x such that for all integers y the equation $xy = y^2$ holds.*

2.4   Indirect proof

If $A \Rightarrow B$ and B is false then only case IV in Table 2.2 is possible (since II is ruled out by the truth of $A \Rightarrow B$) and consequently A is false. If someone says 'If I were rich I would buy a new house', there is a clear implication that the person is not rich. By the same reasoning, if $\neg A$ implies something false then $\neg A$ is false, that is, A is true. This is used in proofs, since quite often the simplest way of proving that proposition A holds is to show that $\neg A$ implies something manifestly wrong. This is known as *indirect proof*. In philosophical debate the method of indirect proof is known as *reductio ad absurdum*. A defence lawyer can show the innocence of the accused by showing that the assumption that the accused committed the crime leads to the inescapable conclusion that the accused was at two distant places at the same time—obviously an untenable proposition.

The next example is a classic example of an indirect proof and goes back to the ancient Greeks. The argument appears in one of Euclid's books (written approximately 300 B.C.). We briefly summarise the prerequisites for the proof. A *prime* is an integer greater than or equal to 2 which has no divisors except 1 and itself. A theorem from elementary number theory says that every positive integer greater than or equal to 2 is divisible by some prime (possibly itself).

Now we prove: *There are infinitely many primes.*

For an indirect proof we assume the contrary. Then we can list all the primes as p_1, p_2, p_3, ..., p_n. Now consider the integer formed by multiplying these together and adding 1: $N = p_1 \cdot p_2 \cdot p_3 \cdots p_n + 1$; this is divisible by some prime, that is, one of the p_i. Consequently 1 is divisible by p_i since $1 = N - p_1 \cdot p_2 \cdot p_3 \cdots p_n$ and both N and $p_1 \cdot p_2 \cdot p_3 \cdots p_n$ are divisible by p_i (to make this rigorous we would also need to show that if a and b are divisible by p_i, then so is $a - b$). However, the proposition that 1 is divisible by a prime (greater than 1) is obviously absurd. We have reached our goal: the theorem in italics has been proved.

We have just given a simple example of a short indirect proof. In mathematics, however, we often encounter situations where an indirect proof requires a number of fairly difficult steps and students are apt to start worrying in the middle of the proof because things look a little strange. If this happens they should realise that in an indirect proof 'the stranger, the better' for as soon as we reach real absurdity we have attained our goal (of course, assuming that we didn't make any errors).

Exercises

Exercise 2.4.1   *Prove that each of the following numbers is not rational.*

(1) $\sqrt{2}$;
(2) $\sqrt{3}$;
(3) $\sqrt{2} + \sqrt{3}$;
(4) $\log_{10} 2$;
(5) $\sqrt[3]{4}$;
(6) $4\sqrt{2}$.

Exercise 2.4.2   *Prove that every prime other than 2 is odd.*

Exercise 2.4.3   *Prove that for every prime $p \geq 5$ there exists a positive integer n such that either $p = 6n - 1$ or $p = 6n + 1$.*

Exercise 2.4.4   *Prove that if $a \in \mathbb{Q}$, $b \in \mathbb{Q}$, $b \neq 0$ then $a + b\sqrt{2}$ is irrational.*

2.5   Comments and supplements

The concept of a set was not introduced precisely in Section 2.1, it was only illustrated by examples. Now we give an example showing the inadequacy of such a simple intuitive approach.

If x is an object then by $\{x\}$ we denote the set consisting of one element x. A set having only one element is called a *singleton*. If X is a set we may ask whether or not $\{X\} \in X$. Clearly, if $X = \emptyset$ then $\{X\} \notin X$ because nothing is in X. On the other hand it seems that the family F of all singletons has the property that $\{F\} \in F$ because $\{F\}$ is a singleton. Now define a set Y by $Y = \{\,\{X\}; \{X\} \notin X\}$. This seems innocent enough. Unfortunately we have a paradox on our hands. Neither $\{Y\} \in Y$ nor $\{Y\} \notin Y$. Indeed, if $\{Y\} \in Y$ then $\{Y\} \notin Y$ by the defining property of the set Y. On the other hand, if $\{Y\} \notin Y$ then again, by the definition of Y, we have $\{Y\} \in Y$. In either case we have both $\{Y\} \in Y$ and $\{Y\} \notin Y$, which is absurd.

This example is a slight modification of a famous paradox, the so-called *Russell's Paradox*, named after its inventor, the British mathematician, logician and philosopher Bertrand Russell.

Any reasonable theory must be free of any paradox. How are we to get rid of the paradoxes of set theory? (There are others besides Russell's.)

Perhaps we may say dogmatically that constructions like the set of all singletons or the set of all sets are not legitimate in set theory. When using a property defining a set it is safe to consider only subsets of a set which we know (like the reals) or which can be constructed from known sets (like the cartesian product $\mathbb{R} \times \mathbb{R}$). Any mathematical discipline should be based on an already established branch of mathematics or founded axiomatically. Since set theory is most basic we have no choice but to build it axiomatically.

An axiomatically based discipline starts with a few basic propositions — axioms—which are not questioned for truth or validity, and develops from these axioms by logical means. Some may feel that axioms must be simple and self-evident. However, what is obvious and simple to a mathematician who has absorbed a great deal of knowledge accumulated over more than two millennia need not be obvious or simple to everybody. The question of the truth of axioms in the modern interpretation of axiomatic theories simply does not arise. Firstly, circular definition must be avoided. Consequently the primitive terms in the axiomatic system cannot be defined. If we set up an axiomatic system, say for set theory, then we simply agree to call the objects which satisfy the axioms sets. It can be objected that with such an interpretation axiomatic mathematics becomes a meaningless game. Mathematics, because of its applicability to other sciences, engineering and its prominence in the history of human thought, is certainly not meaningless. It may be a puzzle why mathematics is at all applicable but this is a rather philosophical question which we will leave aside.[5] To leave the primitive terms undefined has its advantage in that the undefined terms are capable of having several meanings. Imagine that we have proved some theorems on divisibility of integers as logical consequences of a few basic axioms. Then these theorems are applicable to any objects which satisfy the axioms. This actually happens with objects called *Gaussian* integers, and also polynomials with real or rational coefficients. We show this in Subsection 2.5.1 An example of an axiomatic theory which does not require any prerequisites is presented in Eves (1981, Lecture Thirty Five).

The desire for an easy verification that a certain set of objects satisfies a list of axioms leads to the requirement of having as few axioms as possible— within reasonable limits: we don't want to have so few axioms that the system is too weak to do any significant amount of mathematics. If none of the axioms (or parts of the axioms) is deducible from any of the others, the system of axioms is called *independent.*

A fundamentally important requirement of any system of axioms is its *consistency.* A system of axioms is consistent if it is impossible to prove from it two propositions contradicting one another, for example, A and $\neg A$. An inconsistent system is obviously useless for any serious study. Most mathematicians now agree that in principle the only requirement a set of axioms must satisfy is consistency.

An interesting application of sets and axiomatic ideas was developed by the economist Kenneth Arrow. He was interested in the way the preferences of individual members in a society could be combined to produce an overall preference for that society. For example, how do the individual preferences of people for breakfast cereal or clothes styles relate to the cereals or clothes styles most in demand in society. This is of considerable interest to economists and manufacturers. Arrow became famous, and received a Nobel Prize in Economics in 1972, in part for this work. It was realised that his work had much wider implications than just economics—it applies

---

[5]The interested reader is referred to a famous paper by Eugene P. Wigner: The Unreasonable Effectiveness of Mathematics in the Natural Sciences, *Communications in Pure and Applied Mathematics*, vol. XIII (1960), pp. 1–14; or the book *Mathematics and Science*, edited by Ronald E. Mickens, World Scientific Publishing, 1990. The contributors to this book discuss aspects of Wigner's paper.

to any area in which individual preferences need to be combined in some way. In elections, voters (usually) have an order of preference for the candidates. The system of axioms set out by Arrow can be interpreted as a bare minimum statement for a voting system to be regarded as 'democratic'. Arrow proved that the axioms he set out were, in fact, inconsistent, implying that the goal of finding a truly 'democratic' voting system is impossible. His work, which is quite easy to follow, is set out in Arrow (1963). Other people have tried, without success, to find other axioms for a 'democratic' voting system. Some of these, with proofs of inconsistency, can be found in Kelly (1978).

The first scientific discipline which was axiomatized was geometry — about 300 B.C. by Euclid. The geometry he created (or, rather, put on an axiomatic basis) is now called Euclidean geometry. The critical study of the foundations of Euclidean geometry was undertaken towards the end of the 19th century, shortly after discoveries of other geometries. It was found that Euclid used tacitly some propositions unwarranted by his axioms. Euclid's axioms were then augmented and Euclidean geometry perfected. Since their introduction no-one seriously doubted the consistency of the axioms of Euclidean geometry. It was, however, only later, in the 20th century that consistency was proved by the American mathematician of Polish origin, A. Tarski, and then only for a section of Euclidean geometry called 'elementary Euclidean geometry'.

The history of Euclidean geometry seemingly offers a way out of the problems besetting set theory. Set theory ought to be developed from a system of axioms which prevent the occurrence of the paradoxes. This was actually done in the first half of the 20th century. Among the axioms of set theory is one which prohibits constructions like the one in Russell's paradox. Then we should aim at proving consistency for the axioms of set theory. Unfortunately, such efforts are doomed to failure. An Austrian mathematician, K. Gödel, proved that the consistency of the generally accepted axioms of set theory cannot be proved by mathematical means.[6] This creates a problem, in that we do not have an entirely satisfactory philosophy for the foundation of mathematics. The foundation of mathematics is not as rock solid as the layperson believes.

We may add that there is one somewhat controversial axiom in set

---

[6] Actually Gödel proved more than this. He showed that if the axioms of set theory are consistent, then it is possible to formulate a proposition A such that neither A nor $\neg A$ can be proved by mathematical means. 'The axioms are consistent' is one such proposition. These propositions are called *undecidable*.

theory — the axiom of choice mentioned in Chapter 9. It leads to rather surprising results, some of which are contradicted by our intuition.[7] When the foundations of mathematics are studied there is also a need for an analysis of the logical means used in deriving conclusions. Mathematicians are more divided on which arguments are permissible and which are not, rather than which system of axioms should be used. Not all mathematicians are prepared, for example, to accept an unrestricted use of indirect proof.

Set theoretical language plays only an auxiliary, but convenient, role in the mathematics considered here: much of the mathematics in this book was in existence long before set theory was born.

Here we leave axiomatics and set theory. The interested reader is referred to Halmos (1974).

2.5.1 *Divisibility: An example of an axiomatic theory*

This subsection requires more mathematical maturity than the rest of this chapter and uses some concepts explained only later in the book. Skipping this subsection will not affect understanding of the rest of the book.

A set R is called a ring if, for every two elements $x \in R$ and $y \in R$ there is associated a sum $x + y \in R$ and a product $xy \in R$ such that the axioms in Table 2.3 are satisfied.

The set of even integers as well as the set of numbers of the form $9m+15n$ with $m, n \in \mathbb{Z}$ form a ring. The sum and product of even integers is even. The axioms from Table 2.3 are obviously satisfied for even integers, as they are for numbers of the form $9m + 15n$.

A ring R is called Euclidean if it has the following properties:

(1) if $x \neq 0$ and $y \neq 0$ then $xy \neq 0$;
(2) for every element $r \in R$ with $r \neq 0$, there is a positive integer $N(r)$ such that

    (a) $N(xy) \geq N(x)$ if $x \neq 0$ and $y \neq 0$;
    (b) for every two elements x, y with $y \neq 0$ there are q and r in R such that

$$x = qy + r \tag{2.5}$$

    and either $r = 0$ or $N(r) < N(y)$.

---

[7]There are no problems about independence and consistency of the axiom of choice, however. Gödel showed that the axiom of choice is consistent with the other axioms of set theory, and Cohen proved that it is independent of the other axioms.

Table 2.3 Ring axioms

$A_1 : x + y = y + x$ $\qquad$ $M_1 : xy = yx$

$A_2 : x + (y + z) = (x + y) + z$ $\qquad$ $M_2 : x(yz) = (xy)z$

$A_3 :$ There is an element $0 \in R$
such that $0 + x = x$ for all
x in R

$A_4 :$ For every element $x \in R$ there exists an element $(-x) \in R$ such that $(-x) + x = 0$.

$$D : x(y + z) = xy + xz.$$

The integers form a Euclidean ring: q, r are the quotient and the remainder, respectively, by division of x by y; N is the absolute value, $N(r) = |r|$. Polynomials with rational coefficients also form a Euclidean ring. q and r can be found by the long division algorithm (see Section 6.3). For N we take the degree of the polynomial.[8] In a ring[9] x is *divisible* by y if there is $q \in R$ such that $x = qy$. If x is divisible by y we say that y is a *divisor* of x.[10] The *greatest common divisor* d of two elements x, y is defined as a common divisor (that is $x = md$ and $y = nd$ for some m, n in the ring) which is divisible by any other common divisor of x, y. For example, in $\mathbb{Z}$ the numbers -4 and 4 are the greatest common divisors of 8 and 12. If d is a divisor and D the greatest common divisor of two elements in $\mathbb{Z}$ then $D = md$ for some integer m and consequently $|D| \geq |d|$. In $\mathbb{Z}$ the greatest common divisor of two elements has the largest absolute value among all common divisors. The set of all greatest common divisors of a and b is denoted by $\gcd(a, b)$. It is common but perhaps a little confusing to write $d = \gcd(a, b)$ instead of $d \in \gcd(a, b)$. If $x = 0$ or $y = 0$ then the greatest

---

[8] Another example of a Euclidean ring is the Gaussian integers: these are numbers of the form $a + bi$ with a, $b \in \mathbb{Z}$ and $i^2 = -1$. Here it is less obvious what q, r in (2.5) should be. Let $x = a + ib$, $y = c + id$, $x/y = \alpha + i\beta$. Denote by q_1 and q_2 the nearest integer to α and β, respectively. Then $|\alpha - q_1| \leq 1/2$ and $|\beta - q_2| \leq 1/2$. Set $q = q_1 + iq_2$ and $r = x - qy$ with $r = r_1 + ir_2$. We define $N(x) = |x|^2$, that is $N(a + ib) = a^2 + b^2$ and have $N(r) = |r|^2 = |y(q_1 - \alpha) + y(q_2 - \beta)|^2 \leq |y|^2(\frac{1}{4} + \frac{1}{4}) < |y|^2 = N(y)$. Consequently the Gaussian integers form a Euclidean ring.

[9] Not necessarily a Euclidean ring.

[10] 0 is divisible by any element of R and no element distinct from 0 is divisible by 0.

common divisor of x, y is 0. Otherwise two greatest common divisors differ by at most a factor which is a divisor of 1. Indeed, if d, D are two greatest common divisors of x and y then $d = mD$ and $D = nd$ for some m, $n \in \boldsymbol{R}$. Consequently $mn = 1$ and m, n are divisors of 1. In $\mathbb{Z}$ the divisors of 1 are 1 and -1. In the ring of polynomials with rational coefficients the divisors of 1 are all rational numbers distinct from 0.

Theorem 2.1　　*Two elements x, y in a Euclidean ring always have a greatest common divisor.*

Proof.　　The Theorem is obvious if x or y is zero. We may assume for the rest of the proof that $x \neq 0$ and $y \neq 0$. Consider the sets

$$\boldsymbol{M} = \{mx + ny : m \in \boldsymbol{R},\ n \in \boldsymbol{R}\},$$
$$\boldsymbol{P} = \{N(w) : w \in \boldsymbol{M},\ w \neq 0\}.$$

$\boldsymbol{P}$ is a nonempty set of positive integers and has its smallest element $N(d)$ (see Theorem 4.10). Let $w \in \boldsymbol{R}$ be arbitrary. Since $\boldsymbol{R}$ is Euclidean we have $w = dq + r$ and either $r = 0$ or $N(r) < N(d)$. However, the last inequality is impossible since $N(d)$ is the smallest element of $\boldsymbol{P}$. Consequently $r = 0$ and d is a divisor of x and y because both of these elements are in $\boldsymbol{M}$. Every common divisor of x and y divides any element of $\boldsymbol{M}$, and in particular d. Hence d is the greatest common divisor. □

The proof just given is a good example of an existence proof characteristic of modern mathematics. The existence of a mathematical object is established without any laborious process of actually constructing the object. We now show a constructive and effective way of finding the greatest common divisor. This can be done by the so-called Euclid's algorithm. If $y \neq 0$ then we can find q_0 and r_1 such that

$$x = yq_0 + r_1.$$

If $r_1 = 0$ then $\gcd(x, y) = y$. If $r_1 \neq 0$ then again

$$y = r_1 q_1 + r_2.$$

If $r_2 = 0$ then it is easy to see that $\gcd(x, y) = r_1$. If $r_2 \neq 0$ we continue the process and obtain elements $r_1, r_2, r_3 \ldots$ and the corresponding natural numbers $N(r_1) > N(r_2) > N(r_3) > \ldots$. The process must end after at most $N(r_1)$ steps when we obtain

$$r_{n-2} = r_{n-1}q_{n-1} + r_n,$$
$$r_{n-1} = q_n r_n.$$

It can be shown by mathematical induction (see Chapter 5) that $r_n = \gcd(x, y)$. Let us illustrate the Euclid algorithm by a simple example in $\mathbb{Z}$. Start with $x = 133$ and $y = 119$.

$$133 = 119 \times 1 + 14,$$
$$119 = 14 \times 8 + 7,$$
$$14 = 7 \times 2.$$
$$\gcd(133, 119) = 7$$

Maple can be employed to do the computations in the Euclid algorithm. The command for finding the remainder of division of integers x by y, as we know, is irem(x,y). Let us find $\gcd(215441, 149017)$ by using *Maple*.[11]

```
>  irem(215441,149017);
```

$$66424$$

```
>  irem(149017,66424);
```

$$16169$$

```
>  # We can now use the same Maple command repeatedly
>  # to evaluate the successive remainders,
>  # and use these as the divisor in the next step
>  irem(%%,%);
```

$$1748$$

```
>  irem(%%,%);
```

---

[11]The command irem(%%,%); below need not be retyped, you can use the menu of **Edit** to copy and paste.

437

```
>   irem(%%,%);
```

0

Using *Maple* and the Euclid algorithm we found that 437 is the greatest common divisor of 215441 and 149017. One can do even better: *Maple* has an inbuilt facility for determining the greatest common divisor.

The command is `igcd` if we are working in $\mathbb{Z}$ and `gcd` if we are working in the ring of polynomials with rational coefficients. For example

---

```
>   igcd(215441,149017);
```

437

```
>   gcd(x^2-5*x+6,x^2-4*x+4);
```

$x - 2$

```
>   gcd(x^(10)+1,x^(20)+1);
```

1

Exercises

Exercise 2.5.1   *Define the greatest common divisor of n elements. Show that, in a Euclidean ring, n elements always have a greatest common divisor.*

Exercise 2.5.2 $\mathbf{R}$ *is a ring, $w \in \mathbf{R}$, $z \in \mathbf{R}$, $h \in \mathbf{R}$ with $\gcd(w, h) = 1$. Show that $\gcd(w, z) = \gcd(w + hz, z)$. Prove this and apply it to finding a greatest common divisor of the following polynomials:*

(1) $x^{100} + ax^2 + bx + 1, \quad ax^2 + bx + 1$;
(2) $x^6 + x^5 + 3x^4 - 6x^3 - 8x^2 - 2x - 3, \quad x^2 - 2x - 3$.

Exercise 2.5.3   *Use* Maple *to find the greatest common divisors of:*

(1) 108, 144;

(2) 1234567, 7654321;

(3) $5x^8 - 2x^7 + x^6 + 5x^5 + 2x^4 - x^3 + x^2 - x - 1,\ x^5 - x^4 + x^2 - 1.$

Chapter 3

Functions

In this chapter we introduce relations, functions and various notations connected with functions, and study some basic concepts intimately related to functions.

3.1 Relations

In mathematics it is customary to define new concepts by using set theory. To say that two things are related is really the same as saying that the ordered pair (a, b) has some property. This, in turn, can be expressed by saying that the pair (a, b) belongs to some set. We define:

Definition 3.1 (Relation)   A relation is a set of ordered pairs.

This means that if A and B are sets then a relation is a *part* of the cartesian product $A \times B$. If a set R is a relation and $(a, b) \in R$ then the elements a and b are related; we also denote this by writing aRb. For example, if related means that the first person is the father of the second person then this relation consists of all pairs of the form (father, daughter) or (father, son). Another example of a relation is the set $C = \{(1, 2), (2, 3), \ldots, (11, 12), (12, 1)\}$. This relation can be interpreted by saying that m and n are related (mCn) if n is the hour immediately following m on the face of a 12 hour clock.

We define the *domain of the relation R* to be the set of all first elements of pairs in R. Denoting by $\operatorname{dom} R$ the domain of R we have

$$\operatorname{dom} R = \{a \,; \ (a, b) \in R\}\,.$$

The *range of R* is denoted by $\operatorname{rg} R$ and is the set of all second elements of pairs in R, so

$$\operatorname{rg} R = \{b;\ (a,\, b) \in R\}\,.$$

For the relation father-offspring the domain is the set of fathers, the range is the set of daughters and sons. For the 'clock' relation C clearly $\operatorname{dom} C = \operatorname{rg} C = \{1,\, 2,\, \ldots,\, 12\}$.

We can often draw the graph of the relation R as in the diagram below. The graph of R consists of all points P with first coordinate x and second coordinate y such that $P \equiv (x,\, y) \in R$. The domain and the range of a relation is schematically indicated in Figure 3.1.

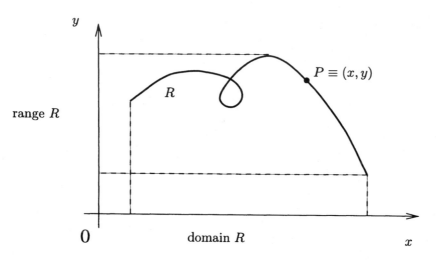

Fig. 3.1   Graph of a relation

The graphs of

(A) $R_1 = \left\{(x,\, y);\ \dfrac{x^2}{4} + y^2 = 1\right\}$,

(B) $R_2 = \left\{(x,\, y);\ \dfrac{x^2}{4} + y^2 \le 1\right\}$, and

(C) $R_3 = \{(x,\, y);\ y > x^2\}$

are shown in Figures 3.2 to 3.4, respectively. The graph of a relation can cover a whole area, as in relations R_2 and R_3. Note that the boundary is included in R_2 but not included in R_3.

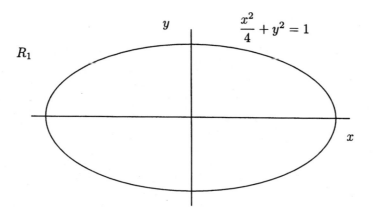

Fig. 3.2   Graph of R_1

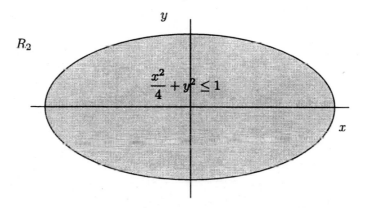

Fig. 3.3   Graph of R_2

The graph of a relation or a geometric picture generally is a telling guide; it helps understanding, and often motivates a proof of a theorem or construction of a counterexample. We shall use geometry freely to motivate or illuminate our theory. For this we presume that the reader is familiar with elementary geometry, including basic analytical geometry. However, we wish to emphasise that our theory itself is independent of any geometry and can stand on its own feet without any support of graphical illustrations. Our ability to draw a graph is restricted by the unavoidable imperfection of the

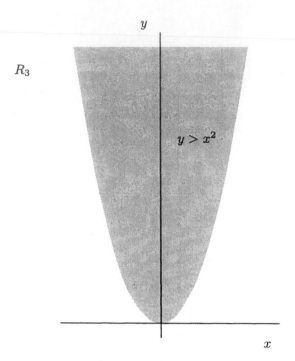

R_3

$y > x^2$

Fig. 3.4   Graph of R_3

drawing instruments. It is perhaps not clear from Figure 3.4 that the points lying *on* the parabola $y = x^2$ do not belong to the graph of R_3. The next example is far more serious. Let $R_4 = \{(x, y);$ either x or y is rational$\}$. Here the imperfection of the drawing instruments makes it impossible to draw the graph of R_4. Either we will blacken the whole plane or the graph of R_4 will be very incomplete. Obviously R_4 cannot be graphed.

We can define a number of properties of relations.

- A relation R is said to be *reflexive* if xRx for every $x \in \operatorname{dom} R$.
- A relation R is said to be *symmetric* if xRy implies yRx for all $x \in \operatorname{dom} R$ and $y \in \operatorname{rg} R$.
- A relation R is said to be *transitive* if xRy and yRz imply xRz.
- A relation which is reflexive, symmetric and transitive is called an *equivalence relation*, or simply an *equivalence*.

Example 3.1   Let $M_k = \{(x, y); \ x \in \mathbb{Z}, \ y \in \mathbb{Z}, \ x - y$ divisible by $k\}$ for

$k \in \mathbb{N}$. Then the relation M_k is reflexive, symmetric and transitive. (Prove this!) The equivalence M_k is often called congruence modulo k, and the notation $x \equiv y \bmod k$ (read x congruent to y modulo k) is used instead of $x M_k y$.

The 'clock' relation is not symmetric, nor reflexive, nor transitive. Examples of relations which are symmetric, reflexive or transitive are given in Exercises 3.1.1–3.1.6.

If R is an equivalence relation on A and $a \in A$ then the set $K = \{b; \, aRb\}$ is called an *equivalence class* and a is a *representative* of K. Note that K depends on a; we can emphasise this by writing K_a instead of K. Two equivalence classes K_a and K_b are either identical or disjoint. For congruence modulo 3 it is easy to see that $K_1 = K_4$ and $K_1 \cap K_2 = \emptyset$.

Remark 3.1 Sometimes all elements in an equivalence class are identified, in other words elements in an equivalence class are considered equal. If this sounds too abstract, it is not. We are used from primary school to equations like $\dfrac{2}{7} = \dfrac{4}{14} = \dfrac{6}{21} = \cdots$. This is based on identification of elements in an equivalence class. A common fraction p/q is just a pair of integers, say (p, q). If two pairs of integers (p, q) and (r, s) with $q \neq 0$, $s \neq 0$ are defined to be equivalent if $ps - qr = 0$, then the pairs $(2, 7) \equiv (4, 14) \equiv (3, 21)$; this is the above equation of common fractions. Similarly, in arithmetic mod 3 we write $1 = 4 = 7 = \cdots$.

Exercises

Exercise 3.1.1 *Graph the following relations from $\mathbb{R}$ into $\mathbb{R}$ and decide which relations are (a) symmetric, (b) reflexive, or (c) transitive.*

$$S_1 = \{(x, y); \ x \leq y\};$$
$$S_2 = \{(x, y); \ x < y\};$$
$$S_3 = \{(x, y); \ x^2 + y^2 = 1\};$$
$$S_4 = \{(x, y); \ x^2 + y^2 \leq 1\};$$
$$S_5 = \{(x, y); \ x^2 + 2x = y^2 + 2y\}.$$

If the relation is symmetric or reflexive, how is this shown by the graph?

Exercise 3.1.2 *Let* $\operatorname{dom} R = \{1, 2, 3, \ldots, 12\}$, *and* aRb *if* a *and* b *differ by* 6. *Write this relation as a set of pairs, and describe the relation using the face of a clock.*

Exercise 3.1.3   *Define separate relations which are:*

(1) *reflexive, but not symmetric or transitive;*
(2) *symmetric, but not reflexive or transitive;*
(3) *transitive, but not symmetric or reflexive.*

(i) **Exercise 3.1.4**   *Let A be a set and B_l with $l \in L$ a family of disjoint sets such that $A = \bigcup_{l \in L} B_l$. Define a relation on A by declaring aRb if a and b lie in the same B_l. Show that this relation is an equivalence.*

Exercise 3.1.5   *Define a relation S for subsets of $\mathbb{R}$ by $S = \{(A, B);\ A \subset \mathbb{R},\ B \subset \mathbb{R},\ (A \setminus B) \cup (B \setminus A)$ is finite $\}$. Prove that S is an equivalence relation, and the family of finite subsets of $\mathbb{R}$ is one equivalence class.*

(i) **Exercise 3.1.6**   *Define a relation R with $\operatorname{dom} R = \mathbb{Z}$ and $\operatorname{rg} R = (\mathbb{Z} \setminus \{0\})$ as follows: $(m, n)R(r, s)$ if $ms = nr$. Prove that R is an equivalence. This equivalence relation is a starting point for defining rational numbers in terms of integers. The fraction $\dfrac{a}{b}$ simply denotes the equivalence class containing the pair (a, b); clearly $\dfrac{2}{3} = \dfrac{4}{6} = \dfrac{-2}{-3} = \dfrac{-4}{-6} = \cdots$.*

3.2   Functions

The concept of a function is the mathematical abstraction of the notion of dependence in ordinary life. The price of a commodity depends on its quality, the average daily temperature depends on the date, the area of a square depends on the length of its side. Using mathematical terminology we say, for example, that the area of a square is a function of the length of its side. A reasonably good description is to say that a function is a rule which associates with every element of a set, called the *domain of a function,* a uniquely determined element y called the *value of the function.* In one of the examples above the rule associates with each date a number equal to the average temperature on that date. However some doubts may arise as to what is meant by the word 'rule'. In this book we define functions using set-theoretical language.

Definition 3.2 (Function)   A function is a relation F such that

$$\big[(x,\, y) \in F \text{ and } (x,\, z) \in F\big] \Rightarrow y = z$$

for all $x \in \operatorname{dom} F$ and $y,\, z \in \operatorname{rg} F$.

A function, then, is a set of ordered pairs such that no distinct pairs have the same first element. If F is a function and $(x,\, y) \in F$ then y is the element associated with x by the rule for F. In the above example, if there are only two qualities of a merchandise, good and bad, for which the prices are \$20 and \$10, respectively, the function is simply $\{(good,\ \$20),\ (bad,\ \$10)\}$. Functions are usually denoted by letters $f,\, g,\, h, \ldots$ or $F,\, G,\, H, \ldots.$. Sometimes letters of the Greek alphabet such as $\phi,\, \psi,\, \ldots$ are also used. Suffices are employed too. For example, we may denote the function $\{(x,\, x/(1 + nx^2))\,;\ x \in \mathbb{R}\}$ by f_n.

Since a function is a relation it is automatically clear what is $\operatorname{dom} f$, $\operatorname{rg} f$ and the graph of f. Definition 3.2 ensures that every line parallel to the y-axis intersects the graph of f in at most one point. Not every relation is a function; the relation father$R$daughter is not a function because there is (at least one) father who has two daughters. The 'clock' relation C is a function because it does not contain two distinct pairs with the same first element. The relation R_1 from the preceding section is not a function because the line $x = 0$ intersects the graph of R_1 in two distinct points (see Figure 3.2).

If f is a function and $(x,\, y) \in f$ then y is denoted by $f(x)$ (read f of x). The symbol $f(x)$ is called the *function value* or the *value of the function* at x. The range of f is simply the set of all function values. It is important to make a clear distinction between f, which is a set of pairs, and $f(x)$, which is an element of $\operatorname{rg} f$.[1] It is possible to encounter in the literature expressions like 'the function x^2'. This is a drastic and undesirable abbreviation for $\{(x,\, x^2)\,;\ x \in \mathbb{R}\}$; undesirable because it confuses function with function value. We shall try to avoid such abuse of terminology.

Instead of writing $f = \{(x,\, y)\,;\ x \in D\}$ we shall often write

$$f : x \mapsto y,\ x \in D.$$

(Read: 'x goes to y' or 'f sends x to y'.)

---

[1] This can be illustrated by reference to using a computer or programmable calculator to evaluate some expression. The computer program is the function, while the output displayed when the program is run with a particular value of x as input is the function value. There is a major difference between the program and the output of the program.

The notation $f : x \mapsto f(x)$ is also used, for instance, $h : x \mapsto x/(1 + x)$. In a situation like this it is understood that the domain of f is the largest subset of $\mathbb{R}$ for which $f(x)$ makes sense. For h it is $\mathbb{R} \setminus \{-1\}$. If there is no need to name a function we may even write $x \mapsto y$ to describe a function, for example, $x \mapsto x/(1 + x)$.

The notation $f : A \mapsto B$ is used to indicate that dom $f = A$ and rg $f \subset B$. This notation is convenient in theoretical discussions, but it has the disadvantage that it does not explicitly describe the values of f. For instance, $f : \mathbb{R} \to \mathbb{P}$ means that f is defined and positive for all real x, but otherwise f can be arbitrary.

It is often convenient to display the function pairs in a table, particularly if we wish to draw the graph of the function. The table for the clock function C is given below.

Table 3.1   The clock function

x	1	2	3	4	5	6	7	8	9	10	11	12
$C(x)$	2	3	4	5	6	7	8	9	10	11	12	1

If dom f is infinite (or simply very large) we may fill in only a few values for x. Such an incomplete table for $h : x \mapsto x/(1 + x)$ is given in Table 3.2.

Table 3.2   The function $h : x \mapsto x/(1 + x)$

x	-3	-2	-3/2	-4/3	-1/2	-1/4	0	1	2
$h(x)$	3/2	2	3	4	-1	-1/3	0	1/2	2/3

The graph of a function usually conveys a good idea of what the function is like. To graph a relatively simple function (by hand) might be laborious, involving a considerable amount of calculation. This is where *Maple* is of great help. Not only does it carry out the required calculations, but it also has excellent graphical capabilities. We deal with graphs in *Maple* in Subsection 3.3.4. Figure 3.5 shows the graph of h obtained by *Maple*.

Two functions are equal if they are equal as sets, that is, if they consist of the same elements (ordered pairs). Equal functions have the same domain

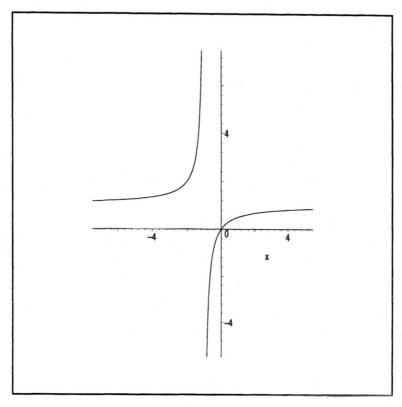

Fig. 3.5   Graph of $\dfrac{x}{1+x}$

and associate the same y with the same x. Of course, they then also have the same range.

Example 3.2   Denote by $\alpha(x)$ the area of a square whose side has length x. Using the arrow notation for functions we can write $\alpha : x \mapsto x^2$, $x \in \mathbb{P}$. Note that α is not the same as $s : x \mapsto x^2$ because s is tacitly defined on all of $\mathbb{R}$ (dom $s = \mathbb{R}$).

So far, the letter x has been used for the first element of a pair belonging to a function. Other letters may be used freely; for example for h above and α from Example 3.2 we could equally write:

$$h : t \mapsto t/(t+1);$$
$$h = \{(v,\, v/(v+1));\ v \in \mathbb{R} - \{-1\}\};$$
$$\alpha : y \mapsto y^2,\ y \in \mathbb{P};$$
$$\alpha = \{(s,\, s^2);\ s \in \mathbb{P}\}.$$

Example 3.3   The function $x \mapsto x$ is perhaps the simplest function. It is called the *identity* and we denote it by id. In other words id $: x \mapsto x$ or $\mathrm{id}(x) = x$. By id_A we denote the function $\{(x,\, x) : x \in A\}$. This notation is used by other authors too but it is by no means standard. It is perhaps a little strange that there is a standard notation for many functions (like sin, cos, log) but not for the very often occurring function id.

Example 3.4   For a set A the *characteristic function* $\mathbf{1}_A$ of A is defined by $\mathbf{1}_A(x) = 1$ if $x \in A$ and $\mathbf{1}_A(x) = 0$ if $x \notin A$. In terms of ordered pairs $\mathbf{1}_A = \{(x,\, 1);\ x \in A\} \cup \{(x,\, 0);\ x \notin A\}$. If we do not specify otherwise we shall assume automatically that $\mathrm{dom}\, \mathbf{1}_A = \mathbb{R}$. However, in mathematics one often encounters a situation in which there is some general set X, fixed during a discourse, like $\mathbb{R}$ in our case, and then it is understood that $\mathrm{dom}\, \mathbf{1}_A = X$.

Clearly, $\mathbf{1}_{\mathbb{R}}(x) = 1$ for every $x \in \mathbb{R}$. We can also write $\mathbf{1}_{\mathbb{R}} : x \mapsto 1$. The function $\mathbf{1}_{\mathbb{P}}$ is sometimes called the Heaviside function in honour of the British mathematician and physicist Oliver Heaviside, who used it when applying mathematics to solve problems in electrical circuit theory.

The function $\mathbf{1}_\emptyset$ is called the *zero function*; clearly $\mathbf{1}_\emptyset : x \mapsto 0$. It is customary to denote $\mathbf{1}_\emptyset$ by 0; however, one should realize that $\mathbf{1}_\emptyset$ and the number 0 are different objects. The context will (usually) make it clear when 0 is the number zero (see Axiom A_3 in Table 4.1 in Chapter 4) and when it is used for the zero function. There is an intimate connection between sets and characteristic functions: see Exercise 4.7.1 and the comments at the end of this chapter.

If f is a function then we define $f(A)$ and $f_{-1}(B)$ by

$$f(A) = \{y;\ y = f(x) \text{ and } x \in A\},$$
$$f_{-1}(B) = \{x;\ f(x) \in B\}.$$

Obviously $f(A) = \{y; (x,\, y) \in f \text{ and } x \in A\}$. The set $f(A)$ is called the *image* of A under f, and $f_{-1}(B)$ is called the *pre-image* of B. Some properties of the symbols $f(A)$ and $f_{-1}(B)$ are established in Exercise 3.2.5. Clearly $\mathrm{rg}\, f = f(\mathrm{dom}\, f)$ and $\mathrm{dom}\, f = f_{-1}(\mathrm{rg}\, f)$. If $s : x \mapsto x^2$ then

$s(\{-1, 1, 2\}) = \{1, 4\}$, $s_{-1}(\{1\}) = \{1, -1\}$, $s_{-1}(\{-1\}) = \emptyset$. There is some harmless ambiguity in this notation. The symbol $f(N)$ may denote the value of f at $N \in$ dom f or it may denote the image of N under f for a set N. Usually it is clear from the context what N is.

If $f : D \mapsto V$ and $A \subset D$ then the function $f|_A$ (read f restricted to A) is defined by

$$f|_A = \{\, (x, f(x)) : x \in A \,\}.$$

$f|_A$ is called the *restriction* of f to A. Obviously, always $f|_A \subset f$ and dom $f|_A = A$. Perhaps we should have written dom$(f|_A)$ but on occasions like this we shall omit the parentheses, since there is little fear of confusion about the meaning. The function α from Example 3.2 is a restriction of $s : x \mapsto x^2$ to $\mathbb{P}$. If g is a restriction of f to some set B, $g = f|_B$, then f is called the *extension* of g from B to D. The function s is the extension of α to $\mathbb{R}$. Given f and B, the function $f|_B$ is uniquely determined. However for a given g and D there may be several extensions of g from dom g to D. For instance, s and $h : x \mapsto x^2.\mathbf{1}_{\mathbb{P}}(x)$ are both extensions of α from $\mathbb{P}$ to $\mathbb{R}$.

It is a slight abuse of notation to denote by the same symbol a function and its restriction or a function and its extension. However we may resort to it in order to conform with widely used notation. For example, there is often no such distinction as we emphasised made between α and s above.

Other words are used with the same meaning as the word 'function'. We may occasionally use the word 'map' or 'mapping' instead of the word 'function', particularly if either dom f or rg f is not a subset of $\mathbb{R}$. Some authors use the word 'function' only if the range is a set of numbers. The terms operator, functional, transformation, transform are also used for some functions. We may occasionally use the word 'operator' for a function whose domain is a set of functions. For example, if $c \in \mathbb{R}$, $f : \mathbb{R} \to \mathbb{R}$ and $g : x \mapsto f(x - c)$ then $T : f \mapsto g$ is called the *translation operator*. Similarly the function $f \mapsto f|_K$ is called the *restriction operator*.

Exercises

Exercise 3.2.1 Let $f : \mathbb{R} \to \mathbb{R}$. *Describe the relation between the graphs of f and g if*

(1) $g : x \mapsto f(2x)$;

(2) $g : x \mapsto 2f(x)$;

(3) $g : x \mapsto f(-x)$;

(4) $g : x \mapsto -f(x)$;

(5) $g : x \mapsto f(x/2)$;

(6) $g : x \mapsto f(x)/2$;

(7) $g : x \mapsto f(x) + c$;

(8) $g : x \mapsto f(x - c)$.

Sketch the graphs for $f = 1_{\mathrm{P}}$ and $f(x) = x^2$.

Exercise 3.2.2   Use the results of Exercise 3.2.1 to graph $x \mapsto 2x^2 + 5x + 3$.

(i) **Exercise 3.2.3**   The function signum is defined by

$$\mathrm{sgn} : x \mapsto 1_{\mathrm{P}} - 1_A$$

where A denotes $\{x;\ x \in \mathbb{R},\ x < 0\}$. Graph the function sgn.

Exercise 3.2.4   Let f be defined as follows: if x is irrational then $f(x) = 0$; if $x = \dfrac{p}{q}$ with $p \in \mathbb{Z}, q \in \mathbb{N}$ and p, q relatively prime, then $f\left(\dfrac{p}{q}\right) = \dfrac{1}{q}$. Determine the values $f(1)$, $f(-1)$, $f(.5)$, $f(\sqrt{2})$, and graph the corresponding points. (*You should realise that f itself cannot be graphed.*)

(i) **Exercise 3.2.5**   Let $f : X \mapsto Y$; A, A_1, A_2, A_l for $l \in L$ be subsets of X, and B, B_1, B_2, B_l for $l \in L$ be subsets of Y. Prove:

(1) $A_1 \subset A_2 \Rightarrow f(A_1) \subset f(A_2)$;

(2) $B_1 \subset B_2 \Rightarrow f_{-1}(B_1) \subset f_{-1}(B_2)$;

(3) $f\left(\bigcup_{l \in L} A_l\right) = \bigcup_{l \in L} f(A_l)$;

(4) $f\left(\bigcap_{l \in L} A_l\right) \subset \bigcap_{l \in L} f(A_l)$; give an example showing that equality need not hold;

(5) $f(A_1) \setminus f(A_2) \subset f(A_1 \setminus A_2)$; give an example showing that equality need not hold;

(6) $f_{-1}\left(\bigcup_{l \in L} B_l\right) = \bigcup_{l \in L} f_{-1}(B_l)$;

(7) $f_{-1}\left(\bigcap_{l \in L} B_l\right) = \bigcap_{l \in L} f_{-1}(B_l)$;

(8) $f_{-1}(B_1 \setminus B_2) = f_{-1}(B_1) \setminus f_{-1}(B_2)$;

(9) $f(f_{-1}(B)) \supset B; \quad f(f_{-1}(B)) = B \Leftrightarrow B \subset f(X);$

(10) $f_{-1}(f(A)) \supset A;$ give an example showing that A can be a proper subset of $f_{-1}(f(A))$.

(11) $A_1 \subset A \Rightarrow f|_A(A_1) = f(A_1);$

(12) $B \subset Y \Rightarrow (f|_A)_{-1}(B) = A \cap f_{-1}(B).$

Exercise 3.2.6   Let rg $f = \{0, 1\}$. Show that there is a set A such that $f(x) = 1_A(x)$ for every $x \in$ dom f. [Hint: $A = f_{-1}(1)$.]

3.3   Functions in *Maple*

3.3.1   *Library of functions*

Maple includes hundreds of mathematical functions. The notation for functions in *Maple* is similar to that used before for functions: these functions operate on one or more *arguments*, which are passed to the function in parentheses. The arguments to the functions usually specify the point at which the function should be evaluated. Some of the more common functions are shown in Table 3.3.

If you want to use a function you must know the name of the function and then you can ask *Maple* for help. You might wonder why the common logarithm (logarithm to base 10) is not in the above table. The reason is simply because mathematicians rarely use it, and in these days of cheap pocket calculators for all sorts of specialised uses, and fairly cheap computers, a negligible number of other people use logarithm tables. The command ?log will tell you all you need about the common logarithm in *Maple*. The following session illustrates use of the common logarithm.

```
>   log10(100);log10(1/100); log10(2);
```

$$\frac{\ln(100)}{\ln(10)}$$

$$-\frac{\ln(100)}{\ln(10)}$$

$$\frac{\ln(2)}{\ln(10)}$$

Table 3.3   Some *Maple* library functions

Mathematical name	Description	*Maple* usage
$\lvert x \rvert$	Absolute value	`abs(x)`
$\sqrt{x}$	Square root	`sqrt(x)`
e^x	Exponential Function	`exp(x)`
$\ln(x)$	Natural logarithm	`log(x)` or `ln(x)`
$\sin(x)$	Trigonometric function	`sin(x)`
$\cos(x)$	Trigonometric function	`cos(x)`
$\tan(x)$	Trigonometric function	`tan(x)`
$\mathrm{Max}(x,y,z)$	Maximum of two or more numbers	`max(x,y,z)`
$\mathrm{Min}(x,y,z)$	Minimum of two or more numbers	`min(x,y,z)`
$\lfloor x \rfloor$	The greatest integer not exceeding x	`floor(x)`
$\lceil x \rceil$	The smallest integer not preceding x	`ceil(x)`

```
>   simplify(%%%);simplify(%%);simplify(%);
```

$$2$$

$$\frac{\ln(2)}{\ln(2)+\ln(5)}$$

$$\frac{\ln(2)}{\ln(2)+\ln(5)}$$

There is an obvious error on the second last line, but we can blame only ourselves for it because we did not take into account the fact that the first command `simplify(%%%)` produces a result, namely 2, which now becomes the most recent result for the next command. The commands should have been

```
>   simplify(%%%);simplify(%%%);simplify(%%%);
```
Now the best way to recover is

```
>   simplify(log10(1/100));evalf(log10(2));
```

$$-2$$

$$.3010299957$$

3.3.2 *Defining functions in* Maple

In *Maple* you can define your own functions. This allows you to evaluate functions at various points, without having to retype the function value. The following example shows how to define a function $f(x) = x^2 + 1$, and how to find the value of this function at $x = 2$ and at $x = 5$. The notation is similar to the one we employed for defining functions: for $\mapsto$, in the *Maple* definition of the function, type -> as a hyphen followed by a greater-than sign.

```
>   # Define f(x) = x^2 + 1
>   f:=x->x^2 + 1;
```

$$f := x \rightarrow x^2 + 1$$

```
>   # Evaluate f at 2 and 5
>   f(2); f(5);
```

$$5$$

$$26$$

In the next example, the variable a is assigned the value 3, then the function $f(x) = ax^2$ is defined, and finally $f(2)$ is evaluated.

```
>   a:=3;
```

$$a := 3$$

```
>   f:=x->a*x^2;
```

$$f := x \rightarrow a\,x^2$$

```
>   f(2);
```

$$12$$

As shown above, when you want to evaluate a function at a particular point, first define the function (say f), and then specify the required point in parentheses (for example, $f(2)$). More generally, you can place variables or expressions in the parentheses, and *Maple* will evaluate the function with respect to those variables or expressions. Look carefully at the following example:

```
>   # Define f(x) = x+2
>   f:=x->x+2;
```

$$f := x \rightarrow x + 2$$

```
>   # Define g(x) = x^2 - 2x + 1
>   g:=x->x^2-2*x+1;
```

$$g := x \rightarrow x^2 - 2\,x + 1$$

```
>   f(x);
```

$$x + 2$$

```
>   g(x);
```

$$x^2 - 2\,x + 1$$

```
>   # Evaluate f(x)*g(x)
>   f(x)*g(x);
```

$$(x+2)(x^2-2x+1)$$

```
>   # Evaluate f((g(x))^2)
>   f(g(x)^2);
```

$$(x^2-2x+1)^2+2$$

In a similar way, you can define functions of two or more variables. Note the slightly different way of typing such functions, as shown below.

```
>   # Define f(x,y) = sqrt(x^2+y^2)
>   f:=(x,y)->sqrt(x^2+y^2);
```

$$f := (x,y) \rightarrow \text{sqrt}(x^2+y^2)$$

```
>   f(3,4);
```

$$5$$

```
>   # Define g(x,y,z) = (x^2+y^2+z^2)/(x+y+z)
>   g:=(x,y,z)->(x^2+y^2+z^2)/(x+y+z);
```

$$g := (x,y,z) \rightarrow \frac{x^2+y^2+z^2}{x+y+z}$$

```
>   g(1,-1,2);
```

$$3$$

Use := and -> to define functions. For example:
$$f(x) = x^2 + 1 \text{ is defined by } \texttt{f:=x->x^2+1;}$$
$$f(x,y) = x^2 + y^2 \text{ is defined by } \texttt{f:=(x,y)->x^2+y^2.}$$

3.3.3   *Boolean functions*

We call a function a Boolean function if its domain of definition is a set of propositions and its range has either two elements, namely *true* and *false*, or three elements, namely *true*, *false* and *fail*.[2] We shall tacitly assume that if the value of a Boolean function at P is true then, indeed, P is a true propostion. A Boolean function is realized in *Maple* by the command is.

For instance

```
>  is(Pi<22/7); is((sqrt(2))^(3/2)>5/2); is(x>0);
```

$$true$$

$$false$$

$$FAIL$$

The last answer is natural, nobody can decide whether x is positive or not unless it is known what x actually represents.

We now define two functions which we shall use in selecting elements from a list.

```
>  big:=x->is(x>2);sqbig:=x->is(x^2>2);
```

$$big := x \to \text{is}(2 < x)$$

$$sqbig := x \to \text{is}(2 < x^2)$$

We have seen earlier in Subsection 2.1.2.1 how to select elements from a list by using the op command, however very often it is needed to select the elements by some property rather than by their position in the list. Let us consider the list

```
>  L:=[Pi,-1,sqrt(3),-2, 0.5,1/3];
```

$$L := [\pi, -1, \sqrt{3}, -2, .5, \frac{1}{3}]$$

from which we wish to create a list of elements greater than 2. The command is select and we use the function big defined above.

```
>  select(big,L);
```

$$[\pi]$$

Similarly, we can create a list of elements with squares greater than 2.

```
>  select(sqbig,L);
```

$$[\pi, \sqrt{3}, -2]$$

---

[2]Sometimes 0, 1 are used instead of *false* and *true*. *Fail* is used in *Maple*; on other occasions the word *undecided* is sometimes more appropriate.

Other properties can also be used for selection or removal. For instance, for removal of rationals from L, we first define a function

```
>   rat:=x->is(x,rational);
```

$$rat := x \rightarrow \mathrm{is}(x, \text{ } rational)$$

and then use the command **remove** .

```
>   remove(rat,L);
```

$$[\pi, \sqrt{3}, .5]$$

Note that for *Maple* the floating point number .5 is not rational. If you do not like .5 in the list you can remove it manually, for example

```
>   [op(1,L),op(3,L)];
```

$$[\pi, \sqrt{3}]$$

Exercises

Exercise 3.3.1 *Which of the following numbers are prime:* 1979, 7919, 131723? [Hint: `is(x,prime)`]

Exercise 3.3.2 *Find all complete squares between* 100 *and* 200. [Hint: Make a list L of these integers, define `sq:=x->is(sqrt(x),integer)` and `select(sq,L)`.]

Exercise 3.3.3 *Find all primes between* 5000 *and* 5100. [Hint: Use the methods of the two previous Exercises.]

3.3.4 *Graphs of functions in* Maple

Maple provides great power and flexibility in graphing various types of functions. The form of the command for graphing a function is

$$\text{plot(name,range,options);} \tag{3.1}$$

The name and range are always compulsory. The word 'range' is used here as in common English meaning the limits (for the first coordinate) within which the graph will be displayed. It should not be confused with range of a function. The 'name' can be a general expression for a function value. This function value can be entered directly, as $x^2 - 3x$ is, or symbolically as $\sin(x)$ is, below. The alternative for the 'name' in (3.1) can be a name of a function, like sin. We deal first with 'name'= function value. There is a great number of options, which are all listed in the **help** file. The ones which are most likely to be needed by readers are explained as we go along.

```
>   plot(x^2-3*x,x=-1..4);plot(sin(x),x=-Pi..Pi);
>
```

The graphs of these two functions are shown in Figures 3.6 and 3.7. Note that the scales on the x- and y-axes are different. We shall see later how to force *Maple* to produce the same scales.

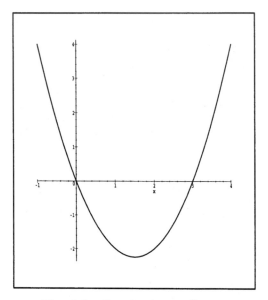

Fig. 3.6   Graph of $y = x^2 - 3x$

The alternative for 'name'=symbol of a function is perhaps more logically consistent, since we graph functions. It also saves typing if the function value is given by a long expression. The difference in 'range' in (3.1) is now in that the form x=a..b must not be used, but just the limits in the form a..b. We will illustrate this by an example with a function which we define ourselves.

```
>   # First define the desired function

>   f:=x->x+1/(x-1)^2;
```

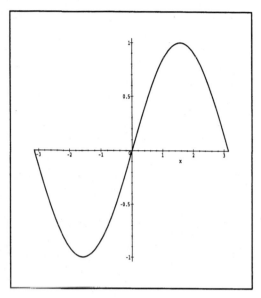

Fig. 3.7   Graph of $\sin x$

$$f := x \rightarrow x + \frac{1}{(x-1)^2}$$

```
>   # Then plot the function;
>   plot(f,-infinity..infinity);
```

The graph, shown in Figure 3.8, is reasonably good but, as must be expected with the infinite range of x, looks a little rough. The use of the infinite range is generally not recommended. It can, however, serve as a guide where to have a closer look at the graph. In our example it is near the dip in the curve. We now restrict the range of x to the interval where we expect the dip to occur.

```
>   # Restrict the range of x for the plot of the function
>   plot(f,3/2..4);
```

The portion of the graph restricted to $1.5 \le x \le 4$ is shown in Figure 3.9.

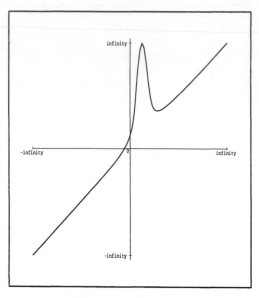

Fig. 3.8   Graph of $x + \dfrac{1}{(x-1)^2}$: unrestricted range

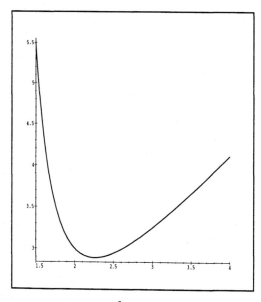

Fig. 3.9   Graph of $x + \dfrac{1}{(x-1)^2}$: restricted horizontal range

If we specify the range of x as -1..5, *Maple* will not be able to cope[3] because f takes very large values near $x = 1$. One possible way to help *Maple* to produce a good graph is to limit the range of the y values as well, for instance between 0 and 6. Note that the range of values for x must also be given, and must precede the range of values for y. As with other computer programs, *Maple* does just what you ask it to. The command plot(f,0..6) will plot the function f for x between 0 and 6 and will not restrict the range of y.

```
>   plot(f,-1..5,0..6);
```

The result of restricting x to the range between -1 and 5 and y to the range between 0 and 6 is shown in Figure 3.10.

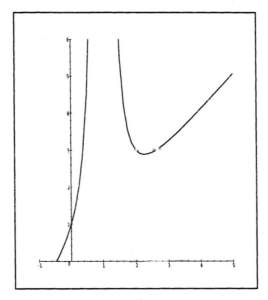

Fig. 3.10 Graph of $x + \dfrac{1}{(x - 1)^2}$: both ranges restricted

The graph of $S : x \mapsto x + \sin 7x$, which was shown in Section 1.1 (see page 3) was not satisfactory. In order to obtain a good graph we can either limit the range of x or use the option numpoints to increase the number of points used to plot the graph. By default *Maple* uses 50 points, and it

---

[3]Try it on your machine.

joins successive points by a straight line. In most cases this is satisfactory. But with functions with a large number of maxima and mimima within the range, more points are needed. With 150 points the graph is fine.

```
>    # First define the function to be graphed
>    S:=x->x+sin(7*x);
>    # Now plot the function:
>    # first with a restricted range,
>    # and the second time over the original range
>    # but using more points
>    plot(S,-5..5);plot(S,-20..20,numpoints=150);
```

The results of these two graphs are shown in Figures 3.11 and 3.12, respectively.

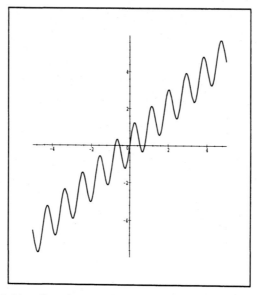

Fig. 3.11   Graph of $x + \sin 7x$: restricted range for x

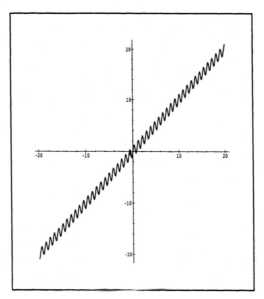

Fig. 3.12   Graph of $x + \sin 7x$: additional points included

Maple adjusts the scales on each axis to fit the picture. This sometimes results in an undesirable distortion as in the next graph of $\sin x$.

```
>  plot(sin,0..Pi);
```

This is shown in Figure 3.13.

The option `scaling=CONSTRAINED` ensures the same scale in both directions. The result of this is shown in Figure 3.14.

```
>  plot(sin,-Pi..Pi,scaling=CONSTRAINED);
```

It is possible to include multiple plots on the same set of axis by using set notation, so that all the function to be plotted are enclosed within braces { and }. Try `plot({sin(x),cos(x),sin(2x)},x=-4..4);`.

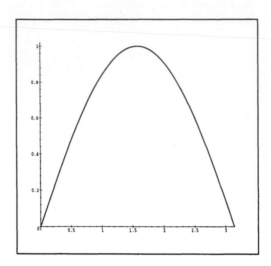

Fig. 3.13   Graph of $\sin x$ over the interval $[0, \pi]$

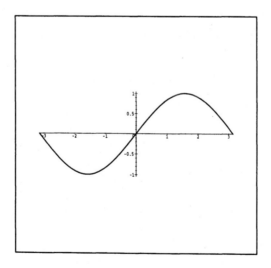

Fig. 3.14   Graph of $\sin x$ using option `scaling=CONSTRAINED`

Exercises

Exercise 3.3.4   *Graph the following functions:*

(1) $x \mapsto \dfrac{x^2 - 5x + 6}{x^2 + x + 1}$;

(2) $x \mapsto \dfrac{x^2 + 2x + 2}{x^2 - x - 2}$;

(3) $x \mapsto x^3 + x - 2$;

(4) $x \mapsto x^3 - 3x$;

(5) $x \mapsto x^3 - 3x^2 - 1$.

Exercise 3.3.5   *In each case, graph each group of three functions in one picture:*

(1) $x \mapsto x^2$, $x \mapsto (x - 2)^2$, $x \mapsto -(x + 2)^2$;

(2) $x \mapsto \cos x$, $x \mapsto \cos(x/2)$, $x \mapsto \cos 2x$.

3.4   Composition of functions

If f and g are two functions then the composition of f and g, denoted by $f \circ g$ (read f circle g) is defined by

$$f \circ g : \quad x \mapsto f\big(g(x)\big).$$

The domain of $f \circ g$ is the set of all $x \in \operatorname{dom} g$ for which $g(x) \in \operatorname{dom} f$. In particular $\operatorname{dom} f \circ g = \operatorname{dom} g$ if $\operatorname{rg} g \subset \operatorname{dom} f$. Generally $f \circ g \neq g \circ f$; see Exercise 3.4.1. The same function can be expressed as a composition of two functions in many different ways; see Exercise 3.4.2. It is easy to see that

$$(f \circ g) \circ h = f \circ (g \circ h)$$

and we shall denote this double composition simply by $f \circ g \circ h$.

Exercises

Exercise 3.4.1   *Find $f \circ g$ and $g \circ f$ if*

(1) $f : u \mapsto u^2$, $g : x \mapsto x + 1$;

(2) $f = \{(1, 3), (2, 45)\}$, $g = \{(1, 2), (0, 0)\}$.

What are $\operatorname{dom} f \circ g$ and $\operatorname{dom} g \circ f$ in (2)?

Exercise 3.4.2   *A given function can generally be expressed as a composition of two functions in many different ways. Let*

$$h : x \mapsto x^2 + 2x,$$
$$f : y \mapsto y^2 - 1,$$
$$g : x \mapsto x + 1,$$
$$F : y \mapsto y^2 + 4y + 3,$$
$$G : x \mapsto x - 1.$$

Show that $h = f \circ g = F \circ G$.

3.5   Bijections

A function f is said to be *one-to-one* (or *injective*) if

$$x_1 \neq x_2 \Rightarrow f(x_1) \neq f(x_2).$$

for every x_1, $x_2 \in \operatorname{dom} f$. An equivalent implication is

$$f(x_1) = f(x_2) \Rightarrow x_1 = x_2.$$

A function f is said to be *one-to-one on a set S* if $f|_S$ is one-to-one.

Example 3.5   In a theatre, associate with each visitor the chair sat on. This association defines a function which is one-to-one: each chair has at most one visitor sitting on it.

Example 3.6   The function $h : x \mapsto x/(1+x)$ is also one-to-one. Indeed, if $\dfrac{x_1}{1+x_1} = \dfrac{x_2}{1+x_2}$ then it follows easily by removing fractions that $x_1 = x_2$.

If $f(C) = D$ then we say that f maps C *onto* D, or that $f : C \mapsto D$ is *surjective*. If the theatre in Example 3.5 is sold-out, then the function defined by the association between people and seats is surjective.

A function $F : X \mapsto Y$ is called a *bijection* of X onto Y if it is one-to-one and onto (or it is both injective and surjective). A bijection $F : X \mapsto Y$ is sometimes described as a one-to-one correspondence between X and Y. The function of Example 3.5 is a bijection of the audience onto the set of chairs in the theatre. The function h from Example 3.6 is a bijection of $\mathbb{R}\backslash\{-1\}$ onto $\mathbb{R}\backslash\{1\}$. Indeed, if $y \neq 1$ there is an x such that $\dfrac{x}{x+1} = y$ and for no x is $\dfrac{x}{x+1} = 1$. For given X and Y there can be several bijections of

X onto Y. For instance, the identity function, the clock function C shown in Table 3.1 and the mapping from one number to the number diametrically opposite on the face of the clock are all bijections of $\{1, 2, 3, \ldots, 12\}$ onto itself.

Exercises

Exercise 3.5.1   *Prove that $f : x \mapsto \frac{1}{10}x^3 + x$ is one-to-one.* [Hint: $\frac{1}{10}(u^3 - v^3) + u - v = (u - v)A$ with $A > 0$.]

3.6   Inverse functions

If S is a relation then the relation $S_{-1} = \{(y, x);\ (x, y) \in S\}$ is called the *inverse relation* to S or simply the inverse of S. Since the point (u, v) is the mirror image of (v, u) across the line $y = x$, the graph of S_{-1} is simply the mirror image of the graph of S across this same line, $y = x$. Every function, as a relation, has an inverse: however this inverse relation need not be a function. If, for a function, the inverse relation f_{-1} is a function then it is called the *inverse function* to f or simply the inverse of f. The inverse function f_{-1} exists if and only if

$$f(x_1) = f(x_2) \Rightarrow x_1 = x_2,$$

that is, if f is one-to-one. The domain of f becomes the range of f_{-1}, and similarly rg $f = \text{dom } f_{-1}$. These results are illustrated in Figure 3.15.

As a concrete example, if $H : x \mapsto 2x$ then H_{-1} exists and

$$\begin{aligned} H_{-1} &= \{(2x, x);\ x \in \mathbb{R}\}, \\ &= \{(y, y/2);\ y \in \mathbb{R}\}, \\ &= \{(x, x/2);\ x \in \mathbb{R}\}. \end{aligned}$$

In other words, $H_{-1} : x \mapsto x/2$.

Viewing the inverse function as a set of (reversed) pairs is useful in graphing. The graph of f_{-1} can be obtained in *Maple* by the command

```
plot([f(x),x,x=a..b]);
```

note the presence of [and] in the command! In Figure 3.16 the graph of the inverse to $x \mapsto \frac{1}{10}x^3 + x$ was obtained in this way.

It follows from the definition of an inverse that $f_{-1}\big(f(x)\big) = x$ for every $x \in \text{dom } f$ and $f\big(f_{-1}(y)\big) = y$ for every $y \in \text{dom } f_{-1} = \text{rg } f$. This can also

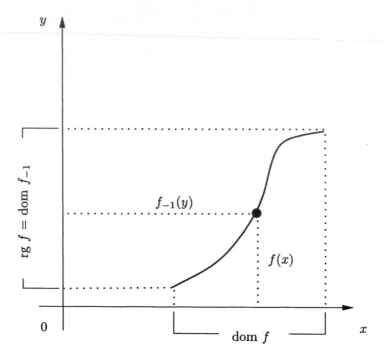

Fig. 3.15   Graph of a one-to-one function

be written in the form $f_{-1} \circ f = \mathrm{id}_A$ and $f \circ f_{-1} = \mathrm{id}_B$ where $A = \mathrm{dom}\, f$ and $B = \mathrm{rg}\, f$.

For finding the inverse of f we simply solve the equation

$$y = f(x) \qquad\qquad (3.2)$$

with respect to x. If there is a unique solution to (3.2) then it is $f_{-1}(y)$. Since we know that it is irrelevant whether x or y or u is used in the notation $f_{-1} : y \mapsto f_{-1}(y)$ we can, after (3.2) has been solved, write x instead of y and y instead of x. This is illustrated in Figure 3.17.

Example 3.7   Let

$$G : \; x \mapsto 1/x, \; x \in \{x;\, 0 < x < 1\}.$$

Find G_{-1}. We have $y = \dfrac{1}{x}$, $x = \dfrac{1}{y}$, so $G_{-1} : x \mapsto \dfrac{1}{x}$. However G and G_{-1}

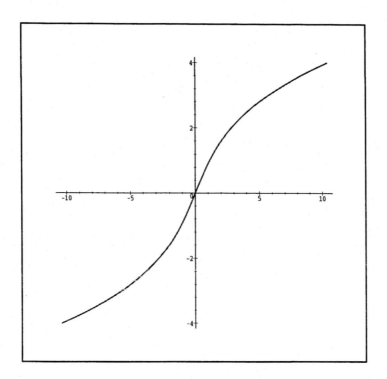

Fig. 3.16   Inverse of $x \mapsto \frac{1}{10}x^3 + x$.

are two distinct functions:

$$\operatorname{dom} G = \{x; \; 0 < x < 1\},$$

while

$$\operatorname{dom} G_{-1} = \{x; \; x > 1\},$$

as can be readily checked. This should also be clear from the graphs of G and G_{-1}.

It may happen that a function f has an inverse but solving the equation $y = f(x)$ is difficult. The existence of f_{-1} is guaranteed if f is one-to-one, and is independent of our ability to find an explicit solution of the equation $y = f(x)$.

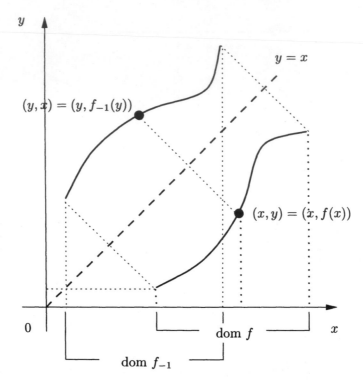

Fig. 3.17   Graph of a one-to-one function and its inverse

Example 3.8   We have seen in Exercise 3.5.1 that $f : x \mapsto \frac{1}{10}x^3 + x$ is one-to-one and has therefore an inverse. Its graph was obtained in Figure 3.16. The equation

$$\frac{1}{10}x^3 + x = y \tag{3.3}$$

is not easy to solve[4]. The *Maple* command fsolve can be used for numerical solution of Equation (3.3) and therefore for approximate evaluation of the inverse. The following *Maple* session illustrates this.

```
>   g:=y->fsolve((1/10)*x^3+x-y=0,x);
```

---

[4]In Chapter 8 we deal with solving equations generally and we also use *Maple* to solve equations like (3.3).

$$g := y \to \text{fsolve}(\frac{1}{10} x^3 + x - y = 0, \, x)$$

> g(0);g(1);g(1.1);g(-2);g(3);

0

.9216989942

1.

−1.594562117

2.088730145

It is important to realize that g is only an approximation to f_{-1}. This happens because the command fsolve gives only the numerical solution. For instance, g(f(sqrt(2*Pi))); gives 2.506628274 instead of $\sqrt{2\pi}$.

Exercises

Exercise 3.6.1   Let $f : x \mapsto \dfrac{1}{2-x}$. Prove that f is one-to-one and find rg f and f_{-1}.

Exercise 3.6.2   Use Maple to graph the inverse to $f : x \mapsto x^5/(1+x^4)$.

Exercise 3.6.3   Find and graph the inverse of $f : x \mapsto x/(1+|x|)$.

3.7   Comments

In our approach we derived the concept of a function from that of a set. It may be argued that the concept of a function is more fundamental than that of a set and should be used as the most primitive notion. It is, indeed, possible to choose the idea of a function as the most basic for all mathematics; such a procedure was conceived by the great mathematician

John von Neumann in the early 1920s. Naturally if we choose functions as the foundation stone their theory has to be established on an axiomatic basis. We cannot *a priori* exclude the possibility that mathematics built on functions would then be distinct from the one based on sets. This does not seem to be the case: set theory and the rest of mathematics as we know it can be recaptured from von Neumann's theory via characteristic functions, that is, functions whose range is $\{0, 1\}$. The fact that functions can be used instead of sets for the foundation of mathematics is interesting but little known, and at our level of sophistication no mathematician would be inclined to adopt it.

Chapter 4

Real Numbers

In this chapter we introduce real numbers on an axiomatic basis, solve inequalities, introduce the absolute value and discuss the least upper bound axiom. In the concluding section we outline an alternative development of the real number system.

4.1 Fields

Real numbers satisfy three groups of axioms—field axioms, order axioms and the least upper bound axiom. We discuss each group separately.

A set F together with two functions $(x, y) \mapsto x + y$, $(x, y) \mapsto xy$ from $F \times F$ into F is called a *field* if the axioms in Table 4.1 are satisfied for all x, y, z in F.

Table 4.1   Field axioms

$A_1 : x + y = y + x$	$M_1 : xy = yx$
$A_2 : x + (y + z) = (x + y) + z$	$M_2 : x(yz) = (xy)z$
$A_3 :$ There is an element $0 \in F$ such that $0 + x = x$ for all x in F	$M_3 :$ There is an element $1 \in F$, $1 \neq 0$, such that $1x = x$ for all x in F;
$A_4 :$ For every element $x \in F$ there exists an element $(-x) \in F$ such that $(-x) + x = 0.$	$M_4 :$ For every element $x \in F$, $x \neq 0$, there exists an element $x^{-1} \in F$ such that $x^{-1}x = 1;$
$D : x(y + z) = xy + xz.$	

The functions $(x, y) \mapsto x + y$ and $(x, y) \mapsto xy$ are called *addition* and *multiplication*, respectively; $x + y$ is the *sum* of x and y, xy is the *product* of x and y. Axioms A_1–A_4 deal with addition, while axioms M_1–M_4 deal with multiplication. Axioms A_1 and M_1 are called the *commutative laws*; axioms A_2 and M_2 are called the *associative laws*; axioms A_3 and M_3 state the existence of a *unit element* under addition and multiplication, respectively, which does not change x; axioms A_4 and M_4 state the existence of an *inverse element* under addition and multiplication, respectively, to produce the respective unit element. Note that the unit element under addition, commonly called the *zero element*, does not have an inverse under multiplication. In more usual language, division by zero is forbidden. Axiom D is called the *distributive law*. In Chapter 2 we defined a ring as a set with addition and multiplication satisfying axioms A_1–A_4, M_1, M_2 and D. Hence every field is a ring, and a ring in which axioms M_3 and M_4 are satisfied is a field.

It follows easily that the element 0 from axiom A_3 is uniquely determined. Indeed, let z be such that $x = z + x$ for every x. Setting $x = 0$ gives

$$
\begin{aligned}
0 &= z + 0 \\
&= 0 + z && \text{by axiom } A_1, \\
&= z && \text{by axiom } A_3.
\end{aligned}
$$

A similar argument shows that there is exactly one element 1 and that $-x$ and x^{-1} are uniquely determined by x. We shall not dwell on such simple consequences of these axioms and hope that the 'usual rules' concerning addition and multiplication are known to the reader from elementary arithmetic or algebra (for example, $-(-a) = a$, $a.0 = 0$, $\left(a^{-1}\right)^{-1} = a$, $(-1)a = -a$, $a^2 - b^2 = (a + b)(a - b)$, etc.). Derivation of some of these rules are to be found in the Exercises at the end of this section. Because of axiom A_2 we write $a + b + c$ instead of either $a + (b + c)$ or $(a + b) + c$; similarly abc denotes either $a(bc)$ or $(ab)c$.

Many other conventions are used, for example, $a - b$ and $a + bc$ are abbreviations for $a + (-b)$ and $a + (bc)$, respectively. A centred dot is sometimes used to indicate multiplication, as in $a \cdot b$ which means the same as ab. The fraction $\dfrac{a}{b}$ or a/b is simply ab^{-1} and $a + b^{-1}$ denotes $a + (b^{-1})$. We trust that the reader is familiar with these and similar conventions from elementary algebra, including the rules for precedence of various operations. In cases where there is the possibility of ambiguity in this book, parentheses will be inserted, or a centred dot used, to clarify the precise meaning.

There are many fields. If we accept reals as known then we may exhibit the rationals also as a field. The structure of a field can be very different from that of the reals or rationals. For example, there exist fields with finitely many elements and these are important, not only theoretically but also in application. Let $F = \{z, u\}$, where $z \neq u$ and addition and multiplication satisfy $z + z = z$, $u + u = z$, $z + u = u + z = u$, $zz = z$, $uu = u$, $uz = zu = z$. It is very easy to check that F is a field, with $z = 0$, $u = 1$. Note that there are no other elements in this field. The addition and multiplication tables for this field, which is known as $GF(2)$, are shown in the following table.

Table 4.2   Addition and multiplication in $GF(2)$

Add				Multiply		
-----	---	---		----------	---	---
+	0	1		×	0	1
0	0	1		0	0	0
1	1	0		1	0	1

It may be thought that any area of mathematics which contained $1 + 1 = 0$ as a basic rule was of little use. However this field is, possibly, more frequently used in the modern world than any other. Every time a computer carries out a parity check $GF(2)$ is invoked. Axiom M_3 guarantees that any field must contain at least two elements, 0 and 1.

Exercises

In Exercises 4.1.1–4.1.3, a, b, c, d, x denote elements of a fixed field F. These Exercises are stated as theorems: your task is to prove them.

ⓘ **Exercise 4.1.1**   The equation $a + x = b$ is uniquely solvable for x. The equation $ax = b$ is uniquely solvable for x if $a \neq 0$.

ⓘ **Exercise 4.1.2**
   (1) $a \cdot 0 = 0 \cdot a = 0$;
   (2) $-(-a) = a$;
   (3) $\left(a^{-1}\right)^{-1} = a$;
   (4) $(-1)a = -a$;
   (5) $-a + (-b) = -(a + b)$;

(6) $(-a)b = a(-b) = -(ab)$;

(7) $(-a)(-b) = ab$;

(8) $\dfrac{a}{b}\dfrac{c}{d} = \dfrac{ac}{bd}$;

(9) $\dfrac{a}{b} + \dfrac{c}{d} = \dfrac{ad + bc}{bd}$;

(10) $\left(\dfrac{a}{b}\right) \Big/ \left(\dfrac{c}{d}\right) = \dfrac{ad}{bc}$;

(11) $a^{-1} \neq 0$.

Exercise 4.1.3

(1) If $ab = 0$ then either $a = 0$ or $b = 0$;

(2) $a + c = b + c \Rightarrow a = b$;

(3) $ac = bc$ and $c \neq 0$ imply $a = b$.

Exercise 4.1.4   Let K be a set containing at least two distinct elements. Assume that addition and multiplication on K satisfy axioms A_1, A_2, M_1, M_2, D and the following:

A_5 : For every a, $b \in K$ the equation $a + x = b$ has a solution.	M_5 : For every a, $b \in K$, $a \neq 0$, the equation $ax = b$ has a solution.

Prove that K is a field.

This means that the axioms A_3, A_4 and M_3, M_4 can be replaced by A_5 and M_5, respectively, and the definition of a field retains the same meaning. [Hint: First solve the equation $a + x = a$, denote it by z and show that $z + b = b$ for every $b \in K$. Show similarly that the solution to $ax = a$ with $a \neq 0$ is the multiplication unit. The existence of $(-a)$ and a^{-1} is obtained by solving the equation $x + a = 0$ or $ax = 1$ respectively. It is obvious that A_5 and M_5 are satisfied in any field.]

4.2   Order axioms

A field F is *ordered* if the axioms in Table 4.3 are satisfied for all a, b, c in F. If $a < b$ we say that a is less (or smaller) than b; if $b < a$ we may also write $a > b$ and say that a is greater than b. The sign of inequality points towards the smaller element. The statement '$a < b$ or $a = b$' is recorded as $a \leq b$. The inequality $a < b$ is called *strict inequality* to distinguish it from the inequality $a \leq b$. If $0 < a$, then a is said to be positive, and negative if $a < 0$. The inequalities $a \leq b$ and $b \leq c$ are shortened to $a \leq b \leq c$ and a similar convention is used if either or both of the inequalities is strict.

Some easy consequences of the axioms are stated in the next Theorem.

Table 4.3   Order axioms

O_1 : There is a relation $<$ on F such that exactly
one of the following possibilities occurs:
$a < b;\ a = b;\ b < a.$
O_2 : $(a < b)$ and $(b < c) \Rightarrow (a < c).$
O_3 : $a < b \Rightarrow (a + c) < (b + c).$
O_4 : If $0 < c$ then
$a < b \Rightarrow ac < bc.$

For the remainder of this section, a, b, c, d denote arbitrary elements in an ordered field F, and x is an element of this field which is (usually) to be found.

Theorem 4.1

(i) $a < b \Rightarrow -b < -a;$

(ii) $(a < b,\ c < 0) \Rightarrow bc < ac;$

(iii) $(a < b,\ c < d) \Rightarrow a + c < b + d;$

(iv) $(a < b,\ c < d) \Rightarrow a - d < b - c;$

(v) $(0 < a < b,\ 0 < c < d) \Rightarrow ac < bd;$

(vi) $a \neq 0 \Rightarrow 0 < a^2;$

(vii) $0 < a \Rightarrow 0 < \dfrac{1}{a};$

(viii) $0 < a < b \Rightarrow 0 < \dfrac{1}{b} < \dfrac{1}{a}.$

Since $bc < ac$ means the same as $ac > bc$ we have the following Memory Aid for (ii): Multiplication of an inequality by a negative number reverses the inequality. A consequence of (vi) is that $1 > 0$.

Proof.   Each of the parts of the theorem is proved separately.

(i) Using O_3 with $c = -(a + b)$ gives
$-b = a - (a + b) < b - (a + b) = -a.$

(ii) If $c < 0$ then by (i) $0 < -c$ and it follows by O_4 that $-ac < -bc$ and by (i) $bc < ac.$

(iii) Successive applications of O_3 give

$$a + c < b + c \quad \text{and} \quad b + c < b + d. \tag{4.1}$$

Applying O_2 to (4.1) (with a replaced by $a + c$, b by $b + c$ and c by $b + d$) gives the result.

(iv) $-d < -c$ by (i) and using (iii) with c replaced by $-d$ and d by $-c$ gives the result.

(v) We have successively $ac < bc$ and $bc < bd$; an appeal to O_4 and then to O_2 completes the proof.

(vi) If $0 < a$ then $0 = 0 \cdot a < a \cdot a = a^2$.
If $a < 0$ then $0 < -a$ and $0 = 0 \cdot (-a) < (-a) \cdot (-a) = a^2$.

(vii) Since $a \dfrac{1}{a} = 1$ it follows that $\dfrac{1}{a} \neq 0$ and by (vi) $\dfrac{1}{a^2} > 0$. Consequently
$$0 = 0 \frac{1}{a^2} < a \frac{1}{a^2} = \frac{1}{a}.$$

(viii) By (vii) we have $\dfrac{1}{a} > 0$, $\dfrac{1}{b} > 0$. Consequently $0 < \dfrac{1}{a} \dfrac{1}{b} = \dfrac{1}{ab}$ and
$$\frac{1}{b} = a \frac{1}{ab} < b \frac{1}{ab} = \frac{1}{a}.$$
$\square$

Theorem 4.1 can advantageously be used to describe the set of $x \in F$ satisfying a given inequality. Consider, for instance,

$$\frac{x}{x - 4} > 1. \tag{4.2}$$

Solving (4.2) means finding all x satisfying (4.2). This inequality only makes sense if $x \neq 4$. For $x \neq 4$ we have $(x - 4)^2 > 0$ and it follows that (4.2) is satisfied if and only if

$$x^2 - 4x = x(x - 4) > (x - 4)^2 = x^2 - 8x + 16\,,$$
$$4x > 16\,,$$
$$x > 4\,.$$

Thus x satisfies (4.2) if and only if $x > 4$. *Maple* can be used with advantage to solve inequalities. We deal with this in Section 4.4.

Example 4.1   We wish to solve

$$x^2 + 6 > 5x. \tag{4.3}$$

We rewrite the inequality in an equivalent form

$$(x - 2)(x - 3) > 0. \tag{4.4}$$

We distinguish two cases:
(a) $x > 2$;

(b) $x < 2$;
(for $x = 2$ the inequality is obviously false).

In case (a) we find by dividing by $x - 2$, that is, multiplying by $\dfrac{1}{x-2}$ that Inequality (4.4) is satisfied if and only if $x - 3 > 0$ and it is not satisfied for $2 < x < 3$.

In case (b) we obtain similarly $x < 3$, and since this provides no further restriction all $x < 2$ satisfy (4.4).

Summarising: Inequality (4.4) is satisfied by x if and only if either $x < 2$ or $x > 3$.

This result is obvious geometrically. Inequality (4.3) is satisfied if and only if the point (x, y) on the parabola $y = x^2 - 5x + 6$ lies above the x-axis. (See Figure 4.1.)

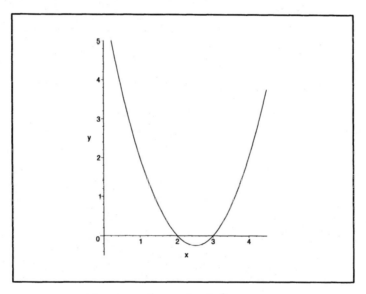

Fig. 4.1   A graph of $x^2 - 5x + 6$

The set P of positive elements in F has the following properties:

(1) For every $x \in F$, $x \neq 0$, either x or $-x$ is in P.
(2) If x and y are in P then so is $x + y$.
(3) If x and y are in P then so is xy.

If, in a field F, a set P is given such that (1), (2) and (3) hold then this field becomes an ordered field by declaring $a < b$ if $b - a \in P$. The proof of this is left as an exercise.

In an ordered field the sets

$$[a, b] = \{x : a \le x \le b\}; \tag{A}$$
$$]a, b[= \{x : a < x < b\}; \tag{B}$$
$$[a, \infty[= \{x : a \le x\}; \tag{C}$$
$$]a, \infty[= \{x : a < x\}; \tag{D}$$
$$]-\infty, a] = \{x : x \le a\}; \tag{E}$$
$$]-\infty, a[= \{x : x < a\}. \tag{F}$$

are called *intervals*. The interval (A) is closed *closed*; the interval (B) is *open*. It is not difficult to guess what the *half-open* (or *half-closed*) interval $[a, b[$ is.

It should be emphasised that we have not introduced any new elements into F; ∞ or $-\infty$ are used for convenience of notation and have no meaning by themselves. This is emphasised by the use of $\infty[$ or $]-\infty$ to indicate the end of the interval, implying that there are no elements in the intervals actually equal to these symbols.

It is also worth pointing out that other notations are used to indicate open intervals. One common notation is $(a, b) = \{x : a < x < b\}$. We shall not use this, since we have used (a, b) to denote an ordered pair of elements.

The length of $[a, b]$ or $]a, b[$ is $b - a$; a and b are called the *end points* of the interval.

Exercises

In these exercises a, b, c, d, x denote elements of an ordered field.

ⓘ **Exercise 4.2.1**   *Prove the following:*

(1) $(a \le b,\ c \ge 0) \Rightarrow ac \le bc$;

(2) $(a \le b,\ c \le 0) \Rightarrow ac \ge bc$;

(3) $(a \le b,\ c \le d) \Rightarrow a - d \le b - c$;

(4) $0 < a \le b \Rightarrow 0 < \dfrac{1}{b} \le \dfrac{1}{a}$;

(5) $0 < a \le b \Rightarrow a^2 \le b^2$.

Exercise 4.2.2   *Solve the following inequalities:*

(1) $3x + 2 \le 2x - 5 \le 4x + 3$;

(2) $2x + 6 \leq x + 10 \leq 2x + 1$;

(3) $x + \dfrac{1}{x} \leq 2$; [Hint: Consider $x > 0$ and $x < 0$ separately.]

(4) $(x + 1)(2x - 1) > 0$;

(5) $(5x - 2)(3x + 1) \leq 0$;

(6) $\dfrac{x + 2}{2x + 3} > 0$;

(7) $\dfrac{x}{x - 4} \geq 1$;

(8) $\dfrac{x + 1}{x + 3} < \dfrac{x + 5}{x + 6}$.

(9) $x^2 + x + 1 > 0$. [Hint: $x^2 + x + 1 = (x + 1/2)^2 + 3/4$.]

(i) **Exercise 4.2.3**   *Prove that $2ab \leq a^2 + b^2$ with equality if and only if $a = b$. [Hint: $(a - b)^2 \geq 0$.]*

4.3   Absolute value

By Max(a, b) we denote the larger of the two numbers a, b and, naturally, if $a = b$ we take Max(a, b) $= a = b$.

Definition 4.1 (Absolute value)   The *absolute value* of a is Max(a, $-a$). Absolute value of a is denoted by $|a|$, that is, $|a| = $ Max(a, $-a$).

Theorem 4.2

   (i) $|a| \geq 0$;

  (ii) $a \leq |a|$;

 (iii) $-a \leq |a|$;

 (iv) $|-a| = |a|$;

  (v) $a \geq 0 \Rightarrow |a| = a$;

 (vi) $a \leq 0 \Rightarrow |a| = -a$;

All assertions (i)–(vi) are easy consequences of the definition. Rewriting (iii) in the form $-|a| \leq a$ we can combine (ii) and (iii) into

$$-|a| \leq a \leq |a|. \tag{4.5}$$

Theorem 4.3

   (i) $|a + b| \leq |a| + |b|$;
   (ii) $\big||a| - |b|\big| \leq |a - b|$;
   (iii) $|ab| = |a||b|$.

Proof.

(i) Combining (4.5) with

$$-|b| \leq b \leq |b|$$

gives

$$-\big(|a| + |b|\big) \leq a + b \leq |a| + |b|.$$

This implies

$$a + b \leq |a| + |b|,$$
$$-(a + b) \leq |a| + |b|.$$

That is, both numbers $a + b$ and $-(a + b)$ do not exceed $|a| + |b|$, and hence

$$|a + b| = \text{Max}\big(a + b, \; -(a + b)\big) \leq |a| + |b|.$$

(ii) Using (i) with a replaced by $a - b$ gives

$$|a| \leq |a - b| + |b|,$$

that is,

$$|a| - |b| \leq |a - b|.$$

Interchanging a and b in this inequality gives

$$|b| - |a| \leq |b - a| = |a - b|.$$

Both numbers $|a| - |b|$ and $-\big(|a| - |b|\big)$ do not exceed $|a - b|$, consequently

$$\big||a| - |b|\big| \leq |a - b|.$$

(iii) The proof is simple when one distinguishes three cases:

   (A) $a \geq 0$, $b \geq 0$;
   (B) $a < 0$, $b < 0$;

(C) one number, say a, is not negative, $a \geq 0$, and the other number, b, is not positive, $b \leq 0$.

In each of the cases (A)–(C) it is easy to verify (iii) of Theorem 4.3 and we leave it to the reader as an exercise. $\square$

The next Theorem will be often used, particularly in Chapter 10.

Theorem 4.4   *If x, a, ε are elements of an ordered field then the inequality*

$$|x - a| < \epsilon \qquad (4.6)$$

is satisfied if and only if

$$a - \epsilon < x < a + \epsilon. \qquad (4.7)$$

Proof.   We split the proof into two parts.

I. Let us assume Inequality (4.6). By Inequality (4.5) we have

$$-|x - a| \leq x - a \leq |x - a|. \qquad (4.8)$$

It follows from Inequality (4.6) that

$$-\epsilon < -|x - a|. \qquad (4.9)$$

Combining (4.6) and (4.9) with (4.8) gives

$$-\epsilon < x - a < \epsilon,$$

which is equivalent to (4.7).

II. The Inequality (4.7) states that

$$x - a < \epsilon,$$

and

$$-(x - a) < \epsilon.$$

Both numbers $x - a$ and $-(x - a)$ are less than ϵ, and consequently $|x - a| < \epsilon$. $\square$

Resorting to geometrical illustrations of elements of F on the number line, we can say $|a|$ is the distance of a from zero, and so $|x - a|$ is the distance of x from a. Theorem 4.4 says that points x having distance from a less than ϵ constitute the interval $]a - \epsilon, a + \epsilon[$. (See Figure 4.2)

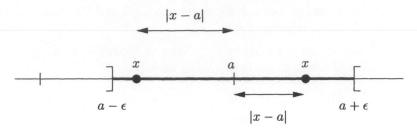

Fig. 4.2   Intervals on the real line

Exercises

Exercise 4.3.1   *Evaluate* $|(x-1)(x-2)|$ *for* $x = 0, 1, -5, 3, 4$.

Exercise 4.3.2   *Solve the following inequalities (describe the set for which each of the inequalities holds):*
  (1) $|x - a| \geq \epsilon$;
  (2) $|x - 1| + x < 2$;
  (3) $0 < |x + 1| < 4$;
  (4) $3x - 1 < |x| < 3x + 1$;
  (5) $|x| \leq |x - 2| + 1$;
  (6) $|x| + |x - s| < 2$;
  (7) $|x^2 - 8x + 10.5| < 2.5$.

4.4   Using *Maple* for solving inequalities

Maple can save a lot of mechanical work when solving inequalities. The form of the command is

```
solve(inequality, unknown);
```

The parameter **unknown** is an option which can be left out if only one variable occurs in the inequality. For instance

---

```
>   solve((x+1)/(x+3)<(x+5)/(x+6));
```

RealRange(Open(−9), Open(−6)), RealRange(Open(−3), ∞)

This answer means that the inequality is satisfied for all x in the intervals $]-9, -6[$ and $]-3, \infty[$. Another example is

---

```
>  solve(abs(2*x^2-16*x+21)<5);
```

$$\text{RealRange}(\text{Open}(4 - 2\sqrt{2}), \text{Open}(4 - \sqrt{3})),$$
$$\text{RealRange}(\text{Open}(4 + \sqrt{3}), \text{Open}(4 + 2\sqrt{2})).$$

The solution set consists again of two intervals. In order to obtain a better idea of what these intervals are we repeat the question, but enter the numbers as floating point numbers.

```
>  solve(abs(x^2-8*x+10.5)<5.0);
```

$$\text{RealRange}(\text{Open}(.7596296508), \text{Open}(3.292893219)),$$
$$\text{RealRange}(\text{Open}(4.707106781), \text{Open}(7.240370349)).$$

If we use the option **unknown**, enclosing the variable in braces, this tells *Maple* to provide the solution as a set of inequalities. We recommend that this option is always used when solving inequalities.

```
>  solve(abs(x^2-8*x+10.5)<5.0,{x});
```

$$\{.7596296508 < x, \, x < 3.292893219\},$$
$$\{x < 7.240370349, \, 4.707106781 < x\}.$$

Most of our readers would find solving the next inequality difficult but *Maple* has no problems. The sign 'less or equal to' is entered into *Maple* as <=. Similarly the sign 'greater than or equal to' is entered as >=. The order in which < and = appear is important! We do not expect a neat result and therefore enter the numbers as floating point numbers.

```
>  solve(x^3+x>=1.0,{x});
```

$$\{.6823278038 \le x\}$$

Solving inequalities involving polynomials of degree three might involve solving an algebraic equation of degree three; this topic is dealt with in Chapter 8. So far *Maple* has provided complete answers. The next example shows that on occasions a small amount of additional work is needed.

```
>   # We are going to solve an inequality which involves a
>   # parameter, so the option 'unknown' is now important.
>   # It is advisable to enclose the unknown in braces.
>   solve(b*x+3<2*b,{x});
```

$$\{\text{signum}(b)\, x < \frac{\text{signum}(b)\,(-3 + 2\,b)}{b}\}$$

From this it is easy to read the result: if b is positive the inequality is satisfied for $x < (-3+2b)/b$, if b is negative it is satisfied for $x > (-3+2b)/b$. (For $b = 0$ it does not make sense to solve the inequality with respect to x.) In this particular case solving with paper and pencil is simpler.

A system of inequalities like $a < b < c$ must be entered in the unabbreviated form as $a < b$ and $b < c$. If there are several inequalities to solve they should be entered as a set. The following example illustrates the result from *Maple* when we solve the system of inequalities $1 < x^2 - 2x - 2 < 3x - 6$.

```
>   ineqs:={1<x^2-2*x-2,x^2-2*x-2<3*x-6}:
>   solve(ineqs,{x});
```

$$\{x < 4,\ 3 < x\}$$

These results mean that the system $1 < x^2 - 2x - 2 < 3x - 6$ is satisfied if and only if $3 < x < 4$. This is illustrated in Figure 4.3 below: the solution set is the set of x where the parabola lies above the line $y = 1$ and below the line $y = 3x - 6$. It is important to understand *Maple's* answer like $\{x < -1\}$. This denotes the set with one element, namely the inequality $x < -1$. The set $\{x < -1\}$ should not be confused with $\{x : x < -1\}$, the set of real x less than -1.

```
>   plot({1,x^2-2*x-2,3*x-6},x=-6..6,y=-6..6);
```

A mathematician cannot solve every and all problems, and neither can *Maple*. It is, however, surprising that *Maple* has difficulties with

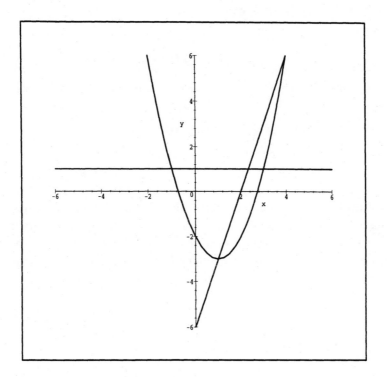

Fig. 4.3   Plot of 1, $x^2 - 2x - 2$, $3x - 6$

solving some very simple problems, for example, $x^2 > a$.[1] In this particular instance this is caused by the presence of the parameter a.

```
>   solve(x^2>a,{x});
```

$$\{-\sqrt{\operatorname{signum}(a)}\, x < -\sqrt{\operatorname{signum}(a)}\, \sqrt{a}\},$$
$$\{\sqrt{\operatorname{signum}(a)}\, x < -\sqrt{\operatorname{signum}(a)}\, \sqrt{a}\}$$

This is an incomprehensible result but if we help *Maple* by making an assumption on a we get the correct answer:

---

[1]This was true at the time of writing this book. Later versions of *Maple* may be more powerful.

```
>  solve(x^2>a,{x}) assuming a>0;
```

$$\{-x < -\sqrt{a}\}, \{x < -\sqrt{a}\}$$

Unfortunately the command **assuming** does not help in the next example.

```
>  solve(x^2>a,{x}) assuming a<0;
```

$$\{-I\,x < \sqrt{-a}\}, \{I\,x < \sqrt{-a}\}$$

The correct answer is that the inequality is always satisfied. At this stage we recommend to our readers not to use *Maple* for solving inequalities which involve some parameter (like a in our case). However the command **assuming** is worth remembering. *Maple* knows also **assume**, the difference between **assuming** and **assume** is that the former applies only to the command which it follows. The latter should be used before the command to which it applies and then it remains valid for the rest of the worksheet.

Exercises

Exercise 4.4.1   *Use* Maple *to solve* Exercises 4.2.2 *and* 4.3.2.

Exercise 4.4.2   *Use* Maple *to find the unions and the intersections of*

(1) $\{x > 0\}$ *and* $\{x < 1\}$,
(2) $\{x > 0\}$ *and* $\{1, 2, 3\}$

4.5   Inductive sets

In this section F denotes an ordered field. A set $S \subset F$ is said to be *inductive* if

(i) $1 \in S$,
(ii) $x \in S \Rightarrow (x + 1) \in S$.

Obviously F itself is inductive. The intersection of all inductive sets in F is denoted by $\mathbb{N}$. We may describe this by saying that $\mathbb{N}$ is the smallest inductive set. Since $1 \in \mathbb{N}$ and $1 + 1 = 2 \in \mathbb{N}$ by (ii), and then again by (ii) $1 + 2 = 3 \in \mathbb{N}$ it is intuitively clear that $\mathbb{N}$ must contain all natural numbers. However, from a strictly logical view natural numbers are defined as elements of $\mathbb{N}$ (in this book). We restate that $\mathbb{N}$ is the smallest inductive set as

---

Theorem 4.5 (Principle of Mathematical Induction) *If M is a subset of $\mathbb{N}$ with the following properties*

(i) $1 \in M$, *and*
(ii) $n \in M \Rightarrow (n+1) \in M$ *for every* $n \in M$,

then $M = \mathbb{N}$.

Although Theorem 4.5 is almost a trivial consequence of the definition of $\mathbb{N}$ it can be useful. As an example of its application we prove

---

Theorem 4.6 $1 \leq n$ *for every* $n \in \mathbb{N}$.

Proof.   We define $M = \{x : x \in \mathbb{N}, 1 \leq x\}$. Obviously $1 \in M$ and if $n \in M$ then $1 \leq n$ and $1+1 \leq n+1$. Since $1 < 1+1$ we have $1 \leq n+1$, that is, $n+1 \in M$. M is inductive and Theorem 4.5 gives $M = \mathbb{N}$. This, in turn, means that $1 \leq n$ for every $n \in \mathbb{N}$. $\square$

Theorem 4.7   *If $n \in \mathbb{N}$, $m \in \mathbb{N}$ and $m < n$ then $n - m \geq 1$.*

In other words, there is no element of $\mathbb{N}$ between m and $m + 1$.

Proof.   We define $S = \{m; m \in \mathbb{N}, m < n$ implies $n - m \geq 1$ for $n \in \mathbb{N}\}$. It suffices to show that $S = \mathbb{N}$. We do this by showing that S is inductive. First we prove that $1 \in S$. For that we consider $M = \{1\} \cup \{n; n \in \mathbb{N}, n - 1 \geq 1\}$. It is easily verified that M is inductive; by Theorem 4.5 then $M = \mathbb{N}$, that is, $1 \in S$. Now let $m \in S$ and $n \in \mathbb{N}$, $m + 1 < n$. This means that $n > 1$ and $m < n - 1$; since $m \in S$ it follows that $n - 1 - m \geq 1$. This last inequality shows that $m + 1 \in S$. $\square$

We are now in a position to start building the theory of natural numbers; however we shall not pursue this, but simply leave some indications in the exercises how this theory can be established in our framework. We proved Theorems 4.5 and 4.6 because they are very important for some fundamental results of the next section (especially Theorem 4.10: The Well Ordering Principle).

The German mathematician L. Kronecker once remarked "God made the natural numbers, everything else is the work of man". This reflects two things. Firstly, the natural numbers are fundamental; we cannot explain what they are. Secondly, after accepting natural numbers (for example, on an axiomatic basis), the negative integers, the rationals and the real numbers themselves can be defined and their respective theories built on the foundation of the theory of natural numbers. The axiomatic theory of natural numbers was established by the Italian mathematician G. Peano and the development indicated above is presented in Landau (1951) and Youse (1972). In this book we have chosen a different starting point, namely the axiomatic system for the reals, created for us by previous generations of mathematicians. However, we briefly outline the Peano approach in Section 4.8.

Having defined $\mathbb{N}$ we can define $\mathbb{Z}$ and $\mathbb{Q}$ as

$$\mathbb{Z} = \mathbb{N} \cup \{0\} \cup \{x;\; -x \in \mathbb{N}\},$$
$$\mathbb{Q} = \{x;\; x = ab^{-1},\; a \in \mathbb{Z},\; b \in \mathbb{N}\}.$$

Recall that we also set $\mathbb{N}_0 = \mathbb{N} \cup \{0\}$ and $\mathbb{P} = \{x;\; x > 0\}$.

Exercises

Exercise 4.5.1   *Prove: $k \in \mathbb{N}$ and $m \in \mathbb{N}$ imply $k + m \in \mathbb{N}$.*
[Hint: Use mathematical induction.]

Exercise 4.5.2   *Prove: $k \in \mathbb{N}$ and $m \in \mathbb{N}$ imply $km \in \mathbb{N}$.*
[Hint: Use mathematical induction.]

Exercise 4.5.3   *Prove: $k \in \mathbb{Z}$ and $m \in \mathbb{Z}$ imply $k + m \in \mathbb{Z}$ and $km \in \mathbb{Z}$.*

Exercise 4.5.4   *Prove: $n \in \mathbb{N} \Rightarrow n \le n^2$.*
[Hint: Use Theorem 4.6.]

4.6   The least upper bound axiom

We start with a definition.

Definition 4.2 (Bounded sets in ordered fields) A set S in an ordered field F is said to be *bounded above* if there exists $K \in F$ such that

$$x \leq K$$

for all $x \in S$. S is said to be *bounded below* if there exists k such that

$$k \leq x$$

for all $x \in S$. S is said to be *bounded* if it is bounded above and bounded below. The element K is called an *upper bound* for S and k is called a *lower bound* for S.

Clearly, set S is bounded if and only if there exists K_1 such that

$$|x| \leq K_1$$

for all $x \in S$. The 'if' part is obvious with $K = K_1$ and $k = -K_1$. For the 'only if' part choose $K_1 = \text{Max}(|K|, |k|)$.

Example 4.2 Let $S = \left\{ x : x = \dfrac{n^2 + 5n + 3}{n^2 + n + 1}, \ n \in \mathbb{N} \right\}$.
M is bounded above since

$$\frac{n^2 + 5n + 3}{n^2 + n + 1} < \frac{n^2 + 5n + 3}{n^2} \leq 1 + 5 + 3 = 9.$$

Thus 9 is *an* upper bound. Is there a smaller upper bound? It is easy to see that the answer is 'yes'. Taking into account that $n \leq n^2$ we obtain

$$\frac{n^2 + 5n + 3}{n^2 + n + 1} = 1 + 2\frac{2n + 1}{n^2 + n + 1} \leq 1 + 2\frac{n^2 + n + 1}{n^2 + n + 1} = 3$$

The natural question now is: Is there an even smaller upper bound? This time the answer is 'no', because for $n = 1$ we have $\dfrac{n^2 + 5n + 3}{n^2 + n + 1} = 3$. The number 3 is the smallest or, in more usual words, the *least upper bound*.

If we illustrate the elements of a set S as points on a vertical number line then we can imagine an upper bound K for S as a kind of obstruction which the elements of S cannot pass; they stay below K. Geometric intuition leads us to believe that K can be moved to its lowest position, thus becoming the least upper bound.

Definition 4.3 (Least upper bound)   An element $s \in F$ is said to be the *least upper bound* for $S \subset F$ if
(i) $x \leq s$ for every $x \in S$;
(ii) if $t < s$ then there exists $a \in S$ such that $t < a(\leq s)$.

Property (i) states that s is an upper bound, while (ii) means that any number strictly less than s is not an upper bound, that is, s is the smallest possible upper bound. A set S can have at most one least upper bound; if there were two, say s_1 and s_2, with $s_1 < s_2$ then by (ii) there will be $a \in S$ such that $s_1 < a(\leq s_2)$. This last inequality contradicts our assumption that s_1 is an upper bound for S.

The Latin word *supremum* is used with the same meaning as least upper bound. The symbol $\sup S$ denotes the least upper bound (supremum) of a set S. The phrase 'least upper bound' is sometimes abbreviated 'lub'.

Definition 4.4 (Complete field)   An ordered field is said to be *complete* if every non-empty set bounded above has a least upper bound.

Definition 4.5 (Real numbers as a complete ordered field) The real number system is a complete ordered field.

Remark 4.1   It can be proved that there is *essentially* only one complete ordered field. Spivak (1967) contains a precise statement of this result as well as the proof on page 507. However, we need not worry unduly now how many complete ordered fields there are, because whatever we prove will be valid in all of them. As agreed previously the set of all reals is denoted by $\mathbb{R}$.

We reformulate Definition 4.5 as

The Least Upper Bound Axiom   Every non-empty set $S \subset \mathbb{R}$ which is bounded above has a least upper bound.

The number 3 was the least upper bound for S in Example 4.2, because it was the largest element in S. Whenever a set M has a largest element then this element is the least upper bound. Also if, for a set S, the least upper bound of S belongs to S then it is the largest element of S. However the least upper bound of S need not belong to S: for the set $S = \{x;\ x \in \mathbb{R},\ x < 0\}$ we obviously have $\sup S = 0$ and $0 \notin S$.

Theorem 4.8   *For any $x \in \mathbb{R}$ there always exists $n \in \mathbb{N}$ such that $x < n$.*

Proof. Assume, contrary to what we want to prove, that there is $K \in \mathbb{R}$ such that $n \le K$ for every $n \in \mathbb{N}$. Then $\mathbb{N}$ has a supremum s. Choose $t = s - 1$; then there exists $k \in \mathbb{N}$ such that $s - 1 < k$ It follows that $s < k + 1 \in \mathbb{N}$, and consequently s is not an upper bound for $\mathbb{N}$, which is a contradiction. □

Theorem 4.8 can be rephrased by saying that $\mathbb{N}$ as a subset of $\mathbb{R}$ is not bounded. This is often referred to as the *Archimedean property*. There are ordered fields which do not have the Archimedean property. Such fields obviously cannot be complete. An example of an ordered field which does not have the Archimedean property is given in Exercise 6.2.10.

Definition 4.6 (Greatest lower bound) An element $m \in F$ is said to be the *greatest lower bound* for $S \subset F$ if
(i) $m \le x$ for every $x \in S$;
(ii) if $m < t$ then there exists $a \in S$ such that $m \le a < t$.

The Latin word *infimum* is used with the same meaning as greatest lower bound. The symbol $\inf S$ denotes the greatest lower bound (infimum) of a set S. The phrase 'greatest lower bound' is sometimes abbreviated 'glb'.

A set $S \subset \mathbb{R}$ can have at most one infimum; the proof is very similar to the above given proof of uniqueness of the least upper bound.

Theorem 4.9 *Every non-empty set $S \subset \mathbb{R}$ bounded below has a greatest lower bound.*

Proof. Define $M = \{x : -x \in S\}$. It is not difficult to see that M is non-empty and bounded above. By the least upper bound axiom there exists $\sup M$. It is now easy to check that $-\sup M$ is the greatest lower bound for S. □

Theorem 4.10 (The Well Ordering Principle) *Every non-empty set $M \subset \mathbb{N}$ has a smallest element.*

Proof. M is bounded below and non-empty, therefore it has a greatest lower bound i. By the second property of the greatest lower bound there is a $k \in M$ such that $i \le k < i + 1$. We now prove indirectly that k is the smallest element of M. If it were not, there would be $n \in M$ such that $n < k$. Hence we would have $i \le n < k < i + 1$ and consequently $k - n < 1$, contradicting Theorem 4.7. □

Theorem 4.11   *For every $x \in \mathbb{R}$ there exists a unique $n \in \mathbb{Z}$ such that $n \le x < n+1$. This number is called the* floor *of x, and is denoted by $\lfloor x \rfloor$.*

We obviously have $\left\lfloor \dfrac{3}{2} \right\rfloor = 1$, $\left\lfloor \dfrac{1}{3} \right\rfloor = 0$, $\left\lfloor -\dfrac{1}{2} \right\rfloor = -1$, $\lfloor 0 \rfloor = 0$, etc.

Proof.   The number n (if it exists) is unique, for if $n \le x < n + 1$ and $\bar{n} \le x < \bar{n} + 1$, with $n < \bar{n}$, we would have $\bar{n} - n \le x - n < 1$. On the other hand, $\bar{n} - n \in \mathbb{N}$, so we have a contradiction with Theorem 4.7.

If $x \in \mathbb{Z}$ there is nothing more to be proved, so we assume $x \notin \mathbb{Z}$. The set of natural numbers strictly greater than $|x|$ has the smallest member k; if $x > 0$ we take $n = k - 1$, if $x < 0$ we take $n = -k$. □

$\lfloor x \rfloor$ is also called the *integer part of* x, it is denoted by `floor(x)` in *Maple*. The function $x \mapsto \lfloor x \rfloor$ is called the *greatest integer function*.

Remark 4.2   Similarly as in Theorem 4.11 it is possible to prove that for every $x \in \mathbb{R}$ there is a unique integer n such that $n - 1 < x \le n$. This number is called the ceiling of x, denoted in *Maple* by `ceil(x)`. The mathematical symbol is $\lceil x \rceil$. For $x \in \mathbb{Z}$ we have $\lfloor x \rfloor = \lceil x \rceil$, otherwise $\lfloor x \rfloor + 1 = \lceil x \rceil$.

Theorem 4.12   *If $a \in \mathbb{R}$, $b \in \mathbb{R}$ and $a < b$ there exists $r \in \mathbb{Q}$ such that $a < r < b$.*

The content of Theorem 4.12 is sometimes described by saying the set of rationals is dense. If we try to illustrate graphically the set $\mathbb{Q}$ on the number line we would not be able to distinguish between $\mathbb{Q}$ and the number line $\mathbb{R}$ itself, because the points for $\mathbb{Q}$ are so densely packed.

Proof.   Since $\mathbb{N}$ is not bounded above there is $k \in \mathbb{N}$ such that $k > \dfrac{2}{b-a}$; that is, $\dfrac{1}{k} < \dfrac{b-a}{2}$. Define $m = \left\lfloor \dfrac{a+b}{2} k \right\rfloor$ and we have $\dfrac{m}{k} \le \dfrac{a+b}{2} < \dfrac{m+1}{k}$. It follows that $a = \dfrac{a+b}{2} - \dfrac{b-a}{2} < \dfrac{m+1}{k} - \dfrac{1}{k} = \dfrac{m}{k} \le \dfrac{a+b}{2} < b$. We have that $r = \dfrac{m}{k} \in \mathbb{Q}$ and $a < r < b$. □

Actually we have proved that there are *infinitely many* rationals lying between any two reals, since we can take for k any of the infinite number of integers greater than $2/(b-a)$.

Exercises

Exercise 4.6.1   *Find* sup S *and* inf S *if*

(1) $S = \left\{ x; \ x = 1 - \dfrac{1}{n}, \ n \in \mathbb{N} \right\};$

(2) $S = \{5\} \cup \{3\} \cup \,]0, 1[;$

(3) $S = \{-5\} \cup \{3\} \cup \,]4, 5[;$

(4) $S = \{x; \ x \in \mathbb{R}, \ x^2 + 5x < 6.$

Exercise 4.6.2   *Show that the set* $S = \left\{ x; \ x = \dfrac{n^2 - 5}{n^4 - 3n^2 - 5}, \ n \in \mathbb{N} \right\}$ *is bounded from above and from below, and find* sup S *and* inf S.

Exercise 4.6.3   *If* S_1 *and* S_2 *are bounded then so is* $S_1 \cup S_2$. *Prove this.*

ⓘ **Exercise 4.6.4**   *If* $S_1 \subset S_2 \subset \mathbb{R}$, $S_1 \neq \emptyset$, S_2 *bounded, then* sup $S_1 \le$ sup S_2 *and* inf $S_1 \ge$ inf S_2. *Prove this. Give an example of an* S_1 *which is a proper subset of* S_2 *but for which* sup S_1 = sup S_2.

ⓘ **Exercise 4.6.5**   *If* X *and* Y *are bounded non-empty subsets of* $\mathbb{R}$ *and for every* $x \in X$ *and every* $y \in Y$ *one has* $x \le y$ *then* sup $X \le$ inf Y. *If, however, for every* $x \in X$ *there exists a* $y \in Y$ *such that* $x \le y$ *then* sup $X \le$ sup Y. *Prove this, and also show that in the second case* sup X *could be greater than* inf Y.

ⓘ **Exercise 4.6.6**   *Prove that if* $a < b$ *there exists an irrational* x *such that* $a < x < b$.
[Hint. Use the same idea as in the proof of Theorem 4.12 and apply the result of Exercise 2.4.4.]

Exercise 4.6.7   *Graph the following functions, preferably without the use of Maple.*

(1) $x \mapsto \lfloor x \rfloor;$

(2) $x \mapsto x - \lfloor x \rfloor;$

(3) $x \mapsto \left(x - \lfloor x \rfloor \right)^2;$

(4) $x \mapsto \lfloor x \rfloor + \lfloor -x \rfloor;$

(5) $x \mapsto \left(x - \lfloor x \rfloor \right)\left(\lfloor x \rfloor + 1 - x \right);$

(6) $x \mapsto \left\lfloor \dfrac{1}{x} \right\rfloor;$

(7) $x \mapsto \dfrac{1}{\left\lfloor \dfrac{1}{x} \right\rfloor}.$

[Hint: For (1)–(3) consider separately the intervals $[n, n + 1[$ with $n \in \mathbb{N}$. The other parts of the exercise should be treated similarly. For instance in part (6), we have $\left\lfloor \dfrac{1}{x} \right\rfloor = n$ if $n \leq \dfrac{1}{x} < n + 1$. Even if you are using *Maple* look at the intervals separately.]

Exercise 4.6.8   *The function* $x \mapsto \operatorname{Min}(x - \lfloor x \rfloor, \lfloor x \rfloor + 1 - x)$ *is called the distance from* x *to the nearest integer, and denoted by* saw. *Graph the following functions:*

(1) $x \mapsto \operatorname{saw}(x)$;

(2) $x \mapsto \operatorname{saw}(2x)$;

(3) $x \mapsto 2 \operatorname{saw} \left(\dfrac{1}{2}x \right)$.

Exercise 4.6.9   *Graph he hump function* h, $h(x) = |x + 1| + |x| + |x - 1|$ *with Maple and also without it.*

4.7   Operation with real valued functions

In this section we assume that the ranges of all functions appearing are subsets of $\mathbb{R}$. It is customary to define the sum, difference, product and quotient of functions as follows.

$$f + g : x \mapsto f(x) + g(x); \; x \in \operatorname{dom} f \cap \operatorname{dom} g,$$

$$f - g : x \mapsto f(x) - g(x); \; x \in \operatorname{dom} f \cap \operatorname{dom} g,$$

$$fg : x \mapsto f(x)g(x); \; x \in \operatorname{dom} f \cap \operatorname{dom} g,$$

$$\frac{f}{g} : x \mapsto \frac{f(x)}{g(x)}; \; x \in \operatorname{dom} f \cap \operatorname{dom} g \cap \{x; \; g(x) \neq 0\}, \qquad (4.10)$$

$$-g : x \mapsto -g(x),$$

$$\frac{1}{g} : x \mapsto \frac{1}{g(x)}; \; x \in \{x; \; g(x) \neq 0\}.$$

Instead of ff we shall write f^2. For example, $\operatorname{id}^2 : x \mapsto x^2$. We shall write $f \leq g$ (or $f < g$) to mean $f(x) \leq g(x)$ (or $f(x) < g(x)$) for all $x \in \operatorname{dom} f \cap \operatorname{dom} g$. If $f < g$ and $\operatorname{dom} f = \operatorname{dom} g$ then the graph of f lies entirely below that of g.

The definitions (4.10) make sense if the ranges of f and g are part of some field, not necessarily of $\mathbb{R}$. If, during a discourse, there is some underlying field, definitions (4.10) are assumed to hold automatically.

Exercises

(i) **Exercise 4.7.1** *Prove:*

(1) $\mathbf{1}_A\mathbf{1}_B = \mathbf{1}_{A\cap B}$;

(2) $A \cap B = \emptyset \Rightarrow \mathbf{1}_A + \mathbf{1}_B = \mathbf{1}_{A\cup B}$;

(3) $\mathbf{1}_A + \mathbf{1}_B - \mathbf{1}_{A\cap B} = \mathbf{1}_{A\cup B}$;

(4) $A \subset B \Rightarrow \mathbf{1}_A \leq \mathbf{1}_B$;

(5) $\mathbf{1}_{A\setminus B} = \mathbf{1}_A - \mathbf{1}_{A\cap B}$.

4.8 Supplement. Peano axioms. Dedekind cuts

The axioms listed earlier in this Chapter, the axioms for a field given in Table 4.1, the order axioms listed in Table 4.3 and the least Upper Bound axiom treated in Section 4.6 are not the only possible system of axioms for the reals. We have already seen in Exercise 4.1.4 that the field axioms can be formulated differently. However the difference between the two sets of axioms there was minor. Different approaches to the real number system are possible. The one we shall consider in this section is, from a philosophical point of view, diametrically opposite to ours: instead of considering the natural numbers as a subset of the reals it starts with the natural numbers and builds up to the reals. Perhaps, from a purely logical viewpoint, this approach may be preferable. We did not employ it because it is long and laborious. It starts with what are called the *Peano axioms* for the natural numbers. We briefly indicate this approach leaving many proofs aside. There are five Peano's axioms.

Peano 1: *There is a nonempty set with a distinguished element 1.*

The set will be henceforth denoted $\mathbb{N}$.

Peano 2: *For each $x \in \mathbb{N}$ there exists one and only one element x'.*

The element x' will be from now on called the *successor* of x: after each element of $\mathbb{N}$ there is one which immediately follows it. This axiom describes in abstract the idea of counting: 2 comes after 1, 3 comes after 2, etc., $n+1$ comes after n, the successor x' comes after x.

Peano 3: $x' \neq 1$ *for every $x \in \mathbb{N}$.*

In words, 1 is not the successor of any element in $\mathbb{N}$.

Peano 4: *If $x' = y'$ then $x = y$.*

This means: there is at most one x for a given x'. In other words the mapping $x \mapsto x'$ is one-to-one.

Peano 5: *If M is a subset of* $\mathbb{N}$ *with the following properties*

   (i) $1 \in M$, *and*
   (ii) $n \in M \Rightarrow (n+1) \in M$ *for every* $n \in M$,

then $M = \mathbb{N}$.

We immediately recognize this axiom as Mathematical Induction: in our approach this is Theorem 4.5.

Note that in our development Mathematical Induction was derived as a theorem, while here it is one of the fundamental axioms. With these five axioms at hand building up the theory of natural numbers proceeds by *defining* addition, multiplication and order in $\mathbb{N}$ and establishing the 'usual' properties of natural numbers. For instance, addition of x and y is defined as follows:

   To every pair of natural numbers x, y there exists a uniquely determined natural number, henceforth denoted $x + y$ such that

   (i) $x + 1 = x'$ *for every* $x \in \mathbb{N}$;
   (ii) $x + y' = (x + y)'$ *for every* $x, y \in \mathbb{N}$.

   This, of course, as well as the definitions of multiplication and order, require proof, which we omit but mention that axiom **Peano 5** is essential for the proof.

   Then the theory moves to extend $\mathbb{N}$ to $\mathbb{Z}$, in other words introducing 0 and the negative integers. We skip over this, noting in passing that it can be shown that the elements of $\mathbb{Z}$ thus defined satisfy the axioms for a ring (see Table 2.3), but pause at the next step, the creation of $\mathbb{Q}$. We denote pairs of elements of $\mathbb{Z}$ as p/q or $\dfrac{p}{q}$ rather than (p, q), though what follows here can be written in terms of ordered pairs if desired. Two such pairs $\hat{p}/\hat{q}$ and p/q with $\hat{q} \neq 0$ and $q \neq 0$ are declared equivalent if

$$\hat{p}q = p\hat{q} \qquad (4.11)$$

This relation is indeed an equivalence relation according to Section 3.1, that is, it is symmetric, reflexive and transitive. As is customary we identify equivalent pairs: two equivalent pairs are simply equal. The set of pairs p/q with $p \in \mathbb{Z}$ and $0 \neq q \in \mathbb{Z}$ for which addition and multiplication is defined below is denoted by $\mathbb{Q}$. The definitions are

addition
$$\frac{p}{q} + \frac{r}{s} = \frac{ps + qr}{qs},$$

multiplication
$$\frac{p}{q} \cdot \frac{r}{s} = \frac{pr}{qs}.$$

These definition are meaningful. Firstly with $q \neq 0$ and $s \neq 0$ we have $qs \neq 0$. Secondly, the definitions are independent of the choice of representation of the pairs. If $p/q = \hat{p}/\hat{q}$ then $\frac{p}{q} + \frac{r}{s} = \frac{\hat{p}}{\hat{q}} + \frac{r}{s}$ and $\frac{p}{q} \cdot \frac{r}{s} = \frac{\hat{p}}{\hat{q}} \cdot \frac{r}{s}$. To prove this, we first note

$$p\hat{q} = \hat{p}q,$$

It follows that

$$\ddot{p}q = p\ddot{q},$$
$$ps\hat{q} = \hat{p}sq,$$
$$(ps + qr)\hat{q}s = (\hat{p}s + \hat{q}r)qs,$$
$$\frac{ps + qr}{qs} = \frac{\hat{p}s + \hat{q}r}{\hat{q}s}.$$

From symmetry arguments we have also

$$\frac{ps + qr}{qs} = \frac{p\hat{s} + q\hat{r}}{q\hat{s}},$$

if $r\hat{s} = \hat{r}s$. This shows the correctness of the definition of addition. The proof for multiplication is similar and left to the readers. It is obvious that addition and multiplication of elements in $\mathbb{Q}$ is commutative and there is no difficulty showing that all the other axioms from the Table 4.1 are also satisfied. As a sample of such proofs let us prove axiom D.

$$\frac{p}{q}\left(\frac{r}{s} + \frac{t}{u}\right) = \frac{p}{q}\frac{ru + ts}{su} = \frac{pru + pts}{qsu} = \frac{pru}{qsu} + \frac{pts}{qsu} = \frac{pr}{qs} + \frac{pt}{qu}.$$

At various places in the above proofs use is made of the equivalence relation among pairs of rational numbers expressed in Equation (4.11). The construction of the field $\mathbb{Q}$ from the ring $\mathbb{Z}$ can be applied more generally: instead of starting with $\mathbb{Z}$ one can take a ring $\mathbf{M}$, proceed in the same manner and the resulting field is called the *quotient field* of $\mathbf{M}$. There is an additional requirement on $\mathbf{M}$ for this construction to work, namely $ab = 0$ must imply that either a or b is 0.[2]

After the field axioms are shown to hold in $\mathbb{Q}$ derived in this fashion, it is time to introduce order in $\mathbb{Q}$. This is done by defining the set P of positive elements:[3] p/q with $p, q \in \mathbb{Z}$ and $q \neq 0$ is defined to be positive if $pq \in \mathbb{N}$. The relation a is less than b is then defined by declaring $a < b$ if and only if $(b - a) \in P$. It is not difficult to prove that this relation satisfies the order axioms in Table 4.3.[4] We have now almost reached our goal: we have in $\mathbb{Q}$ an ordered field, and it remains to extend it to a complete ordered field. This can be accomplished by what are known as *Dedekind cuts*, which later will be renamed real numbers. A pair of subsets $A, \bar{A}$ of $\mathbb{Q}$ is called a (Dedekind) cut if the following conditions are satisfied:

C1 Both sets are nonempty and $\mathbb{Q} = A \cup \bar{A}$.
C2 If $a \in A$ and $\bar{a} \in \bar{A}$ then $a < \bar{a}$.
C3 If $a \in A$ then there exists $a' \in A$ satisfying $a < a'$.

Such a cut is denoted by $A|\bar{A}$, the symbol $|$ denoting the 'cut' in $\mathbb{Q}$ between A and $\bar{A}$. The set A is the called the lower section of the cut, and $\bar{A}$ the upper section. Property C3 simply means that A does not have a largest element: $\bar{A}$ may or may not have a smallest element. For cuts an order relation, using the symbol $\prec$, is *defined* as follows:

Order

$A|\bar{A} \prec B|\bar{B}$ if and only if $B \cap \bar{A} \neq \emptyset$. This is fairly intuitive: $B|\bar{B}$ is bigger than $A|\bar{A}$ if B reaches over A into $\bar{A}$. Interpreting this may be easier if the reader makes a sketch. This order has the trichotomy property, Axiom O_1 in Table 4.3: for two cuts $A|\bar{A}$ and $B|\bar{B}$, one and only one of the following is true:[5]

$$A|\bar{A} \prec B|\bar{B}, \qquad A|\bar{A} = B|\bar{B}, \qquad B|\bar{B} \prec A|\bar{A}.$$

If $B \cap \bar{A} \neq \emptyset$ then the first order relation holds. If $B \cap \bar{A} = \emptyset$ then $A \subset B$. If $A = B$ then we have equality, otherwise $\bar{B} \cap A \neq \emptyset$ and then $B|\bar{B} \prec A|\bar{A}$.

In our development the existence of the least upper bound was guaranteed by an axiom: here it becomes a provable theorem. We now prove the least upper bound theorem for cuts.

---

[2] This is, of course, satisfied in $\mathbb{Z}$.
[3] See page 103.
[4] With F replaced by $\mathbb{Q}$ in O_1.
[5] $A|\bar{A} = B|\bar{B}$ means naturally that $A = B$ and $\bar{A} = \bar{B}$

Theorem 4.13 (The Least Upper Bound Theorem for cuts)
*If $\mathfrak{S}$ is a non-empty set of cuts and $M|\bar{M}$ is such that it exceeds every
cut in $\mathfrak{S}$, that is $X|\bar{X} \prec M|\bar{M}$ for every $X|X \in \mathfrak{S}$, then there exists a
cut $S|\bar{S}$ such that*

(i) *if $X|\bar{X} \in \mathfrak{S}$ and $X|\bar{X} \neq S|\bar{S}$ then $X|\bar{X} \prec S|\bar{S}$,*
(ii) *if $T|\bar{T} \prec S|\bar{S}$ then there exists $X|\bar{X} \in \mathfrak{S}$ such that $T|\bar{T} \prec X|\bar{X}$.*

Proof. To make the proof easily readable we divide it into several easy
steps.

Step 1. Definition of $S|\bar{S}$. The set S consists of all rationals which
belong to a lower section of a cut in $\mathfrak{S}$. The complement of S, that is $\mathbb{Q}\setminus S$,
is $\bar{S}$.
The next three steps prove that $S|\bar{S}$ is a cut.

Step 2. Condition C1. It suffices to show that $\bar{M} \cap S = \emptyset$. If this were
not true then $\bar{M} \cap X_0 \neq \emptyset$ for some $X_0|\bar{X}_0 \in \mathfrak{S}$ but then $M|\bar{M} \prec X_0|\bar{X}_0$
contradicting the assumption that $X|\bar{X} \prec M|\bar{M}$ for every $X|\bar{X} \in \mathfrak{S}$.

Step 3. Condition C2. If $s \in S$ and $\bar{s} \in \bar{S}$ then $s \in X_1$ for some
$X_1|\bar{X}_1 \in \mathfrak{S}$ and $\bar{s}$ is in no X for $X|\bar{X} \in \mathfrak{S}$. Consequently $\bar{s} \in \bar{X}_1$ and
therefore $s < \bar{s}$.

Step 4. Condition C3. If $s \in S$ then $s \in X_2$ for some $X_2|\bar{X}_2 \in \mathfrak{S}$ and
there is some $\breve{s} \in X_2 \subset S$ with $s < \breve{s}$.

Step 5. Proof of (i). Let $X|\bar{X} \in \mathfrak{S}$ and $X|\bar{X} \neq S|\bar{S}$. Assume contrary
to what we want to prove that $S|\bar{S} \prec X|\bar{X}$. Then, by the definition of
order in $\mathfrak{S}$, we have $X \cap \bar{S} \neq \emptyset$ and this is impossible, since $X \subset S$.

Step 6. Proof of (ii). By assumption $S \cap \bar{T} \neq \emptyset$ which means there is
an $X|\bar{X} \in \mathfrak{S}$ with $X \cap \bar{T} \neq \emptyset$. By the definition of order in $\mathfrak{S}$ this means
$T|\bar{T} \prec X|\bar{X}$. $\qquad\square$

We now define addition and multiplication for cuts.

Addition
Let $Y = \{a + b; a \in A, b \in B\}$ and $\bar{Y} = \mathbb{Q}\setminus Y$. Then $Y|\bar{Y}$ is a cut[6] and we
define

$$Y|\bar{Y} \overset{\text{def}}{=} A|\bar{A} + B|\bar{B}.$$

We denote by $\hat{0}$ the cut $Z|\bar{Z}$ with $Z = \{x; x \in \mathbb{Q}, x < 0\}$ and $\bar{Z} = \mathbb{Q}\setminus Z$.
The cut $\hat{0}$ is the zero element for addition: for every cut $A|\bar{A}$ we have

---

[6]This, of course, must be proved, but we omit the proof as we did with many other
proofs in this section.

$A|\bar{A} + \hat{0} = A|\bar{A}$. We say that a cut $A|\bar{A}$ is positive if $\hat{0} \prec A|\bar{A}$, which simply means that all elements of $\bar{A}$ are positive. The definition of addition of cuts was simple enough, but unfortunately multiplication is not as easy. It is defined in stages, first for positive cuts and then generally.

Multiplication: positive cuts

Let $A|\bar{A}$ and $B|\bar{B}$ be positive cuts. Let X be the set consisting of negative rationals, zero and all numbers of the form ab with $a \in A$, $b \in B$ and both numbers a, b positive. Define $\bar{X} = \mathbb{Q} \setminus X$. Then $X|\bar{X}$ is a cut[7], and we define multiplication by

$$X|\bar{X} \overset{\text{def}}{=} (A|\bar{A}) \cdot (B|\bar{B}).$$

Clearly the restriction to positive cuts was necessary, otherwise the definition of $X|\bar{X}$ as a cut would not be meaningful.

Multiplication: general cuts

If x, y are cuts then xy is defined by

$$
\begin{aligned}
xy &= \hat{0} && \text{if } x = \hat{0} \text{ or } y = \hat{0}; \\
xy &= -\big((-x)y\big) && \text{if } x \prec \hat{0} \text{ and } \hat{0} \prec y; \\
xy &= -\big(x(-y)\big) && \text{if } \hat{0} \prec x \text{ and } y \prec \hat{0}; \\
xy &= (-x)(-y) && \text{if } x \prec \hat{0} \text{ and } y \prec \hat{0}.
\end{aligned}
$$

With these definitions, it is possible to prove that the cuts form a complete ordered field, that is they satisfy Axioms[8] O_2–O_4 (see Table 4.3) and all the field axioms (see Table 4.1). This is not very difficult but could be tedious: it also involves defining the cut $-x$ (used in the definitions of general cuts) and the unit cut for multiplication, $\hat{1}$.

When is a cut a rational number?

If $\bar{X}$ contains a smallest element x^* then the mapping $X|\bar{X} \mapsto x^*$ is clearly one-to-one and onto $\mathbb{Q}$, so it is a bijection. Moreover, it can be shown that

$$X|\bar{X} + Y|\bar{Y} \mapsto x^* + y^*, \tag{4.12}$$

$$X|\bar{X} \cdot Y|\bar{Y} \mapsto x^* \cdot y^*, \tag{4.13}$$

$$X|\bar{X} \prec Y|\bar{Y} \Leftrightarrow x^* < y^*. \tag{4.14}$$

Relations (4.12) to (4.14) show that the mapping $X|\bar{X} \mapsto x^*$ preserves addition, multiplication and order. Two fields F and F_1 are said to be

---

[7] Again, this needs to be proved

[8] We have already established Axiom O_1.

isomorphic if there exists a bijection of F onto F_1 which preserves addition and multiplication. Two ordered fields are isomorphic if there is a bijection of one onto the other which preserves addition, multiplication and order. The bijection is called an isomorphism. Frequently in mathematics isomorphic structures are identified. This is natural: in two isomorphic structures, elements act the same way in both structures. They may be different in appearance, but not in substance. In accordance with this, we say that those cuts for which the upper section contains a smallest element *are* the rationals.

Cuts and real numbers

There remains the case where, for the cut $X|\bar{X}$, the set $\bar{X}$ does not contain a smallest element. We then use this cut to *define* a real number (in fact, an irrational real number). Our outline of an alternative route to reals has come to an end. We have constructed a complete, ordered field which contains the rationals as an ordered subfield, so we now 'have' the reals.

Summary of Peano's axioms

It is not possible to establish the consistency of Peano's axioms but, as Gödel showed, neither is it possible to establish the consistency of the axioms of set theory. However, the construction of the reals by cuts shows the relative consistency of the axioms for the reals. If Peano's axioms are consistent then so are the axioms for reals. From a philosophical view this is the main difference between the two approaches. From a mathematical point of view they are equally valid.

Chapter 5

Mathematical Induction

In this chapter we study proof by induction and prove some important inequalities, particularly the arithmetic-geometric mean inequality. In order to employ induction for defining new objects we prove the so called recursion theorems. Basic properties of powers with rational exponents are also established in this chapter.

5.1 Inductive reasoning

The process of deriving general conclusions from particular facts is called induction. It is often used in the natural sciences. For example, an ornithologist watches birds of a certain species and then draws conclusions about the behaviour of all members of that species. General laws of motion were discovered from the motion of planets in the solar system. The following example shows that we encounter inductive reasoning also in mathematics.

Example 5.1   Let us consider the numbers $n^5 - n$ for the first few natural numbers:

n	$n^5 - n$
1	0
2	30
3	240
4	1020
5	3120
6	7770
7	16800

It seems likely that for every $n \in \mathbb{N}$ the number $n^5 - n$ is a multiple of 10.

The reasoning in the above example does not give us the feeling of cast-iron certainty which mathematical arguments usually have. It may not be true for $n = 8$, though you can easily check that it is. Even if you have used a computer to check the first billion natural numbers, that does not prove that it is true for *all* natural numbers. Indeed, basing arguments on a finite number of examples is an uncertain procedure, and it can lead to serious mistakes, as we shall shortly see in Example 5.2.

In everyday life, and in the natural sciences, our conclusions are subject to further observations and experiments (devised to check the conclusions). In mathematics this additional check is missing, and there is yet another important difference. In Example 5.1 we observed a few particular cases, but we made conclusions about the validity of the formula for infinitely many cases.

Example 5.2   Consider the inequality

$$\frac{5n + \frac{1}{2}}{5n - \frac{1}{2}} > 1 + \frac{1}{5^5}. \tag{5.1}$$

If we substitute for n quite a few of the early natural numbers we see that the inequality holds. However it would be wrong to conclude that (5.1) holds for all natural numbers n. For natural numbers n the (rational) number $5n - \frac{1}{2}$ is positive, and therefore Inequality (5.1) holds if and only if

$$5n + \frac{1}{2} > \left(5n - \frac{1}{2}\right)\left(1 + \frac{1}{5^5}\right),$$

and this inequality holds if and only if

$$\frac{1}{2} + \frac{1}{2}\left(1 + \frac{1}{5^5}\right) > \frac{1}{5^4}\, n.$$

The last inequality holds if and only if $n < 625.1$. We summarise this as: Inequality (5.1) holds for the natural number n if and only if $1 \le n \le 625$. Inequality (5.1) represents a statement valid for the first 625 natural numbers but not for all natural numbers.

If we considered the inequality

$$\frac{100n + \frac{1}{2}}{100n - \frac{1}{2}} > 1 + \frac{1}{100^{100}} \tag{5.2}$$

instead of (5.1) we could spend all our life (or even all the life of the universe) testing successive particular cases for n and we would never discover that there are natural numbers n for which (5.2) is not true.

Something more than testing the first few (or the first few million) natural numbers is need to prove that $n^5 - n$ is a multiple of 10. The key is Theorem 4.5, the Principle of Mathematical Induction. Let $M = \{n; n \in \mathbb{N}, n^5 - n$ is a multiple of 10 $\}$. Clearly $1 \in M$. Now we want to show that if $n^5 - n$ is a multiple of 10, then so is $(n+1)^5 - (n+1)$. By the Principle of Mathematical Induction it will follow then that $M = \mathbb{N}$, that is, $n^5 - n$ is a multiple of 10 for every $n \in \mathbb{N}$. By expanding $(n+1)^5$ we have

$$(n+1)^5 - (n+1) = n^5 - n + 10(n^3 + n^2) + 5n(n^3 + 1).$$

We are assuming $n^5 - n$ is a multiple of 10, and $10(n^3 + n^2)$ is obviously a multiple of 10, it suffices to show that $5n(n^3 + n)$ is a multiple of 10. One of the numbers n and $n^3 + 1$ must be even, and so $5n(n^3 + 1)$ is indeed a multiple of 10.

We can now summarise: A proof by mathematical induction has two important parts:

(1) *A check that the proposition is valid for the natural number 1.*
(2) *A proof that if the proposition holds for any natural number, then it also holds for the next natural number.*

The second step is often called *inference from n to $n+1$*. The assumption made in the second part, namely that the statement holds for some natural number is often referred to as the *induction hypothesis*.

We have seen in Example 5.2 that the first part by itself is insufficient, even if we consider a lot of particular cases. It is natural to concentrate on the second part, particularly because it is usually the more difficult one. However, neglect of the first part can also spell disaster.

Example 5.3   Let us 'prove' by induction that for every $n \in \mathbb{N}$ the number $2n + 1$ is even. Assuming that $2n + 1$ is even, we obtain then $2(n+1) + 1 = (2n + 1) + 2$ is even because it is the sum of an even number $2n + 1$ (by the induction hypothesis) and 2.

Of course, $2n + 1$ is never even. It is not likely that somebody would make such a blunder as asserting that $2n + 1$ is even for some particular $n \in \mathbb{N}$, but example 5.3 clearly demonstrates the necessity of step 1.

The particular starting value in the first step need not always be 1 as the next example shows. See also exercises 5.1.8, 5.1.9 and 5.1.10.

Example 5.4   We wish to prove that the equation

$$2x + 5y = n \tag{5.3}$$

has a solution for every $n \in \mathbb{N}$, $n \geq 4$ such that x and y belong to $\mathbb{N}_0$. An everyday interpretation of (5.3) is that any amount of whole dollars of at least $4 can be paid with a mixture of $2 and $5 notes (or coins).

Step 1: For $n = 4$ Equation (5.3) has a solution $x = 2$, $y = 0$, and both x and y belong to $\mathbb{N}_0$.

Step 2: Assume that (5.3) has a solution x_0, y_0 in non-negative integers for a particular value of n and consider the equation

$$2x + 5y = n + 1. \tag{5.4}$$

If $y_0 \geq 1$ then it is easy to check that $(x_0 + 3, y_0 - 1)$ is a solution of (5.4) and both $x_0 + 3$ and $y_0 - 1$ are non-negative integers. If, on the other hand, $y_0 = 0$, then $2x_0 = n \geq 4$, and so n must be even. Consequently $x_0 \geq 2$, leading to $x_0 - 2 \geq 0$, $y_0 + 1 > 0$ and $(x_0 - 2, y_0 + 1)$ satisfy (5.4) and both belong to $\mathbb{N}_0$.

We now wish to emphasize the phrase *any natural number* in step 2.

Example 5.5   We 'prove' by induction that all natural numbers are equal. Let S be the set containing 1 and all natural numbers n such that $n = n + 1$. Clearly $1 \in S$ by the definition of S. Now let $n \in S$, that is, $n = n + 1$. Then $n + 2 = (n + 1) + 1 = n + 1$, and consequently $(n + 1) \in S$. So, by the Principle of Mathematical Induction, $S = \mathbb{N}$, and this, in turn, implies that $n = n + 1$ for every $n \in \mathbb{N}$, and so all natural numbers are equal.

The error was that the inference from n to $n + 1$ was incorrect: if $n \in S$ then *either* $n = n + 1$ *or* $n = 1$. However, if $n = 1$ there is no way of showing that $n + 1 = 2 \in S$.

Another amusing example with a similar twist is given in Exercise 5.1.12.

Exercises

Exercise 5.1.1   *Show that $n^{13} - n$ is divisible by 13 for every positive integer n.*

Exercise 5.1.2   *Prove that if p is a prime then $n^p - n$ is divisible by p for every positive integer n.*

Exercise 5.1.3   *Prove the following formulae by mathematical induction:*

(1) $1^2 + 2^2 + 3^2 + \cdots + n^2 = \dfrac{n(n+1)(2n+1)}{6}$;

(2) $1^3 + 2^3 + 3^3 + \cdots + n^3 = \left[\dfrac{n(n+1)}{2}\right]^2$;

(3) $1^4 + 2^4 + 3^4 + \cdots + n^4 = \dfrac{n(n+1)(6n^3 + 9n^2 + n - 1)}{30}$;

(4) $1^5 + 2^5 + 3^5 + \cdots + n^5 = \dfrac{n^2(n+1)^2(2n^2 + 2n - 1)}{12}$;

(5) $1^2 + 3^2 + 5^2 + \cdots + (2n-1)^2 = \dfrac{n(2n-1)(2n+1)}{3}$;

(6) $2^2 + 4^2 + 6^2 + \cdots + (2n)^2 = \dfrac{2n(n+1)(2n+1)}{3}$;

(7) $1^3 + 3^3 + 5^3 + \cdots + (2n-1)^3 = n^2(2n^2 - 1)$;

(8) $2^3 + 4^3 + 6^3 + \cdots + (2n)^3 = 2n^2(n+1)^2$;

(9) $1 \cdot 2 + 2 \cdot 3 + 3 \cdot 4 + \cdots + n(n+1) = \dfrac{n(n+1)(n+2)}{3}$;

(10) $1 \cdot 2 \cdot 3 + 2 \cdot 3 \cdot 4 + 3 \cdot 4 \cdot 5 + \cdots + n(n+1)(n+2) = \dfrac{n(n+1)(n+2)(n+3)}{4}$;

(11) Suggest a generalisation of (9) and (10), and then prove the suggested formula, by using induction.

Exercise 5.1.4 *Derive formulae (5)–(8) in the previous Exercise from formulae (1) and (2).* [Hint:
$1^2 + 2^2 + \cdots + (2n)^2 = 1^2 + 3^2 + \cdots (2n-1)^2 + 4(1^2 + 2^2 + \cdots + n^2).$]

(i) **Exercise 5.1.5** *If $a_1, a_2, \ldots, a_n$ are real numbers then*

$$(a_1 + a_2 + \cdots a_n)^2 = a_1^2 + a_2^2 + \cdots + a_n^2$$
$$+ \ 2(a_1 a_2 + a_1 a_3 + \cdots + a_1 a_n)$$
$$+ \quad a_2 a_3 + a_2 a_4 + \cdots + a_2 a_n$$
$$+ \quad \cdots$$
$$+ \quad a_{n-1} a_n)$$

Prove this by using mathematical induction.

Exercise 5.1.6 *Prove, by mathematical induction, for $x \in \mathbb{R}$*
$(1 + x)\left(1 + x^2\right)\left(1 + x^4\right) \cdots \left(1 + x^{2^n}\right) = 1 + x + x^2 + x^3 + \cdots + x^{2^{n+1} - 1}$,
where the right hand side contains all powers of x up to the highest shown.

Exercise 5.1.7 *Let k be a natural number. Prove that for every $n \in \mathbb{N}$*

$$\left(1 - \frac{1}{k}\right)\left(1 - \frac{2}{k}\right)\left(1 - \frac{3}{k}\right) \cdots \left(1 - \frac{n}{k}\right) \geq 1 - \frac{n(n+1)}{2k}.$$

(i) **Exercise 5.1.8**   *The formula*

$$1 + 2 + 3 + \cdots n = \frac{n(n+1)}{2} + (n-1)(n-2)(n-3)(n-4)(n-5)$$

is valid for $n = 1, 2, 3, 4, 5$ but not for $n = 6$. Construct a formula valid for the first million natural numbers but not valid for all $n \in \mathbb{N}$.

(i) **Exercise 5.1.9**   *Let $M \subset \mathbb{Z}$ with the following properties:*

(1) *M contains an integer r;*
(2) *$k \in M$ implies $k + 1 \in M$.*

Then M contains all integers greater than or equal to r, $M = \{x; \ x \in \mathbb{Z}, \ x \geq r\}$. Prove this.

Exercise 5.1.10   *Use induction to decide for what values of $n \in \mathbb{N}$ the inequality $n + 12 < n^2$ is correct.*

Exercise 5.1.11   *If $m \in \mathbb{N}$, $m > 1$ then $n \leq m^n$. Prove this.*

(i) **Exercise 5.1.12**   *This puzzle is attributed to the Hungarian born American mathematician G. Polya.*

Theorem: *If a finite set of blonde girls contains one girl with blue eyes then all girls in the set have blue eyes.*

The 'proof' proceeds by induction on the number of girls in the set. The statement is obviously true if the set contains only one girl. Consider now a set of $k + 1$ blonde girls $\{G_1, G_2, \ldots, G_{k+1}\}$, and assume that at least one of them, say G_1, has blue eyes. Now taking the set $\{G_1, G_2, \ldots, G_k\}$, and using the induction hypothesis, we deduce that all the girls in this set have blue eyes. Taking now the set $\{G_2, G_3, \ldots, G_{k+1}\}$, which has only k elements and contains a girl with blue eyes, namely G_2, the induction hypothesis ensures that all these girls, including G_{k+1}, have blue eyes. Combining this with the earlier result we see that all $k + 1$ girls have blue eyes, which completes the proof.

Explain this paradox, that is, find the logical error on the above 'proof'.

(i) **Exercise 5.1.13**   *Prove: if $n \in \mathbb{N}_0$ and $m \in \mathbb{N}$ then there exist unique numbers q and r such that*

$$n = mq + r,$$
$$0 \leq r < m,$$

and $q \in \mathbb{N}_0$, $r \in \mathbb{N}_0$. (This theorem expresses what you know from school as division of n by m, with quotient q and remainder r.)

$\textcircled{i}$ **Exercise 5.1.14**   *Prove that if $n \in \mathbb{N}$ then there exist numbers $k \in \mathbb{N}_0$ and a_0, a_1, a_2, ..., a_k, such that*

$$n = 10^k a_k + 10^{k-1} a_{k-1} + \cdots + 10 a_1 + a_0\,,$$

with $a_i \in \{0, 1, 2, 3, \ldots, 9\}$, $i = 0, 1, 2, \ldots, k$.
[Hint: use induction and Exercise 5.1.13.]

5.2   Aim high!

When proving theorems by mathematical induction, we often encounter a seemingly paradoxical situation that a stronger theorem is easier to prove.
   If we try to prove the inequality

$$\frac{1}{1 \cdot 2} + \frac{1}{2 \cdot 3} + \frac{1}{3 \cdot 4} + \cdots + \frac{1}{n(n+1)} < 1 \tag{5.5}$$

by mathematical induction, then (5.5) is of no use for the proof that

$$\frac{1}{1 \cdot 2} + \frac{1}{2 \cdot 3} + \frac{1}{3 \cdot 4} + \cdots + \frac{1}{n(n+1)} + \frac{1}{(n+1)(n+2)} < 1. \tag{5.6}$$

On the other hand, the stronger statement that

$$\frac{1}{1 \cdot 2} + \frac{1}{2 \cdot 3} + \frac{1}{3 \cdot 4} + \cdots + \frac{1}{n(n+1)} = 1 - \frac{1}{n+1} \tag{5.7}$$

can easily be proved by induction. Let $M = \{n;\ n \in \mathbb{N}$, and (5.7) holds$\}$. Clearly $1 \in M$. Let $n \in M$, that is, let (5.7) be true, then

$$\frac{1}{1 \cdot 2} + \frac{1}{2 \cdot 3} + \frac{1}{3 \cdot 4} + \cdots + \frac{1}{n(n+1)} + \frac{1}{(n+1)(n+2)} =$$
$$1 - \frac{1}{n+1} + \frac{1}{(n+1)(n+2)} = 1 - \frac{1}{n+2}.$$

This proves that $n + 1 \in M$, so by the principle of mathematical induction $M = \mathbb{N}$ and (5.7) holds for every $n \in \mathbb{N}$.

5.3   Notation for sums and products

Let $n \in \mathbb{N}$ and a_1, a_2, ..., a_n be elements of a field. The sum and product of the numbers a_i are denoted respectively by

$$a_1 + a_2 + a_3 + \cdots + a_n = \sum_{i=1}^{n} a_i \,,$$

$$a_1 a_2 a_3 \cdots a_n = \prod_{i=1}^{n} a_i \,.$$

The index i is called the summation or product index, respectively, and the numbers below and above the signs $\sum$ and $\prod$ are called the limits of the sum or the product. The upper limit is written simply n, but is to be interpreted as $i = n$. The index i can be replaced by any other letter (or symbol) without altering the value of the sum or product, and is sometimes referred to as a 'dummy index'. Thus

$$\sum_{i=1}^{n} a_i = \sum_{j=1}^{n} a_j = \sum_{t=1}^{n} a_t \,.$$

If $m \in \mathbb{N}$, $m < n$ we can first sum the numbers a_1, a_2, ..., a_m and then the rest, so we have

$$\sum_{i=1}^{m} a_i + \sum_{i=m+1}^{n} a_i = \sum_{i=1}^{n} a_i \,. \tag{5.8}$$

The second sum on the left hand side of (5.8) can also be rewritten in the form

$$\sum_{i=m+1}^{n} a_i = \sum_{j=1}^{n-m} a_{m+j} \,. \tag{5.9}$$

We obtain the right hand side of (5.9) from the left by setting $i = m + j$ and by changing the limits of summation accordingly.

The symbols $\sum$ and $\prod$ also represent parentheses including any of the following terms which involve the summation or product index, so that we can write

$$c \Big(\sum_{i=1}^{n} a_i b_i \Big) d_j = c \sum_{i=1}^{n} a_i b_i d_j \,. \tag{5.10}$$

It follows from the distributive law that

$$c \sum_{i=1}^{n} a_i = \sum_{i=1}^{n} ca_i \, . \tag{5.11}$$

If $b_1, b_2, \ldots, b_n$ are elements of the same field then

$$\sum_{i=1}^{n} a_i + \sum_{i=1}^{n} b_i = \sum_{i=1}^{n} (a_i + b_i) \, . \tag{5.12}$$

Consider now the mn numbers a_{ij}, $i = 1, 2, \ldots, n$, $j = 1, 2, \ldots, m$. For these numbers we can first form the sum

$$b_j = \sum_{i=1}^{n} a_{ij} \, ,$$

for $j = 1, 2, \ldots, m$, and then sum these to obtain

$$\sum_{j=1}^{m} b_j = \sum_{j=1}^{m} \sum_{i=1}^{n} a_{ij} \, , \tag{5.13}$$

or we can first form the sum

$$\sum_{j=1}^{m} a_{ij} \, ,$$

for $i = 1, 2, \ldots, n$ and then sum these numbers and obtain

$$\sum_{i=1}^{n} \sum_{j=1}^{m} a_{ij} \, . \tag{5.14}$$

Since both (5.13) and (5.14) are the sum of the mn numbers a_{ij} it is obvious that

$$\sum_{j=1}^{m} \sum_{i=1}^{n} a_{ij} = \sum_{i=1}^{n} \sum_{j=1}^{m} a_{ij} \, , \tag{5.15}$$

and either sum is often written

$$\sum_{\substack{1 \le i \le n \\ 1 \le j \le m}} a_{ij} \, . \tag{5.16}$$

If $m = n$ then (5.16) can be further abbreviated to

$$\sum_{i,j=1}^{n} a_{ij} \,. \tag{5.17}$$

When dealing with sums like (5.13) or (5.14) care should be taken to use distinctive letters for the summation index. For instance

$$\sum_{i=1}^{4} a_i b_i = a_1 b_1 + a_2 b_2 + a_3 b_3 + a_4 b_4 \,,$$

whereas

$$\sum_{i,j=1}^{4} a_i b_j = a_1 b_1 + a_1 b_2 + a_1 b_3 + a_1 b_4 +$$
$$a_2 b_1 + a_2 b_2 + a_2 b_3 + a_2 b_4 +$$
$$a_3 b_1 + a_3 b_2 + a_3 b_3 + a_3 b_4 + \tag{5.18}$$
$$a_4 b_1 + a_4 b_2 + a_4 b_3 + a_4 b_4 \,.$$

The inner summation limit in a sum like (5.14) can depend on the outer summation index. Thus we have, for example,

$$\sum_{i=1}^{n} \sum_{j=1}^{i} a_{ij} = a_{11} +$$
$$a_{21} + a_{22} +$$
$$a_{31} + a_{32} + a_{33} +$$
$$a_{41} + a_{42} + a_{43} + a_{44} +$$
$$\ldots +$$
$$a_{n1} + a_{n2} + a_{n3} + a_{n4} + \cdots + a_{nn}$$

If we sum the columns in this expansion first and add these sums together we obtain

$$\sum_{i=1}^{n} \sum_{j=1}^{i} a_{ij} = \sum_{j=1}^{n} \sum_{i=j}^{n} a_{ij} \,. \tag{5.19}$$

Formula (5.19) can be proved without explicitly writing out the terms as in (5.18) by realising that each term in (5.19) contains a_{ij} with i and $j \in \mathbb{N}$ such that $1 \leq j \leq i \leq n$. Denoting the summation indices by distinct symbols is also necessary when, for example, two sums need to be multiplied, as in

$$\sum_{i=1}^{n} a_i \sum_{j=1}^{n} b_j = \sum_{i,j=1}^{n} a_i b_j.$$

Formulae (5.8), (5.11) and (5.12) are consequences of the field axioms, and should, in fact, be proved by using mathematical induction. We do not wish to dwell on simple proofs such as these, but, as an example of how these proofs could be arranged, we prove the next theorem, and leave a few similar proofs as exercises.

Theorem 5.1   *If $n \in \mathbb{N}$ and $a_1, a_2, \ldots, a_n$ are real numbers then*

$$\left| \sum_{k=1}^{n} a_k \right| \le \sum_{k=1}^{n} \left| a_k \right|. \tag{5.20}$$

Proof.   If $n = 1$ then (5.20) is true with the equality sign. Now assume (5.20) holds for some $n \in \mathbb{N}$, and consider $n + 1$ real numbers a_k, $k = 1, 2, \ldots, n, n + 1$. Since

$$\sum_{k=1}^{n+1} a_k = \sum_{k=1}^{n} a_k + a_{n+1}$$

we have, by Theorem 4.2 and by the induction hypothesis,

$$\begin{aligned}
\left| \sum_{k=1}^{n+1} a_k \right| &= \left| \sum_{k=1}^{n} a_k + a_{n+1} \right| \\
&\le \left| \sum_{k=1}^{n} a_k \right| + \left| a_{n+1} \right| \\
&\le \sum_{k=1}^{n} \left| a_k \right| + \left| a_{n+1} \right| \\
&= \sum_{k=1}^{n+1} \left| a_k \right|.
\end{aligned}$$

$\square$

There are formulae for products similar to those for sums in (5.8), (5.9),

(5.12) and (5.15).

$$\prod_{i=1}^{m} a_i \prod_{i=m+1}^{n} a_i = \prod_{i=1}^{n} a_i \, ; \tag{5.21}$$

$$\prod_{i=m+1}^{n} a_i = \prod_{i=1}^{n-m} a_{i+m} \, ; \tag{5.22}$$

$$\prod_{i=1}^{n} a_i \prod_{i=1}^{n} b_i = \prod_{i=1}^{n} a_i b_i \, ; \tag{5.23}$$

$$\prod_{i=1}^{n} \prod_{j=1}^{m} a_{ij} = \prod_{j=1}^{m} \prod_{i=1}^{n} a_{ij} \, . \tag{5.24}$$

Care is needed in handling sums and products if the terms following the signs $\sum$ or $\prod$ are independent of the summation or product index, respectively. If, for instance, $a_i = c$ for $i = 1, 2, \dots, n$ then

$$\sum_{i=1}^{n} a_i = \sum_{i=1}^{n} c = nc \, . \tag{5.25}$$

Similarly

$$\prod_{i=1}^{n} c b_i = c^n \prod_{i=1}^{n} b_i \, . \tag{5.26}$$

In order to preserve some formulae in limiting cases, and to avoid continually having to consider these cases separately, it is convenient to introduce the void sum and product by defining

$$\sum_{i=1}^{0} a_i = 0 \, , \tag{5.27}$$

and

$$\prod_{i=1}^{0} a_i = 1 \, . \tag{5.28}$$

These formulae are purely conventional and are devoid of any deeper meaning.

5.3.1 *Sums in* Maple

The `sum()` function can be used to calculate sums. In the form `sum(expr,variable=m..n)`, the specified expression is summed over the given range for the specified variable.

```
>   # Evaluate the sum of i from 1 to 10
>   sum(i,i=1..10);
```

$$55$$

```
>   # Evaluate the sum of i^2+i from 1 to 5
>   sum(i^2+i,i=1..5);
```

$$70$$

Recall that *Maple* gives exact solutions. Often when using the `sum()` command, you will need to use either `evalf()` to obtain a floating point result, or a decimal point somewhere in the sum. In the following examples, `sum()` is being used on fractions.

```
>   # Evaluate the sum of 1/i from 1 to 50
>   sum(1/i,i=1..50);
```

$$\frac{13943237577224054960759}{3099044504245996706400}$$

```
>   # Use evalf() to evaluate the same sum
>   # giving a decimal result
>   evalf(sum(1/i,i=1..50));
```

$$4.499205338$$

```
>   # Use a decimal point to evaluate
>   # the sum of 1/i from 1 to 100,000
>   # giving a decimal result
>   sum(1./i,i=1..100000);
```

12.09014612

Rather than specifying numbers as the range for summation, you can also use variables. In the following examples, the indefinite sum is calculated for $i = 1 \ldots N$).

```
>  # Evaluate the sum of i from 1 to N
>  sum(i,i=1..N);
```

$$\frac{1}{2}(N+1)^2 - \frac{1}{2}N - \frac{1}{2}$$

The above result is not in the form you will find quoted in textbooks or other reference works such as Gradshtein and Ryzhik (1996), so we try to simplify it.

```
>  simplify(%);
```

$$\frac{1}{2}N^2 + \frac{1}{2}N$$

```
>  # Evaluate the sum of i^2 from 1 to N
>  # in simplified form
>  simplify(sum(i^2,i=1..N));
```

$$\frac{1}{3}N^3 + \frac{1}{2}N^2 + \frac{1}{6}N$$

Maple honours our convention on void sums (5.27), at least in cases where the second index is 1 less than the first index. For instance

```
>  sum(i,i=1..0); sum(j^2-1,j=7..6);
```

$$0$$

$$0$$

Maple can sum elements of a list. This enables evaluation of sums which would be difficult to form. The next few lines of *Maple* show how to evaluate a sum of numbers of the form $1/p$ where p runs through the first 100 primes.

```
>  L:=[seq(ithprime(k),k=1..100)]:
>  evalf(sum(1/L[i],i=1..100));
```

$$2.106342121$$

5.3.2 *Products in* Maple

The `product()` function can be used to calculate products. Its usage is very similar to that of the `sum()` function described above.

```
>  # Evaluate the product of 1+1/i^2 from 1 to 10
>  product(1+1/i^2,i=1..10);
```

$$\frac{2200962205}{658409472}$$

Now turn this result into a decimal number.

```
>  evalf(%);
```

$$3.342847117$$

```
>  # Evaluate the product of (i+1)/(i-1) from 2 to 20
>  product((i+1)/(i-1),i=2..20);
```

$$210$$

```
>  # Evaluate the product of (i+1)/(i-1)
>  # for i ranging from 2 to N and reduce it
>  # to what Maple thinks is the simplest form
>  simplify(product((i+1)/(i-1),i=2..N));
```

$$\frac{1}{2} N (N+1)$$

Maple can produce sums and products which would not be so easy to produce otherwise. Assume we want to sum terms of the form $a_i a_j a_k$ where i, j, k are all distinct and run from 1 to 4. Subscripts are created in *Maple* by brackets: a_i is entered into *Maple* as a[i]. The first step is to produce a list of all possible combinations of three elements from $\{1, 2, 3, 4\}$. We use the combinatorics package called combinat and the command is choose.

```
> chs:=combinat[choose](4,3);
```

$$chs := [[1,\ 2,\ 3],\ [1,\ 2,\ 4],\ [1,\ 3,\ 4],\ [2,\ 3,\ 4]]$$

The second element in the third list [1,3,4] can be recalled by

```
> chs[3][2];
```

$$3$$

The required sum is now easily created

```
> sum(product(a[chs[i][j]],j=1..3),i=1..4);
```

$$a_1\, a_2\, a_3 + a_1\, a_2\, a_4 + a_1\, a_3\, a_4 + a_2\, a_3\, a_4$$

Exercises

(i) **Exercise 5.3.1**   *Prove, by induction, for $n \in \mathbb{N}$, $q \in \mathbb{R}$*

(1) $2 \displaystyle\sum_{i=1}^{n} i = n(n+1)$;

(2) $\displaystyle\sum_{i=1}^{n}(a_i + c) = \sum_{i=1}^{n} a_i + nc$;

(3) $(q-1) \displaystyle\sum_{i=1}^{n} q^i = q^{n+1} - 1$.

Exercise 5.3.2   *Rewrite the formulae in Exercise 5.1.3 using the $\sum$ notation.*

Exercise 5.3.3   *Let a_i, b_i be real numbers for $i = 1, 2, 3, \ldots, n$. Prove*

(1) $0 \le a_i < b_i$ *for $i = 1, 2, \ldots, n$ implies* $\displaystyle\prod_{i=1}^{n} a_i < \prod_{i=1}^{n} b_i$;

(2) $0 \le a_i \le b_i$ *for $i = 1, 2, \ldots, n$ implies* $\displaystyle\prod_{i=1}^{n} a_i \le \prod_{i=1}^{n} b_i$;

(3) $\left| \prod_{k=1}^{n} a_k \right| = \prod_{k=1}^{n} |a_k|$;

(4) $\left(\sum_{i=1}^{n} a_i \right)^2 = \sum_{i,j=1}^{n} a_i a_j$;

(5) $\left(\prod_{i=1}^{n} a_i \right)^k = \prod_{i=1}^{n} a_i^k$.

(i) **Exercise 5.3.4**   *Prove that among n real numbers $a_1, a_2, \ldots, a_n$ there is a largest and a smallest. The largest number is denoted by $\mathrm{Max}(a_1, a_2, \ldots, a_n)$ and the smallest by $\mathrm{Min}(a_1, a_2, \ldots, a_n)$.*

Exercise 5.3.5   *Let $a_1, a_2, \ldots, a_n$ be real numbers, and $b_1, b_2, \ldots, b_n$ be positive real numbers. Prove that*

$$\mathrm{Min}\left(\frac{a_1}{b_1}, \frac{a_2}{b_2}, \ldots, \frac{a_n}{b_n} \right) \leq \frac{a_1 + a_2 + \cdots + a_n}{b_1 + b_2 + \cdots + b_n} \leq \mathrm{Max}\left(\frac{a_1}{b_1}, \frac{a_2}{b_2}, \ldots, \frac{a_n}{b_n} \right) .$$

[Hint: Start with $n = 2$. Let the minimum on the left-hand side be a_1/b_1. Derive the inequality $a_1 b_2 + a_1 b_1 \leq a_2 b_1 + a_1 b_1$ from which the left-hand inequality follows. Then use induction to prove the left-hand inequality. The argument for the other inequality is similar.]

(!) **Exercise 5.3.6**   *Show that*

(1) $\displaystyle\sum_{i=0}^{n} \sum_{j=i}^{n} r^j = \sum_{j=0}^{n} (j+1) r^j$;

(2) $\displaystyle\sum_{j=0}^{n} (j+1) x^j = \frac{1 - x^{n+1}}{(1-x)^2} - \frac{(n+1) x^{n+1}}{1-x}$.

[Hint: (1) Use (5.19) and (5.25). (2) Simplify the double sum in (1) by using Exercise 5.3.1 (3).]

5.4   Sequences

A function whose domain is $\mathbb{N}$ is called a *sequence*. Letters a, b, x, y, z, u are often used to denote sequences. The function value of a sequence u at n is called the nth *term of the sequence* and is denoted by u_n. However, we may occasionally use the notation $u(n)$ also. Displaying the first few terms often gives a good idea of the behaviour of the sequence although one must realize that the first few terms alone do not determine the sequence.

Exercises

Exercise 5.4.1   *For a given k define two different sequences which have the same first k terms.*

5.5   Inductive definitions

Sequences can often be defined inductively. For example, the arithmetic sequence with first term c and difference d can be so defined, by setting $a_1 = c$ and $a_{n+1} = a_n + d$ for $n \in \mathbb{N}$. Using the last relation successively for $n = 1, 2, 3, \ldots$ we obtain

$$a_2 = a_1 + d = c + d \,,$$
$$a_3 = a_2 + d = c + 2d \,,$$
$$a_4 = a_3 + d = c + 3d \,,$$
$$\ldots$$

There seems to be no difficulty in finding a_n for every $n \in \mathbb{N}$. Is it really certain that a_n is defined for every n? The answer is 'yes', in this case. However this process of defining a sequence has some similarity with inductive reasoning discussed in Section 5.1 and we saw there that such a process needed to be made quite specific in mathematics. Consider the following.

Example 5.6   Let $\alpha = \dfrac{99}{100}$ and define $f(1) = 1$, $f\left(\lfloor (n+1)\alpha \rfloor + 1\right) = f(n) + 1$. Setting n successively to 1, 2, 3, ... we obtain

$$f(2) = f\left(\left\lfloor \frac{198}{100} \right\rfloor + 1\right) = f(1) + 1 = 2 \,,$$
$$f(3) = f\left(\left\lfloor \frac{297}{100} \right\rfloor + 1\right) = f(2) + 1 = 3 \,,$$
$$f(4) = f\left(\left\lfloor \frac{396}{100} \right\rfloor + 1\right) = f(3) + 1 = 4 \,,$$
$$\ldots =$$

There seems to be no difficulty in finding $f(n)$ successively for every $n \in \mathbb{N}$. However this time there is no sequence f such that

$$f(1) = 1 \,,$$

and

$$f\left(\lfloor (n+1)\alpha \rfloor + 1\right) = f(n) + 1$$

for every $n \in \mathbb{N}$. Indeed, for $n = 100$ we have $\lfloor (n+1)\alpha \rfloor + 1 = 100$, leading to $f(100) = f(100) + 1$. This is clearly impossible.

Something more is needed to define a sequence than just checking that there is no difficulty in calculating the first few (or the first few billion) terms of the sequence. The solution is found in the Recursion Theorem.

Theorem 5.2 (Recursion Theorem) *Let S be a set, $a \in S$ and $g : S \to S$. Then there exists a unique sequence f such that*

$$f(1) = a \,,$$

and

$$f(n+1) = g(f(n)) \,.$$

Proof. Here we prove only the uniqueness and postpone the proof of existence to Section 5.9 at the end of this chapter. Let F satisfy

$$F(1) = a \,,$$
$$F(n+1) = g(F(n)) \,.$$

Define $M = \{n;\ n \in \mathbb{N} \text{ and } F(n) = f(n)\}$; clearly $1 \in M$. Assume now that $n \in M$. Then $F(n+1) = g(F(n)) = g(f(n)) = f(n+1)$. This means that $n+1 \in M$. Consequently $M = \mathbb{N}$, $f(n) = F(n)$ for every $n \in \mathbb{N}$ and therefore $f = F$. $\qquad\qquad \square$

We now use the Recursion Theorem to define the arithmetic sequence. It is sufficient to take $g : x \mapsto x + d$ and $f(1) = a_1 = c$. Theorems concerned with sequences defined inductively (recursively) are usually proved by mathematical induction.

It is also possible to start with earlier or later terms than with $n = 1$. For instance, if $a \in \mathbb{R}$, $n \in \mathbb{N}_0$ then we define a^n recursively as follows: $a^0 = 1$, $a^{n+1} = a \cdot a^n$.

Theorem 5.3   *If $a \in \mathbb{R}$, $b \in \mathbb{R}$, $n \in \mathbb{N}_0$, $m \in \mathbb{N}_0$ then*

   (i) $a^n b^n = (ab)^n$;

   (ii) $a^n a^m = a^{n+m}$;

   (iii) $(a^m)^n = a^{mn}$;

   (iv) $0 \le a < b \Rightarrow a^n < b^n$;

   (v) $m < n$, $a > 1 \Rightarrow a^m < a^n$.

This theorem is elementary and we leave the proof, generalised to the case where $m \in \mathbb{Z}$, $n \in \mathbb{Z}$ as Exercises 5.5.1, 5.5.5 and 5.5.7. As a guide, here we prove item (ii) by induction.

Proof.   The statement is true for $n = 0$ and arbitrary m. Assume (ii) and consider $a^{n+1} \cdot a^m$. By the recursive definition $a^{n+1} = a \cdot a^n$ and by the induction hypothesis $a^n a^m = a^{n+m}$. Consequently $a^{n+1} \cdot a^m = a \cdot a^n \cdot a^m = a \cdot a^{n+m} = a^{n+m+1}$. $\qquad\qquad\square$

Define $f_1 = 1$, $f_2 = 1$ and $f_{n+2} = f_{n+1} + f_n$. The first few terms are

$$1, \, 1, \, 2, \, 3, \, 5, \, 8, \, 13, \, 21, \, \ldots$$

The sequence f is well defined and is called the *Fibonacci sequence* after the mathematician who first investigated it. It occurs in a wide range of seemingly unrelated areas, such as the arrangements of branches on a tree, the size of breeding populations of animals and in efficient procedures for sorting large amounts of data. However Theorem 5.2 is not directly applicable to this definition, and the required modification is stated in Exercise 5.5.4. If a similar recursive definition is needed relating more than two previous terms in the sequence, this will also need to be justified. Exercise 5.5.4 can serve as a guide to more general recursive definitions. A very general recursion theorem is proved later as Theorem 5.13.

Exercises

(i) **Exercise 5.5.1**   For $a \ne 0$, $a \in \mathbb{R}$ and $n \in \mathbb{N}$ define $a^{-n} = (a^{-1})^n$. For $a \ne 0$, $b \ne 0$, $m \in \mathbb{Z}$ and $n \in \mathbb{Z}$ prove:

   (1) $a^{n+m} = a^n a^m$;

   (2) $(a^m)^n = a^{nm}$;

   (3) $(ab)^n = a^n b^n$.

Exercise 5.5.2   Define the geometric sequence (progression) inductively and prove the formula for the nth term and for the sum of n terms.

(i) **Exercise 5.5.3**   Prove that

(1) $a^n - b^n = (a - b) \displaystyle\sum_{k=1}^{n} a^{n-k} b^{k-1}$;

(2) $a^{2n+1} + b^{2n+1} = (a + b) \displaystyle\sum_{k=0}^{2n} (-1)^k a^{2n-k} b^k$;

for $n \in \mathbb{N}$ and a, b elements of a field.

Exercise 5.5.4   Let S be a set and $g : S \times S \to S$. Then there exists exactly one sequence f such that $f(1) = a \in S$, $f(2) = b \in S$ and $f(n+2) = g(f(n), f(n + 1))$ for $n \in \mathbb{N}$. Prove the uniqueness of f.

(i) **Exercise 5.5.5** $n \in \mathbb{N}$ and $0 \leq a < b$ imply $a^n < b^n$. Prove this.

(i) **Exercise 5.5.6** $n \in \mathbb{N}$ and $a < b$ imply $a^{2n+1} < b^{2n+1}$. Prove this.

(i) **Exercise 5.5.7** $m \in \mathbb{Z}$, $n \in \mathbb{Z}$, $n > m$ and $a > 1$ imply $a^m < a^n$. Prove this.

(i) **Exercise 5.5.8** $-n \in \mathbb{N}$ and $0 \leq a < b$ imply $b^n < a^n$. Prove this.

(i) **Exercise 5.5.9** $m \in \mathbb{Z}$, $n \in \mathbb{Z}$, $n > m$ and $0 < a < 1$ imply $a^m > a^n$. Prove this.

Exercise 5.5.10   Let $x_1 = -\dfrac{1}{3}$, $x_{n+1} = \dfrac{2}{1 + x_n}$. Show that x_n is not defined for all n, and explain why this does not contradict Theorem 5.2! [Hint: Search for S.]

5.6   The binomial theorem

Definition 5.1 (Factorial and Binomial Coefficient) $n!$ (read n factorial) is defined for $n \in \mathbb{N}_0$ by $n! = \displaystyle\prod_{i=1}^{n} i$. Consequently we have $0! = 1$, $1! = 1$, $2! = 2$, $3! = 1 \cdot 2 \cdot 3 = 6$, etc. The binomial coefficient $\dbinom{n}{k}$ (read 'n by k' or 'n choose k') is defined by $\dbinom{n}{k} = \dfrac{n!}{k!(n - k)!}$.

It is a matter of simple calculation to check that $\binom{n}{k} = \binom{n}{n-k}$ and $\binom{n}{k} + \binom{n}{k+1} = \binom{n+1}{k+1}$. The last relation enables us to calculate $\binom{n}{k}$ successively as shown in the following table.

$$
\begin{array}{ccccccccccc}
 & & & & & 1 & & & & & \\
 & & & & 1 & & 1 & & & & \\
 & & & 1 & & 2 & & 1 & & & \\
 & & 1 & & 3 & & 3 & & 1 & & \\
 & 1 & & 4 & & 6 & & 4 & & 1 & \\
1 & & 5 & & 10 & & 10 & & 5 & & 1
\end{array}
\tag{5.29}
$$

Each number in this table is the sum of the two numbers which are situated to the left and the right in the row directly above it. The triangular table (5.29) is called *Pascal's Triangle*, and the kth element in the nth row is $\binom{n-1}{k-1}$. The Pascal triangle shows that $\binom{n}{k}$ is an integer for n, $k \in \mathbb{N}_0$ with $k \leq n$.

Theorem 5.4 (Newton's Binomial Theorem)   *If a, b are elements of a ring and $n \in \mathbb{N}$ then*

$$
(a + b)^n = \sum_{k=0}^{n} \binom{n}{k} \times a^{n-k} b^k.
\tag{5.30}
$$

Remark 5.1   A ring need not contain all the natural numbers: for instance the ring of even numbers does not. In such a ring the product na may not be defined for a in the ring and $n \in \mathbb{N}$. Usually multiplication by a natural number means repeated addition: by analogy we define

$$
n \times a \overset{\text{def}}{=} \underbrace{a + a + \cdots a}_{n \text{ summands}}
$$

for a in a ring and $n \in \mathbb{N}$. For the proof of the Newton Theorem we also need the identity $n \times a + m \times a = (n + m) \times a$.

Proof.   The theorem is obviously true if $n = 1$. Assume (5.30) and consider $(a + b)^{n+1}$.

$$(a+b)^{n+1} = (a+b)(a+b)^n$$

$$= (a+b) \sum_{k=0}^{n} \binom{n}{k} \times a^{n-k} b^k$$

$$= \sum_{k=0}^{n} \binom{n}{k} \times a^{n+1-k} b^k + \sum_{k=0}^{n} \binom{n}{k} \times a^{n-k} b^{k+1}$$

$$= \sum_{k=0}^{n} \binom{n}{k} \times a^{n+1-k} b^k + \sum_{j=1}^{n+1} \binom{n}{j-1} \times a^{n-j+1} b^j$$

$$= a^{n+1} + \sum_{k=1}^{n} \left[\binom{n}{k} + \binom{n}{k-1} \right] \times a^{n+1-k} b^k + b^{k+1}$$

$$= \sum_{k=0}^{n+1} \binom{n+1}{k} \times a^{n+1-k} b^k$$

$\square$

Two easy consequences of (5.30) are

$$2^n = \sum_{k=0}^{n} \binom{n}{k}, \qquad (5.31)$$

and

$$0 = \sum_{k=0}^{n} (-1)^k \binom{n}{k}. \qquad (5.32)$$

The binomial coefficient $\binom{n}{k}$ is produced in *Maple* by `binomial(n,k)`.

Exercises

Exercise 5.6.1   *Prove the following formulae for the binomial coefficients:*

(1) $\displaystyle \sum_{k=0}^{N} \binom{n}{2k} = 2^{n-1}$ *where* $N = \left\lfloor \dfrac{n+1}{2} \right\rfloor$;

(2) $\displaystyle \sum_{k=0}^{n} \binom{n}{k}^2 = \binom{2n}{n}$.

[Hint: (1) Add (5.31) and (5.32). (2) Expand $(1+x)^{2n}$ and $(1+x)^n(1+x)^n$ and compare the coefficient of x^n.]

Exercise 5.6.2   *For integers k, n, $0 \le k \le n$ prove*

$$\sum_{j=k}^{n} \binom{j}{k} = \binom{n+1}{k+1}.$$

[Hint: Use induction on $m = n - k$.]

Exercise 5.6.3   *For natural numbers k, n, $k \le n$ prove*

$$\left(1 + \frac{1}{n}\right)^k < 1 + \frac{k}{n} + \frac{k^2}{n^2}.$$

(i) **Exercise 5.6.4**   *Define the polynomial coefficients* $\begin{bmatrix} n \\ k_1, \, k_2, \, \ldots, \, k_r \end{bmatrix}$ *by*

$$\begin{bmatrix} n \\ k_1, \, k_2, \, \ldots, \, k_r \end{bmatrix} = \frac{n!}{k_1! \, k_2! \ldots k_r!}$$

and prove the polynomial theorem

$$\left(\sum_{i=0}^{r} a_i\right)^n = \sum_{\substack{k_1+k_2+\cdots+k_r=n \\ k_i \ge 0}} \begin{bmatrix} n \\ k_1, \, k_2, \, \ldots, \, k_r \end{bmatrix} \times a_1^{k_1} \, a_2^{k_2} \ldots a_r^{k_r}.$$

for a_i in some ring.

(i) **Exercise 5.6.5**   *Prove: $0 < h < 1$, $n \in \mathbb{N}$ implies $(1 + h)^n < 1 + 2^n h$.*
[Hint:$(1 + h)^n < 1 + h \displaystyle\sum_{k=0}^{n} \binom{n}{k}$ and use (5.31).]

Exercise 5.6.6   *Prove: $(a + b)^5 \equiv a^5 + b^5 \mod 5$*

5.7   Roots and powers with rational exponents

A basic result of this section is the following:

> **Theorem 5.5**   *If $a > 0$, $a \in \mathbb{R}$, $n \in \mathbb{N}$ then there exists exactly one positive number x such that $x^n = a$.*

Definition 5.2 (n^{th} root of a)   The number x from Theorem 5.5 is called the nth root of a and is denoted by $\sqrt[n]{a}$. For $n = 2$ it is customary to write $\sqrt{a}$ instead of $\sqrt[2]{a}$. For $a = 0$ we set $\sqrt[n]{a} = 0$.

For the proof of Theorem 5.5 we need the following Lemma.

Lemma 5.1   Let $a > 0$, $n \in \mathbb{N}$, $x > 0$.

(i) *if $x^n < a$ there exists $y \in \mathbb{R}$ such that $x < y$, $y^n < a$.*
(ii) *if $a < x^n$ there exists $z \in \mathbb{R}$ such that $0 < z < x$, $a < z^n$.*

Proof.

(i) We find h with $0 < h < 1$ such that $x^n(1+h)^n < a$ and set $y = x(1+h)$. Since $(1+h)^n < 1 + 2^n h$ by Exercise 5.6.5 we have $x^n(1+h)^n < x^n + 2^n x^n h$, and it is now sufficient to choose h such that $0 < h < 1$ and $x^n + 2^n x^n h \le a$. $h = \text{Min}\left(\dfrac{1}{2}, \dfrac{a - x^n}{2^n x^n}\right)$ satisfies both of these requirements.

(ii) We note that $\dfrac{1}{x} > 0$, $\dfrac{1}{a} > 0$, $\left(\dfrac{1}{x}\right) < \dfrac{1}{a}$. By (i) there is $y > \dfrac{1}{x}$ such that $y^n < \dfrac{1}{a}$. We can now take $z = \dfrac{1}{y}$. $\qquad\square$

We now take up the proof of Theorem 5.5.

Proof.   Since $0 < u < v$ implies $u^n < v^n$ there is at most one x satisfying $x^n = a$. For the existence proof we assume $n > 1$ since the case $n = 1$ is trivial. We now distinguish three cases.

(A) $a = 1$ then $x = 1$.
(B) $a > 1$. We define $S = \{u;\ u > 0,\ u^n < a\}$. The set S is non-empty since $1 \in S$. It is also bounded above by a, since $v \ge a$ implies $v^n \ge a^n > a$ (see Exercises 5.5.5 and 5.5.7). By the least upper bound axiom there exists $x = \sup S$. We prove that $x^n = a$ by showing that each of $x^n < a$ and $a < x^n$ is impossible.

  1. If $x^n < a$ then by (i) of Lemma 5.1 there exists $y \in S$, $y > x$, contradicting the definition of x as the supremum of S.
  2. If $a < x^n$ then by (ii) of the same Lemma there exists z, $z < x$ with $a < z^n$. Since x is the least upper bound of S there exists $u \in S$, with $z < u \le x$. This implies that $a < z^n < u^n$, so that $u \notin S$, another contradiction.

(C) $0 < a < 1$. Then $b = \dfrac{1}{a} > 1$ and by II there is $w > 0$, $w^n = b$.

Clearly $\left(\dfrac{1}{w}\right)^n = a$.

□

It is worth emphasizing that $\sqrt[n]{a}$ is, by its very definition, positive or zero. Consequently $\sqrt{a^2}$ need not be a (it is not if $a < 0$); however $|a|^2 = a^2$ and it follows from the uniqueness part of Theorem 5.5 that $\sqrt{a^2} = |a|$ (always). Further, if $a > 0$, $b > 0$, since $(\sqrt[n]{a}\,\sqrt[n]{b})^n = ab$ we have (again by the uniqueness part of Theorem 5.5)

$$\sqrt[n]{ab} = \sqrt[n]{a}\,\sqrt[n]{b}. \tag{5.33}$$

If n is odd then for every $a \in \mathbb{R}$ there exists an $x \in \mathbb{R}$ such that $x^n = a$. Indeed, for $a < 0$ we have $a = -|a|$, so that $x^n = a = -|a|$, from which it follows that $x = -\sqrt[n]{|a|}$. It is customary to denote this x also by $\sqrt[n]{a}$. Hence for an odd n and $a < 0$ we have $\sqrt[n]{a} = -\sqrt[n]{-a}$.

Theorem 5.6 *If $a > 0$, $b > 0$ and $m \in \mathbb{N}$ then*

    (i) $\sqrt[n]{ab} = \sqrt[n]{a}\,\sqrt[n]{b}$;

    (ii) $\sqrt[n]{\dfrac{a}{b}} = \dfrac{\sqrt[n]{a}}{\sqrt[n]{b}}$;

    (iii) $\sqrt[n]{\dfrac{1}{b}} = \dfrac{1}{\sqrt[n]{b}}$;

    (iv) $a < b \Rightarrow \sqrt[n]{a} < \sqrt[n]{b}$.

Proof.

    (i) is just a restatement of (5.33).

    (ii) $\sqrt[n]{b}\,\sqrt[n]{\dfrac{a}{b}} = \sqrt[n]{a}$ by (i).

(iii) This follows from (ii) by setting $a = 1$.

(iv) $\sqrt[n]{a} = \sqrt[n]{b}$ and $\sqrt[n]{a} > \sqrt[n]{b}$ are both impossible, since it would follow that $a = b$ in the first case and $a > b$ in the second. Hence we must have $\sqrt[n]{a} < \sqrt[n]{b}$.

□

Every rational number r can be expressed in many different ways as $r = \dfrac{p}{q}$ with $p \in \mathbb{Z}$ and $q \in \mathbb{N}$. We wish to define a^r as $\sqrt[q]{a^p}$ but before we do that we prove

Theorem 5.7 Let p, m be integers, q and n natural numbers and $a > 0$. If $\dfrac{p}{q} = \dfrac{m}{n}$ then

$$\sqrt[q]{a^p} = \sqrt[n]{a^m}. \tag{5.34}$$

Proof. Let $\sqrt[q]{a^p} = x$, $\sqrt[n]{a^m} = y$. Then $a^{pn} = x^{qn}$ and $a^{mq} = y^{nq}$. It follows that $x^{nq} = y^{nq}$ and by the uniqueness part of Theorem 5.5 we have $x = y$. $\qquad\square$

Definition 5.3 (a^r for rational r) For $a > 0$, $r \in \mathbb{Q}$, $r = \dfrac{p}{q}$, $q \in \mathbb{N}$ we define

$$a^r = \sqrt[q]{a^p}. \tag{5.35}$$

Remark 5.2 If $r \in \mathbb{Z}$ then a^r defined by (5.35) is the same as defined in Section 5.5.

Theorem 5.8 If a, b are positive, $r \in \mathbb{Q}$, $s \in \mathbb{Q}$ then

(i) $1^r = 1$, $a^r > 0$, $a^0 = 1$;

(ii) $a^r b^r = (ab)^r$;

(iii) $\left(\dfrac{a}{b}\right)^r = \dfrac{a^r}{b^r}$;

(iv) $\left(\dfrac{1}{b}\right)^r = \dfrac{1}{b^r}$;

(v) $a^r a^s = a^{r+s}$;

(vi) $\dfrac{a^r}{a^s} = a^{r-s}$;

(vii) $a^{-s} = \dfrac{1}{a^s}$;

(viii) $(a < b,\ r > 0) \Rightarrow a^r < b^r$;

(ix) $(1 < a,\ s < r) \Rightarrow a^s < a^r$.

Proof.

(i) is obvious.

For the rest of the proof we put $r = \dfrac{p}{q}$, $s = \dfrac{m}{q}$ with $q \in \mathbb{N}$ and $p \in \mathbb{Z}$, $m \in \mathbb{Z}$ (it is a trivial exercise to show that r and s can be written with the same denominator).

(ii)
$$a^r b^r = \sqrt[q]{a^p} \sqrt[q]{b^p}$$
$$= \sqrt[q]{a^p b^p} \qquad \text{(by Theorem 5.6 (i))}$$
$$= \sqrt[q]{(ab)^p} \qquad \text{(by Exercise 5.5.1)}$$
$$= (ab)^r.$$

(iii) $b^r \left(\dfrac{a}{b}\right)^r = \left(b\dfrac{a}{b}\right)^r$ by (ii), and therefore $b^r \left(\dfrac{a}{b}\right)^r = a^r$;

(iv) This follows from (iii) by setting $a = 1$.

(v)
$$a^r a^s = \sqrt[q]{a^p} \sqrt[q]{a^m}$$
$$= \sqrt[q]{a^p a^m} \qquad \text{(by Theorem 5.6 (i))}$$
$$= \sqrt[q]{a^{p+m}} \qquad \text{(by Exercise 5.5.1)}$$
$$= a^{(p+m)/q}$$
$$= a^{r+s}.$$

(vi) $a^s a^{r-s} = a^r$ by (v).

(vii) Put $r = 0$ in (vi).

(viii) Since $r > 0$, $q \in \mathbb{N}$ we have $p \in \mathbb{N}$. Consequently $a^p < b^p$ (by Theorem 5.3 (iv)), and by (iv) of Theorem 5.6 (with a replaced by a^p, b replaced by b^p, etc.) we have $\sqrt[q]{a^p} < \sqrt[q]{b^p}$;

(ix) We have $r - s > 0$ and by using (viii) we obtain $1^{r-s} < a^{r-s} = \dfrac{a^r}{a^s}$, the last equation following from (vi).

$\square$

Exercises

ⓘ **Exercise 5.7.1**   Prove that Theorem 5.6 remains valid for arbitrary real a, b provided n is odd and $b \neq 0$ in (ii) and (iii).

ⓘ **Exercise 5.7.2**   Prove by induction: if a is positive, $n \in \mathbb{N}_0$ and $r \in \mathbb{Q}$ then $(a^r)^n = a^{rn}$.

ⓘ **Exercise 5.7.3**   Prove: if a is positive, $r \in \mathbb{Q}$ and $k \in \mathbb{Z}$ then $(a^r)^k = a^{rk}$.

ⓘ **Exercise 5.7.4**   Prove: if a is positive, $r \in \mathbb{Q}$ and $s \in \mathbb{Q}$ then $a^{rs} = (a^r)^s$. [Hint: $r = \dfrac{p}{q}$, $s = \dfrac{m}{n}$, $x = a^{p/q}, a^p = x^q, a^{pm} = x^{qm} = (x^m)^q = (x^{m/n})^n q$.]

ⓘ **Exercise 5.7.5**   *We have* $\sqrt{\sqrt[3]{-1}\ \sqrt[3]{-1}} = \sqrt{(-1)\cdot(-1)} = \sqrt{1} = 1$. *But we also have* $\sqrt{\sqrt[3]{-1}\ \sqrt[3]{-1}} = \sqrt{\sqrt[3]{-1}}\ \sqrt{\sqrt[3]{-1}} = \left(\sqrt{\sqrt[3]{-1}}\right)^2 = \sqrt[3]{-1} = -1$. *Hence* $1 = -1$. *Explain this paradox.*

Exercise 5.7.6   *Let* $g : x \mapsto \sqrt[3]{1+\sqrt{x}} + \sqrt[3]{1-\sqrt{x}}$; $\operatorname{dom} g = [0,1]$. *By showing that the equation* $y = g(x)$ *has a unique solution for* $y \in [\sqrt[3]{2}, 2]$ *prove the existence of* g_{-1} *and find it.*

Exercise 5.7.7   *Let* $h : x \mapsto x^2$, $\operatorname{dom} h =]-\infty, -1[$. *Show that* h *is one-to-one and find* h_{-1}.

Exercise 5.7.8   *Let* $F : x \mapsto x^2$ *for* $x < -1$ *and* $x \mapsto -x$ *for* $x \geq -1$. *Graph* F. *Prove that* F *is one-to-one, and find* $\operatorname{rg} F$ *and* F_{-1}.

5.8   Some important inequalities

Mathematical induction is useful in proving general inequalities, as the next three theorems show.

Theorem 5.9 (Bernoulli's Inequality)   *If* $x \in \mathbb{R}$, $x \geq -1$, $n \in \mathbb{N}$ *then*

$$(1+x)^n \geq 1 + nx. \tag{5.36}$$

Remark 5.3   Inequality (5.36) is a trivial consequence of the Binomial Theorem 5.4 if $x \geq 0$.

Proof.   If $n = 1$, equality holds in (5.36). Assume (5.36) holds for n and multiply it by $1 + x \geq 0$.

$$(1+x)^{n+1} \geq (1+nx)(1+x),$$
$$= 1 + (n+1)x + nx^2,$$
$$\geq 1 + (n+1)x.$$

$\square$

If $a_1, a_2, \ldots, a_n$ are non-negative numbers then

$$G = \sqrt[n]{a_1\, a_2 \ldots a_n}$$

is called the *geometric mean* of a_1, a_2, ..., a_n and

$$A = \frac{1}{n}(a_1 + a_2 + \ldots + a_n)$$

is called the *arithmetic mean* of a_1, a_2, ..., a_n.

Theorem 5.10 (Arithmetic-Geometric Mean Inequality)   *If* $a_i \geq 0$ *for* $i = 1, 2, \ldots, n$ *then*

$$\sqrt[n]{\prod_{i=1}^{n} a_i} \leq \frac{1}{n} \sum_{i=1}^{n} a_i. \tag{5.37}$$

Proof.   Equality holds in (5.37) for $n = 1$. Clearly, rearranging the numbers doesn't affect the result, so we can arrange them in order of increasing size

$$0 \leq a_1 \leq a_2 \leq \ldots \leq a_n \leq a_{n+1}.$$

Let A be the arithmetic mean of these $n + 1$ numbers, $A = \dfrac{1}{n+1} \sum_{i=1}^{n+1} a_i$. It is clear that $a_1 + a_{n+1} - A \geq 0$. The arithmetic mean of the n non-negative numbers a_2, a_3, ..., a_n, $a_1 + a_{n+1} - A$ is A, and applying the induction hypothesis to these numbers we have

$$(a_1 + a_{n+1} - A) \prod_{i=1}^{n-1} a_{i+1} \leq A^n,$$

and hence

$$A(a_1 + a_{n+1} - A) \prod_{i=1}^{n-1} a_{i+1} \leq A^{n+1}. \tag{5.38}$$

From the obvious inequality $a_1 \leq A \leq a_{n+1}$ it follows that $(A - a_1)(a_{n+1} - A) \geq 0$, and this can be expanded to give $A(a_1 + a_{n+1} - A) - a_1 a_{n+1} \geq 0$, or

$$a_1 a_{n+1} \leq A(a_1 + a_{n+1} - A). \tag{5.39}$$

If this is combined with Inequality (5.38) we get

$$a_1 a_{n+1} \prod_{i=1}^{n-1} a_{i+1} \leq A^{n+1}. \tag{5.40}$$

This is (5.37) with n replaced by $n + 1$. $\square$

Corollary 5.10.1   *Equality holds in (5.37) if and only if all a_i are equal.*

Proof.   The 'if' part is obvious. We prove 'the only' part by showing that if not all a_i are equal then

$$\prod_{i=1}^{n+1} a_i < A^{n+1}. \tag{5.41}$$

This is obvious if $a_1 = 0$, so let $0 < a_1 < a_{n+1}$ With notation as before, we now have $a_1 < A < a_{n+1}$, so that $(A - a_1)(a_{n+1} - A) > 0$, leading to strict inequality in (5.39). Multiplying this inequality with the positive number $\prod_{i=1}^{n-1} a_{i+1}$ and taking into account Inequality (5.38) gives (5.41). $\square$

For $n = 2$, $a = \sqrt{a_1}$, $b = \sqrt{a_2}$ Inequality (5.37) becomes

$$2ab \le a^2 + b^2. \tag{5.42}$$

This was established directly and easily in Exercise 4.2.3 by considering the inequality $(a - b)^2 \ge 0$.

Theorem 5.11 (Schwarz's Inequality)   *If a_k and b_k are in $\mathbb{R}$ for $k = 1, 2, \ldots, n$ then*

$$\left(\sum_{k=1}^{n} a_k b_k \right)^2 \le \sum_{k=1}^{n} a_k^2 \sum_{k=1}^{n} b_k^2. \tag{5.43}$$

Proof.   If $n = 1$ equality holds in (5.43). Assume now that the theorem is true for $n \in \mathbb{N}$. Writing $\sum_{k=1}^{n} a_k b_k = s_n$, $\left(\sum_{k=1}^{n} a_k^2 \right)^{1/2} = r_n$, $\left(\sum_{k=1}^{n} b_k^2 \right)^{1/2} = t_n$, we have

$$s_{n+1}^2 = (s_n + a_{n+1} b_{n+1})^2$$
$$\le s_n^2 + 2|s_n a_{n+1} b_{n+1}| + a_{n+1}^2 b_{n+1}^2.$$

The induction hypothesis

$$s_n^2 \le r_n^2 t_n^2, \tag{5.44}$$

implies

$$s_{n+1}^2 \leq r_n^2 t_n^2 + 2r_n t_n |a_{n+1} b_{n+1}| + a_{n+1}^2 b_{n+1}^2. \tag{5.45}$$

Since by (5.42)

$$2r_n t_n |a_{n+1} b_{n+1}| \leq t_n^2 a_{n+1}^2 + r_n^2 b_{n+1}^2, \tag{5.46}$$

we have, substituting this in (5.45),

$$s_{n+1}^2 \leq (r_n^2 + a_{n+1}^2)(t_n^2 + b_{n+1}^2). \tag{5.47}$$

$\square$

Corollary 5.11.1   *There is equality in* (5.43) *if and only if for* $i = 1, 2, \ldots, n$ *either* $b_i = 0$ *or there exists* $c \in \mathbb{R}$ *such that* $a_i = cb_i$. *($a_i = 0$ is covered by the special case $c = 0$.)*

Proof.   See Exercise 5.8.7. $\square$

Exercises

Exercise 5.8.1   *Let $S \subset \mathbb{N}$ with the following properties:*

(1) $2^n \in S$ *whenever* $n \in \mathbb{N}$;
(2) $m \in S$ *implies* $k \in S$ *for all* $k < m$, $k \in \mathbb{N}$.

Show that $S = \mathbb{N}$.

Exercise 5.8.2   *Prove the arithmetic-geometric mean inequality for $n = 2^m$ by induction on m, and then use Exercise 5.8.1 to prove it for general n.*

Exercise 5.8.3   *Let $x_1, x_2, \ldots, x_n$ be non-negative.*
If $\prod_{i=1}^{n} x_i = 1$, then $\sum_{i=1}^{n} x_i \geq n$; if $\sum_{i=1}^{n} x_i = n$, then $\prod_{i=1}^{n} x_i \leq 1$.
Prove these results.

Exercise 5.8.4   *Use Theorem 5.10 to prove that among all rectangles of perimeter s the square has the largest area.*

ⓘ **Exercise 5.8.5**   *For $x \geq -1$ and $n = r/s$ with $r, s \in \mathbb{N}$ and $r < s$ prove that $\sqrt[n]{1+x} \leq 1 + \dfrac{x}{n}$. [Hint: Use Theorem 5.10 on $a_1 = a_2 = \cdots = a_r = 1 + x$, $a_{r+1} = 1, \ldots, a_s = 1$.]*

Exercise 5.8.6 For x and n as in the previous exercise prove that $1 + \dfrac{x}{n \sqrt[n]{(1+x)^{n-1}}} \leq \sqrt[n]{1+x}$.

[Hint: Use Exercise 5.8.5 to obtain $x/n + \sqrt[n]{(1+x)^{n-1}} \leq 1 + x$ and then divide by $\sqrt[n]{(1+x)^{n-1}}$.]

Exercise 5.8.7 Prove Corollary 5.11.1.

[Hint: The 'if' part is easy. The 'only if' part is obvious if $n = 1$. Proceed by induction. If equality holds in (5.47) then from 5.45 and 5.46 we must have

$$\begin{aligned}
s_{n+1}^2 &\geq r_n^2 t_n^2 + 2 r_n t_n |a_{n+1} b_{n+1}| + a_{n+1}^2 b_{n+1}^2, \\
&= \left(r_n t_n + |a_{n+1} b_{n+1}| \right)^2, \\
&\geq \left(|s_n| + |a_{n+1} b_{n+1}| \right)^2, \\
&\geq s_{n+1}^2.
\end{aligned}$$

Consequently equality holds throughout and $|s_n| = r_n t_n$. Now use the induction hypothesis to complete the proof. Distinguish three cases:

(a) $t_n = 0$, $b_{n+1} = 0$;
(b) $t_n = 0$, $b_{n+1} \neq 0$;
(c) $a_i = c b_i$ for $i = 1, 2, \ldots, n$.]

5.9 Complete induction

In some inductive proofs the assumption that the assertion is valid for $n - 1$ is not strong enough to conclude that the assertion is valid for n. A stronger induction hypothesis is employed, namely, that the assertion is valid for all natural k less than n. The justification for such proofs is based on the following theorem.

Theorem 5.12 *Let S be a subset of $\mathbb{N}$ with the following properties:*

(i) $1 \in S$;
(ii) *for every $n \in \mathbb{N}$, the following implication holds:*

$$(k \in S \text{ for all } k < n) \Rightarrow n \in S$$

Then $S = \mathbb{N}$.

Proof. Assume that $S \neq \mathbb{N}$, that is, $\mathbb{N} \setminus S \neq \emptyset$. By the well ordering principle (Theorem 4.10) there is a smallest $n \in \mathbb{N} \setminus S$, say m. Clearly $m \neq 1$ since $1 \in S$ by (i). For all $k < m$, $k \in \mathbb{N}$ we therefore have $k \in S$, and by (ii) we have $m \in S$, which is a contradiction, so our initial assumption is wrong, and $S = \mathbb{N}$.   □

Theorem 5.12 is often useful as the next example shows.

Example 5.7   For every $n \in \mathbb{N}$ there exists $s \in \mathbb{N}_0$ and $m \in \mathbb{N}$ such that $n = 2^s(2m - 1)$.

Let $S = \{n; \ n \in \mathbb{N}, \ n = 2^s(2m - 1) \text{ with } s \in \mathbb{N}_0 \text{ and } m \in \mathbb{N}\}$. We have $1 \in S$; indeed every odd $n \in S$. Assume now that n is even and $k \in S$ for all $k \in \mathbb{N}$, $k < n$. Since n is even we have $n = 2k$, $k \in \mathbb{N}$, $k < n$ and by the induction hypothesis $k = 2^t(2m - 1)$, $n = 2^{t+1}(2m - 1)$ and $n \in S$. By Theorem 5.12 $S = \mathbb{N}$.

Complete induction can also be used for recursive definitions. The next example shows this.

Example 5.8   Let $x \in \mathbb{R}$, $x > 0$. There exists the largest integer not exceeding x, and we denote this[1] by a_0. Then find the largest integer a_1 such that

$$a_0 + \frac{a_1}{10} \leq x,$$

and proceed indefinitely. At the nth step find the largest integer a_n such that

$$a_0 + \frac{a_1}{10} + \frac{a_2}{10^2} + \cdots + \frac{a_n}{10^n} \leq x.$$

This defines a sequence a_n and we say that x is expressed as a decimal fraction $a_0.a_1a_2a_3 \ldots$. This notation should not be confused with any product: it simply expresses the fact that for every n the number x satisfies

$$\sum_{i=0}^{n} \frac{a_i}{10^i} \leq x < \sum_{i=0}^{n-1} \frac{a_i}{10^i} + \frac{a_n + 1}{10^n}.$$

If $x > 0$ and is expressed as a decimal fraction $a_0.a_1a_2a_3 \ldots$ then

$$x = \sup \left\{ u : u = a_0 + \frac{a_1}{10} + \frac{a_2}{10^2} + \cdots + \frac{a_n}{10^n} \leq x \right\}.$$

---

[1] In other words $a_0 = \lfloor x \rfloor$.

This, together with the proof that $a_i \in \mathbb{N}_0$ and, in particular, $0 \leq a_i \leq 9$ is postponed to Exercises 5.9.2 and 5.9.3.

The decimal fraction $0.000\ldots$ is associated with the number 0. If $y < 0$ we write $x = -y$, find the decimal expansion of x in the form $a_0.a_1a_2a_3\ldots$ and then associate the decimal fraction $-a_0.a_1a_2a_3\ldots$ with y. If x is a positive rational and of the form

$$x = b_0 + \frac{b_1}{10} + \frac{b_2}{10^2} + \cdots + \frac{b_n}{10^n}$$

then the decimal fraction for x is $b_0.b_1b_2\ldots b_n000\ldots$.

Any other number $b \in \mathbb{N}$, $b > 1$ can be used instead of 10 in the above construction. For $b = 2$ we obtain what are called binary fractions, and for $b = 3$ ternary fractions.

The above construction of decimal fractions or, more precisely, of the sequence with n^{th} term a_n can be made precise by using the following theorem.

Theorem 5.13 (Recursion Theorem) *Let S be a set, $a \in S$ and $g_n : S^{n-1} \mapsto S$ for every $n \in \mathbb{N}$, $n > 1$. Then there exists a unique function $f : \mathbb{N} \to S$ such that*

(i) $f(1) = a;$
(ii) $f(n) = g_n(f(1),\ f(2),\ \ldots,\ f(n-1)).$

Proof. The proof of uniqueness is rather similar to the uniqueness proof of Theorem 5.2. The proof of existence is postponed to the last section of this chapter. $\quad\square$

The definitions of a_i above can now be formally made:

$$a_0 = \lfloor x \rfloor$$

and

$$a_n = \left\lfloor 10^n x - \sum_{i=0}^{n-1} 10^{n-i} a_i \right\rfloor.$$

S is simply $\mathbb{R}$ and g_n of $(u_1,\ u_2,\ \ldots,\ u_{n-1})$ is defined by

$$g(u_1,\ u_2,\ \ldots,\ u_{n-1}) = \left\lfloor 10^n x - \sum_{i=0}^{n-1} 10^{n-i} u_i \right\rfloor.$$

Exercises

ⓘ **Exercise 5.9.1** *Every natural number is divisible by a prime. We used this theorem as self-evident without proof in Chapter 2. Prove it using complete induction!* [Hint: If n is a prime there is nothing to prove, otherwise it is divisible by a number smaller than n.]

Exercise 5.9.2 *Let a real number $x > 0$ be expressed as a decimal fraction $a_0.a_1 a_2 a_3 \ldots$. Prove that $0 \le a_n \le 9$ for every $n \in \mathbb{N}$. Also prove that for binary and ternary fractions one has $a_n \in \{0, 1\}$ and $a \in \{0, 1, 2\}$, respectively.*

ⓘ **Exercise 5.9.3** *Let $x > 0$ and $a_0.a_1 a_2 a_3 \ldots$ be its decimal fraction. Let $A_n = \sum_{i=0}^{n} \dfrac{a_i}{10^i}$. Prove that $x = \sup\{y;\ y = A_n\}$.*

ⓘ **Exercise 5.9.4** *Let x and y be positive with the same decimal fraction $a_0.a_1 a_2 a_3 \ldots$. Prove that $x = y$.*
[Hint: The proof is immediate from Exercise 5.9.3.]

ⓘ **Exercise 5.9.5** *Our definition of decimal fractions is slightly different from the conventional one. We shall rectify this in Chapter 11. Prove that with our definition $0.999\ldots$ is not a decimal fraction for any $x \in \mathbb{R}$. Also show that the decimal fraction corresponding to $\dfrac{1}{3}$ is $0.3333\ldots$.*

ⓘ **Exercise 5.9.6** *Prove that the numbers s and m from Example 5.7 are uniquely determined by n.*

Exercise 5.9.7 *Prove that if two positive real numbers x and y have distinct decimal fractions then they are distinct.*

Exercise 5.9.8 *Prove that two distinct real numbers have distinct decimal fractions.*

5.10 Proof of the recursion theorem

To complete our treatment of the Recursion Theorems 5.2 and 5.13 we are going to prove the existence part of Theorem 5.13. We need the following Lemma.

Lemma 5.2   *For every $K \in \mathbb{N}$ there exists a unique function f_K : $\{1, 2, \ldots, K\} \mapsto S$ such that*

$$f_K(1) = a \qquad\qquad (5.48)$$
$$f_K(n) = g_n(f_K(1), f_K(2), \ldots, f_K(n-1)) \qquad (5.49)$$

for all $n \le K$, $n \in \mathbb{N}$.

Proof.   First we prove that f_K, if it exists, is uniquely determined. The proof is by induction. The function f_1 is uniquely determined, $f_1 : 1 \mapsto a$. Assume now f_{K-1} is unique, then

$$f_K(n) = f_{K-1}(n) \qquad\qquad (5.50)$$

for $n \le K - 1$ by the induction hypothesis and

$$f_K(K) = g_K(f_K(1), \ldots, f_K(K-1)) \qquad (5.51)$$

by (5.49). In view of (5.50) we have

$$f_K(n) = g_n(f_{K-1}(1), \ldots, f_{K-1}(n-1)), \qquad (5.52)$$

which means that $f_K(n)$ is uniquely determined for $n = 1, 2, \ldots, K$.

If f_K exists it must satisfy (5.52). We use this piece of knowledge to define f_K. More precisely we define first $f_1(1) = a$. Then we proceed by induction again. Assume f_{K-1} exists and define $f_K(n)$ by (5.50) and by (5.52). It remains to show that f_K so defined satisfies (5.48) and (5.49). The first equation is easy, $f_K(1) = f_{K-1}(1) = a$. Then we have

$$
\begin{aligned}
f_K(n) &= f_{K-1}(n), \\
&= g_n(f_{K-1}(1), \ldots, f_{K-1}(n-1)), \\
&= g_n(f_K(1), \ldots, f_K(n-1)),
\end{aligned}
$$

for $n \le K - 1$ and

$$
\begin{aligned}
f_K(n) &= g_n(f_{K-1}(1), \ldots, f_{K-1}(n-1)), \\
&= g_n(f_K(1), \ldots, f_K(n-1)),
\end{aligned}
$$

for $n = K$ by (5.52) and by (5.50). $\qquad\square$

We now take up the proof of Theorem 5.13.

Proof. We define $f = \bigcup\limits_{K=1}^{\infty} f_K$ and emphasize that each f_K is a set of pairs $(n, f_K(n))$. Obviously f is a relation, and it is sufficient to prove that it is a function. Let $(s, u) \in f$ and $(s, v) \in f$. Then $u = f_p(s)$ and $v = f_q(s)$ for some natural numbers p, q. Since both the restriction of f_p and the restriction of f_q to $\{1, 2, \ldots, s\}$ must, by uniqueness, be equal to f_s, we have $f_p(s) = f_q(s) = f_s(s)$, that is, $u = v$. Then $f(1) = a$ since $f(1) = f_1(1)$. Similarly $f(k) = f_n(k)$ for all $k \leq n$, $k \in \mathbb{N}$. Consequently

$$
\begin{aligned}
f(n) &= f_n(n), \\
&= g_n(f_n(1), \, f_n(2), \, \ldots, \, f_n(n-1)), \\
&= g_n(f(1), \, f(2), \, \ldots, \, f(n-1)),
\end{aligned}
$$

for every $n \in \mathbb{N}$. $\qquad\qquad\square$

5.11 Comments

We devoted this chapter to mathematical induction and its applications and to recursive definitions. The foundation for this was laid in the previous chapter in Theorems 4.5 and 4.10. These theorems were established as consequences of the axioms for the reals, that is, axioms of a complete ordered field. In Peano's axiomatic approach, which we sketched in Section 4.8, mathematical induction was one of the principal axioms. However the recursion theorem still must be proved and it plays an even more fundamental and central role than in our approach. In Peano's approach, the recursion theorem or some modification of it is needed for the definition of addition and multiplication of natural numbers.

Chapter 6

Polynomials

Polynomial functions have always been important, if for nothing else than because, in the past, they were the only functions which could be readily evaluated. In this chapter we define polynomials as algebraic entities rather than functions, establish the long division algorithm in an abstract setting, we also look briefly at zeros of polynomials and prove the Taylor Theorem for polynomials in a generality which cannot be obtained by using methods of calculus.

6.1   Polynomial functions

If M is a ring and $a_0, a_1, a_2, \ldots, a_n \in M$ then a function of the form

$$A : x \mapsto a_n x^n + a_{n-1} x^{n-1} + \cdots + a_1 x + a_0 \tag{6.1}$$

is called a *polynomial*, or sometimes more explicitly, a polynomial with coefficients in M. Obviously, one can add any number of zero coefficients, or rewrite Equation (6.1) in ascending order of powers of x without changing the polynomial. The domain of definition of the polynomial is naturally M, but the definition of $A(x)$ makes sense for any x in a ring which contains M. This natural extension of the domain of definition is often understood without explicitly saying so. If A and B are two polynomials then the polynomials $A + B$, $-A$ and AB are defined in the obvious way as

$$A + B : x \mapsto A(x) + B(x)$$
$$-A : x \mapsto -A(x)$$
$$AB : x \mapsto A(x)B(x)$$

The coefficients of $A+B$ are obvious; they are the sums of the corresponding coefficients of A and B. The zero polynomial function is the zero function, that is $\mathbf{1}_{\emptyset} : x \mapsto 0$. Similarly, the coefficients of $-A$ have opposite signs to the coefficients of A. The coefficients of AB are obtained by multiplying through, collecting terms with the same power of x and sorting them in descending (or ascending) powers of x. If

$$B(x) = b_m x^m + b_{m-1} x^{m-1} + \ldots + b_1 x + b_0,$$

and

$$P = AB,$$

then

$$P(x) = p_{m+n} x^{m+n} + p_{m+n-1} x^{m+n-1} + \cdots + p_1 x + p_0,$$

where

$$
\begin{aligned}
p_0 &= a_0 b_0 \\
p_1 &= a_1 b_0 + a_0 b_1 \\
p_2 &= a_2 b_0 + a_1 b_1 + a_0 b_2 \\
&\ \ \vdots \\
p_{n+m-1} &= a_n b_{m-1} + a_{n-1} b_m \\
p_{n+m} &= a_n b_m
\end{aligned}
\tag{6.2}
$$

There is a clear pattern to the formulae (6.2). In order to subsume them in a compact formula we set $a_k = 0$ for $k > n$ and $b_k = 0$ for $k > m$. Then we can rewrite Equations (6.2) as

$$p_k = \sum_{j=0}^{k} a_{k-j} b_j \tag{6.3}$$

for[1] $k = 1, 2, \ldots$.

With these definitions of addition and multiplication polynomials themselves form a ring. In modern mathematics there is a need to regard polynomials as algebraic expressions. The precise meaning of this will be made clear in the next section. Treating polynomials as algebraic expressions is needed because the coefficients of a polynomial function, generally speaking, are not uniquely determined by the polynomial function. For instance, the polynomial $x \mapsto x^2 + x$ over $GF(2)$ and the zero polynomial $x \mapsto 0$

---

[1]For $k > n + m$, all $p_k = 0$.

are equal as functions but they do not have the same coefficients. Later in this chapter we prove that if the ring M is a field which contains the field of rationals, then indeed, equal polynomial functions have the same coefficients.

6.2 Algebraic viewpoint

A sequence of elements from M constitutes an *unending polynomial*, more precisely an *unending polynomial over M*. To indicate that we regard a sequence $n \mapsto a_n$, $n \in \mathbb{N}_0$ as an unending polynomial rather than as a mere sequence we write $\langle a_n \rangle$, or more explicitly

$$\langle a_0, a_1, a_2, \ldots, a_n, \ldots \rangle.$$

Two unending polynomials $\langle a_n \rangle$ and $\langle b_n \rangle$ are equal, by definition, if and only if $a_k = b_k$ for all $k \in \mathbb{N}_0$. For an unending polynomial $\langle a_n \rangle$ the elements a_k are referred to as coefficients. The reason for this name becomes apparent later in this section. The sum and product of two unending polynomials are defined as

$$\langle a_n \rangle + \langle b_n \rangle = \langle a_n + b_n \rangle \tag{6.4}$$
$$\langle a_n \rangle \langle b_n \rangle = \langle p_k \rangle \tag{6.5}$$

where p_k are defined as in (6.3).

Unending polynomials are also (more commonly) called *formal power series*, and a different notation is used too. Instead of $\langle a_n \rangle$ one writes

$$a_0 + a_1 X + \cdots + a_n X^n + \cdots. \tag{6.6}$$

One problem with this notation is that X^n is not a power of an element from M, it is a purely formal symbol indicating term number $n + 1$ of the sequence. An advantage of this notation is that it facilitates multiplication. To multiply two unending polynomials, we multiply term by term, replace $X^k X^l$ by X^{k+l} and collect terms with the same 'power' of X. For instance

$$(1 + 2X + 3X^2 + \cdots + (n+1)X^n + \ldots)$$
$$\cdot (1 - 2X + 3X^2 + \cdots + (-1)^n(n+1)X^n + \cdots)$$
$$= 1 + 2X^2 + 3X^4 + \cdots + (n+1)X^{2n} + \cdots$$

Some examples on multiplication of unending polynomials are given in Exercises 6.2.1 and 6.2.2. Addition of unending polynomials satisfies axioms

A_1 to A_4 from Table 4.1. The role of the zero element is played by the zero unending polynomial $0 + 0X + \cdots + 0X^n + \cdots$, which we denote by 0. Axiom M_1 is also obviously satisfied, since

$$\sum_{j=0}^{k} a_{k-j} b_j = \sum_{j=0}^{k} b_{k-j} a_j.$$

Now we prove that multiplication of unending polynomials is associative. Using formula (6.3) for multiplication of unending polynomials we have

$$(\langle a_n \rangle \langle b_n \rangle) \langle c_n \rangle = \left\langle \sum_{j=0}^{n} \sum_{i=1}^{n-j} a_{n-j-i} b_i c_j \right\rangle \tag{6.7}$$

$$\langle a_n \rangle (\langle b_n \rangle \langle c_n \rangle) = \left\langle \sum_{j=0}^{n} \sum_{i=1}^{n-j} a_j b_{n-j-i} c_i \right\rangle \tag{6.8}$$

Both sums in (6.7) and (6.8) consist of terms of the form $a_\alpha b_\beta c_\gamma$, where α, β and γ are non-negative integers with $\alpha + \beta + \gamma = n$, and each such term appears in the sum exactly once. Consequently the sums are equal and the multiplication is associative. It is easy to see that Axiom D from Table 4.1 is also satisfied. It follows that unending polynomials form a ring.

A *polynomial* is a formal power series which has only a finite number of coefficients distinct from zero. In other words $\langle a_n \rangle$ is a polynomial if there is a natural number N such that $a_k = 0$ for $k > N$. Similarly to the notation (6.6) we write

$$a_0 + a_1 X + \cdots a_N X^N \quad \text{or} \quad a_N X^N + a_{N-1} X^{N-1} + \cdots a_0. \tag{6.9}$$

We shall denote the polynomial in (6.9) briefly as $A\langle X \rangle$. The coefficient a_0 is called the absolute term, the coefficient a_N is said to be the leading coefficient. The polynomials over a ring M themselves form a ring which is denoted by $M[X]$. Polynomials are a special class of unending polynomials: therefore to show that they form a ring it suffices to show that the sum and product of two polynomials have only a finite number of coefficients distinct from zero. This is obvious for the sum. For the product of the polynomials, $\langle a_n \rangle$ with $a_k = 0$ for $k > n$ and $\langle b_n \rangle$ with $b_k = 0$ for $k > m$, the general term is $\sum_{i=0}^{N} a_{N-i} b_i$. If $N > n + m$ then (at least one of) a_{N-i} or b_i is zero. This completes the proof that polynomials form a ring.

If $a_n \neq 0$ and $a_k = 0$ for $k > n$ then the *degree* of $A\langle X \rangle$ is defined to be n; in symbols $\deg A\langle X \rangle = n$. However, the zero polynomial which has all coefficients zero and which is denoted by 0 has no degree. The non-zero elements of M can be regarded as polynomials of degree zero. Indeed, as far as addition or multiplication is concerned, polynomials of degree zero behave exactly like elements of M.

$$\langle a, 0, 0, \ldots \rangle + \langle b, 0, 0, \ldots \rangle = \langle a + b, 0, 0, \ldots \rangle,$$
$$\langle a, 0, 0, \ldots \rangle \cdot \langle b, 0, 0, \ldots \rangle = \langle a \cdot b, 0, 0, \ldots \rangle,$$
$$\langle a, 0, 0, \ldots \rangle \cdot \langle b_0, b_1, \ldots \rangle = \langle a \cdot b_0, a \cdot b_1, \ldots \rangle.$$

If we have two algebraic structures, such as two fields or two rings, and if the elements of these structures behave, as far as addition and multiplication are concerned, in the same way, we say that these structures are *isomorphic*. More precisely, if M and M' are two rings (fields) we say that they are isomorphic if there is a bijection $I : M \mapsto M'$ such that $I(a + b) = I(a) + I(b)$ and $I(a \cdot b) = I(a) \cdot I(b)$. The bijection I is called an *isomorphism*. In the situation above, for $a \in M$ we have $I(a) = \langle a, 0, 0, \ldots \rangle$. In abstract algebra and in this book we shall consider two isomorphic structures as identical. In this spirit we say that the set of polynomials of degree zero together with the zero element *is* M.

It is easy to see that

$$\deg\left(A\langle X \rangle + B\langle X \rangle\right) \leq \mathrm{Max}\left(\deg A\langle X \rangle, \deg B\langle X \rangle\right) \qquad (6.10)$$
$$\deg\left(A\langle X \rangle \cdot B\langle X \rangle\right) \leq \deg A\langle X \rangle + \deg B\langle X \rangle, \qquad (6.11)$$

provided $\deg\left(A\langle X \rangle + B\langle X \rangle\right)$, $\deg A\langle X \rangle$ and $\deg B\langle X \rangle$ are defined. The case $B\langle X \rangle = 1 - A\langle X \rangle$ shows that strict inequality can occur in (6.10). Exercise 6.2.5 gives an example showing that strict inequality can occur in (6.11). If, however, M is a field then the equality holds in (6.11): indeed, the coefficient of X^{n+m} in this case must be different from zero. Hence we have

$$\deg(A\langle X \rangle \cdot B\langle X \rangle) = \deg A\langle X \rangle + \deg B\langle X \rangle, \qquad (6.12)$$

For more details see Exercises 6.2.4–6.2.6. The ring M need not have an element 1 (the unit element for multiplication) of Axiom M_3 from Table 4.1; for example, the ring of even integers has no unit element. If M does have a unit element then we can denote the polynomial

$$\langle 0, 1, 0, \ldots 0, \ldots \rangle$$

by X and then $X.X = X^2$ is the polynomial

$$\langle 0, 0, 1, 0, \ldots 0, \ldots \rangle.$$

Generally

$$X^n = \underbrace{\langle 0, 0, \ldots,}_{n \text{ zeros}} 1, 0, \ldots \rangle.$$

X is called the indeterminate and it behaves, with respect to addition and multiplication, like an element of M or an element of some ring containing M. The notation

$$a_n X^n + a_{n-1} X^{n-1} + \cdots + a_1 X + a_0$$

now becomes less formal and more meaningful. Each term $a_k X^k$ is a product of a polynomial a_k of degree zero with the polynomial X^k, and the original polynomial itself is a sum of polynomials $a_n X^n$, $a_{n-1} X^{n-1}, \ldots, a_1 X, a_0$.

The following rule looks obvious but still requires proof[2].

Cancellation rule:
If F is a field and $A\langle X \rangle$, $B\langle X \rangle$, $C\langle X \rangle \in F[X]$ with $A\langle X \rangle$ not a zero polynomial and if

$$A\langle X \rangle B\langle X \rangle = A\langle X \rangle C\langle X \rangle$$

then $B\langle X \rangle = C\langle X \rangle$.

Proof. Since $F[X]$ is a ring we have

$$A\langle X \rangle \big(B\langle X \rangle - C\langle X \rangle \big) = 0. \tag{6.13}$$

If $B\langle X \rangle - C\langle X \rangle$ were not the zero polynomial then the leading coefficient on the right hand side of (6.13) would not be zero and Equation (6.13) could not hold. Consequently $B\langle X \rangle = C\langle X \rangle$. $\qquad\square$

With the polynomial $A\langle X \rangle$ associate the polynomial function A defined on M

$$A : x \mapsto a_n x^n + a_{n-1} x^{n-1} + \cdots + a_0. \tag{6.14}$$

Let us make the notation very clear: A is the function defined in (6.14), $A(x)$ is the value of this function at x and $A\langle X \rangle$ is the polynomial (6.9). This notation is convenient because we have the following rule:

---

[2]This rule is false if $F[X]$ is replaced by $M[X]$.

Substitution rule:

If $A\langle X\rangle$ and $B\langle X\rangle$ are two equal polynomials over M and z is an element of M or an element of a ring containing M then $A(z) = B(z)$.

Proof. $A\langle X\rangle$ and $B\langle X\rangle$ are equal polynomials which means that when both are written in order of ascending powers of X, and each term of the form $c_k X^k$ appears exactly once, they have the same coefficients. Consequently the function values of A and B are evaluated in exactly the same way and must be equal. $\qquad\square$

Exercises

Exercise 6.2.1 *Multiply the following unending polynomials*

(1) $(1 + X + \cdots + X^n + \cdots) \cdot (1 - X + X^2 + \cdots + (-1)^n X^n + \cdots)$
(2) $(1 + 2X + X^2 + 0X^3 + \cdots + 0X^n + \cdots) \cdot (1 - X + X^2 + \cdots + (-1)^n X^n + \cdots)$

!) (i) **Exercise 6.2.2** *If M has a multiplicative unit element 1 then $1 + 0X + \cdots + 0X^n + \cdots$ is the multiplicative unit in the ring of unending polynomials: this unit is also denoted by 1. If $A\langle X\rangle = a_0 + a_1 X + \cdots + a_n X^n + \cdots$ with $a_0 = 1$ then there exists an unending polynomial $B\langle X\rangle$ such that $A\langle X\rangle \cdot B\langle X\rangle = 1$. Prove this. Show also that if M is a field then the polynomial $B\langle X\rangle$ exists if merely $a_0 \neq 0$.*

(i) **Exercise 6.2.3** *An easy way to multiply two polynomials $A(x)$ and $B(x)$ of degree m and n, respectively, is to use the Maple command[3]*

```
taylor(A(x)B(x),m+n);convert(%,polynom):
```
Verify this for $A(x) = x^2 - 1$ and $B(x) = x^{100} + x^{99} + \cdots + 1$.

Exercise 6.2.4 *The relation $x \equiv y \bmod m$ was introduced in Example 3.1. Denote the equivalence class containing an element x as $\tilde{x}$. Define $\tilde{x} + \tilde{y} = \widetilde{x + y}$ and $\tilde{x} \cdot \tilde{y} = \widetilde{x \cdot y}$. Show that these definitions are correct, that is, if $x \equiv x' \bmod m$ and $y \equiv y' \bmod m$ then $x + y \equiv x' + y' \bmod m$, with the corresponding relationship for the product. Show that the equivalence classes modulo m with addition and multiplication just defined form a ring. Show also that this ring has a unit element satisfying axiom M_3 from Table 4.1. For greater clarity we denoted the classes with a tilde sign. It is however common to omit the tilde sign, and denote the class by the number which it contains.*

---

[3]The command `convert` is needed only for further computations: if you want only the result of multiplication you can omit it.

(i) **Exercise 6.2.5**   *If M is the ring of integers modulo 4 then the product of two polynomials $\tilde{1} + \tilde{2}X$ and $\tilde{2}X$ of degree one is a polynomial of degree one. Verify this. Give your own example of M and polynomials of second and third degree for which the product (a) is of third degree; and (b) is the zero polynomial.*

(i) **Exercise 6.2.6**   *A ring M is called an integral domain if*

(a) *it has a multiplicative unit element,[4] as in axiom M_3.*
(b) *if $x, y \in M$ then $x \neq 0$, $xy = 0 \Rightarrow y = 0$.*

Show that

(1) *a field is an integral domain;*
(2) *if M is an integral domain then Equation (6.12) holds;*
(3) *if M is an integral domain then so is $M[X]$.*

(i) **Exercise 6.2.7**   *Prove that an integral domain having only a finite number of elements is a field.*

(i) **Exercise 6.2.8**   *Let p be a prime. Prove that the equivalence classes mod p with addition and multiplication defined in Exercise 6.2.4 form a field.*

Exercise 6.2.9   *Algebraic expressions of the form*

$$\frac{a_{-k}}{X^k} + \frac{a_{-k+1}}{X^{k-1}} + \cdots + a_0 + a_1 X + \cdots + a_n X^n + \cdots \qquad (6.15)$$

can be introduced as pairs $(k, A\langle X \rangle)$ where $k \in \mathbb{N}$ and $A\langle X \rangle = a_{-k} + a_{-k+1}X + \cdots + a_0 X^k + \cdots$ is an unending polynomial. If the coefficients $a_i \in \mathbb{R}$, we call such pairs formal L-series. We say that two L-series $(k, A\langle X \rangle) = (l, B\langle X \rangle)$ are equal if and only if $X^l A\langle X \rangle = X^k B\langle X \rangle$ as unending polynomials. We define addition and multiplication as follows:

$$(k, A\langle X \rangle) + (l, B\langle X \rangle) \stackrel{\text{def}}{=} (k + l, X^l A\langle X \rangle + X^k B\langle X \rangle),$$

$$(k, A\langle X \rangle) \cdot (l, B\langle X \rangle) \stackrel{\text{def}}{=} (k + l, A\langle X \rangle \cdot B\langle X \rangle).$$

Show that with these definitions of addition and multiplication the formal L-series constitute a field. The set of elements of the form $(1, X \cdot A\langle X \rangle)$ with $A\langle X \rangle = a + 0 \cdot X + \cdots$ is isomorphic to $\mathbb{R}$. [Hint: The unit element of multiplication is (m, X^m). To find a multiplicative inverse use Exercise 6.2.2.]

---

[4]Some authors do not include this axiom in the definition of an integral domain.

Exercise 6.2.10   Let a be the first non-zero coefficient of an unending polynomial $A\langle X \rangle$, that is $A\langle X \rangle = aX^m + a_{m+1}X^{m+1} + \cdots$. In the field of formal L-series (see the previous exercise) define $(k, A\langle X \rangle)$ to be positive if and only if $a > 0$. Show that the positive formal L-series have the properties of positive elements mentioned on page 103 and conclude that formal L-series constitute an ordered field. Show that this field lacks the Archimedean property.

[Hint: For $A\langle X \rangle = 1 + 0X \ldots + 0X^n + \cdots$ and $n \in \mathbb{N}$ prove $(1, A\langle X \rangle) > n$.]

ⓘ **Exercise 6.2.11**   Let S be a non-empty set and $\mathfrak{F}$ the set of functions $f : S \mapsto \mathbb{R}$. Define $(f + g)(x) = f(x) + g(x)$ and $(f \cdot g)(x) = f(x) \cdot g(x)$. Show that with these definitions $\mathfrak{F}$ becomes a ring which has a multiplicative unit. Show that if S contains at least two distinct elements then $\mathfrak{F}$ is not an integral domain.

6.3   Long division algorithm

The next theorem describes so-called long division. We have already used it in Section 1.5.2 and readers have probably encountered it many times before, however, most likely only for $F = \mathbb{R}$ and without a proper proof.

Theorem 6.1   If $P\langle X \rangle$, $H\langle X \rangle$ are polynomials over a field F and $H\langle X \rangle$ is not the zero polynomial then there exist polynomials $Q\langle X \rangle \in F[X]$ and $R\langle X \rangle \in F[X]$ such that

$$P\langle X \rangle = H\langle X \rangle Q\langle X \rangle + R\langle X \rangle \qquad (6.16)$$

and either $\deg R\langle X \rangle < \deg H\langle X \rangle$ or $R\langle X \rangle$ is the zero polynomial. The polynomials $Q\langle X \rangle$, $R\langle X \rangle$ are uniquely determined.

Proof.   First we prove uniqueness. If besides $Q\langle X \rangle$ and $R\langle X \rangle$ there are $Q_1\langle X \rangle$ and $R_1\langle X \rangle$ with the same properties then

$$H\langle X \rangle[Q\langle X \rangle - Q_1\langle X \rangle] = R_1\langle X \rangle - R\langle X \rangle. \qquad (6.17)$$

If $Q\langle X \rangle - Q_1\langle X \rangle$ were not the zero polynomial then neither would $R_1\langle X \rangle - R\langle X \rangle$ be and by (6.10) and (6.12)

$$\deg H\langle X \rangle(Q\langle X \rangle - Q_1\langle X \rangle) \geq \deg H\langle X \rangle$$
$$> \mathrm{Max}(\deg R_1\langle X \rangle, \deg R\langle X \rangle) \geq \deg(R_1\langle X \rangle - R\langle X \rangle). \qquad (6.18)$$

This contradicts (6.17) and consequently $Q\langle X \rangle = Q_1 \langle X \rangle$ and then $R\langle X \rangle = R_1 \langle X \rangle$.

Now we prove existence. Let

$$P\langle X \rangle = p_n X^n + p_{n-1} X^{n-1} + \cdots + p_0,$$
$$H\langle X \rangle = h_m X^m + h_{m-1} X^{m-1} + \cdots + h_0,$$

with $p_n \neq 0$ and $h_m \neq 0$. If $m > n$ or $P\langle X \rangle = 0$ then the theorem is true, with $Q\langle X \rangle = 0$ and $R\langle X \rangle = P\langle X \rangle$. In the remaining case we proceed by complete induction. The assertion is true if $\deg P\langle X \rangle = 0$ with $R\langle X \rangle = 0$ and $Q\langle X \rangle = p_0 h_0^{-1} = P\langle X \rangle / H\langle X \rangle$. Let us now consider the polynomial $p\langle X \rangle = P\langle X \rangle - p_n h_m^{-1} X^{n-m} H\langle X \rangle$. The polynomial $p\langle X \rangle$ has been chosen so that $\deg p\langle X \rangle < \deg P\langle X \rangle$ or possibly $p\langle X \rangle = 0$. In the latter case $R\langle X \rangle = 0$ and $Q\langle X \rangle = a_n h_m^{-1} X^{n-m}$. In the former case either by induction hypothesis or because $\deg p\langle X \rangle > \deg H\langle X \rangle$ there exist polynomials $G\langle X \rangle$ and $R\langle X \rangle$ such that $p\langle X \rangle = H\langle X \rangle G\langle X \rangle + R\langle X \rangle$ with $\deg R\langle X \rangle < \deg H\langle X \rangle$ or $R\langle X \rangle = 0$. It follows that $P\langle X \rangle = H\langle X \rangle [p_n h_m^{-1} X^{n-m} + G\langle X \rangle] + R\langle X \rangle$, as required. $\qquad \square$

Maple can find the polynomials $Q\langle X \rangle$ and $R\langle X \rangle$. The commands are quo(P(X),H(X),X) for the quotient and rem(P(X),H(X),X) for the remainder. *Maple* works in the field which is implied by the coefficients, this means, in the smallest field which contains the coefficients of $P\langle X \rangle$ and $H\langle X \rangle$. If, for example, $P\langle X \rangle$ and $H\langle X \rangle$ have integer coefficients then $Q\langle X \rangle$ and $R\langle X \rangle$ are found in $\mathbb{Q}[X]$. In the following example the field is formed by numbers of the form $a + b\sqrt{2}$ with $a, b \in \mathbb{Q}$.

```
>    quo(X^2+sqrt(2)*X+sqrt(2),3*X+1,X);
```

$$\frac{1}{3}\sqrt{2} - \frac{1}{9} + \frac{1}{3}X$$

```
>    rem(X^2+sqrt(2)*X+sqrt(2),3*X+1,X);
```

$$\frac{2}{3}\sqrt{2} + \frac{1}{9}$$

One can obtain both the remainder and the quotient by one command by adding an additional fourth argument, either q or r. However this argument must be enclosed in ' '. For instance

---

```
>   rem(x^2+3,x-1,x,'q');
```

$$4$$

```
>   q;
```

$$x + 1$$

---

or

```
>   quo(x^2+3,x-1,x,'r');r;
```

$$x + 1$$

$$4$$

The capitalized commands `Quo` and `Rem` work the same way, but the coefficients can belong to the field of equivalence classes mod p, where p is a prime. This is illustrated in the next example.

```
>   Quo(x^10,6*x-1,x,'r') mod 7;r;
```

$$6\,x^9 + x^8 + 6\,x^7 + x^6 + 6\,x^5 + x^4 + 6\,x^3 + x^2 + 6\,x + 1$$

$$1$$

For polynomials of zero degree the `Quo` command can be used to find the multiplicative inverse of an element in the field of integers mod p. The next example shows how to find the multiplicative inverse of 91 mod 97.[5] `irem` is then used to check the result.

```
>   Quo(1,91,X) mod 97;
```

---

[5] Of course, there is no remainder.

16

```
>  irem(16*91,97);
```

1

Exercises

ⓘ **Exercise 6.3.1**   *If F is replaced by $\mathbb{Z}$ in Theorem 6.1 then polynomials $Q\langle X \rangle$ and $R\langle X \rangle$ need not exist. Consider the following example: $P\langle X \rangle = X^2 + 1$ and $H\langle X \rangle = 2X$. However, polynomials in $\mathbb{Z}[X]$ lie also in $\mathbb{Q}[X]$ and the theorem can be applied to $\mathbb{Q}[X]$. Obviously, the required polynomials do not have integer coefficients. Prove all these assertions.*

ⓘ **Exercise 6.3.2**   *Prove that numbers of the form $a + b\sqrt{2}$ with $a \in \mathbb{Q}$ and $b \in \mathbb{Q}$ form a field.*
[Hint: Only Axiom M_4 is not obvious.]

Exercise 6.3.3   *Let M be the ring of equivalence classes modulo 4. (See Exercise 6.2.4.) Considering $P\langle X \rangle = \tilde{2}X^2$, $Q_1\langle X \rangle = X$ and $Q_2\langle X \rangle = \tilde{3}X$ show that Equation (6.16) is satisfied with two distinct polynomials $Q\langle X \rangle$.*

ⓘ **Exercise 6.3.4**   *If $H\langle X \rangle = X^m + h_{m-1}X^{m-1} + \cdots + h_0$ and F is merely an integral domain then the assertion of Theorem 6.1 remains valid. Prove it!*

Exercise 6.3.5   *Find the multiplicative inverse of 2 mod 101 using the Quo command.*

6.4   Roots of polynomials

If F is a field, $P\langle X \rangle \in F[X]$ then a belonging to some field containing F is a root of the polynomial $P\langle X \rangle$ or of its associated function P if $P(a) = 0$. A root of $P\langle X \rangle$ or of P is also called a zero of $P\langle X \rangle$ or of P, respectively. Theorem 6.1 is often used for $H\langle X \rangle = X - a$, in which case Equation (6.16) reads

$$P\langle X \rangle = (X - a)Q\langle X \rangle + P(a). \tag{6.19}$$

It follows that if a is a zero of $P\langle X \rangle$ then $X - a$ divides $P\langle X \rangle$. From this we have

---

Theorem 6.2   *If F is a field then the polynomial $P\langle X \rangle \in F[X]$ of degree n has at most n distinct roots in any field which contains F.*

Proof.   We proceed by induction. The theorem is obviously correct for polynomials of degree zero. If $P\langle X \rangle$ of degree n has no roots then there is nothing more to prove. If it has a root a then by Equation (6.19) we have $P\langle X \rangle = (X - a)Q\langle X \rangle$, where $Q\langle X \rangle$ is of degree $n - 1$. By the induction hypothesis $Q\langle X \rangle$ has at most $n - 1$ roots and consequently $P\langle X \rangle$ has at most n roots. $\square$

If M is merely a ring then $P\langle X \rangle \in M[X]$ of degree n can have more roots than n: one such example is given in Exercise 6.4.1. Obviously a polynomial of degree n can have fewer than n roots, for example, the polynomial of the second degree $X^2 - 2X + 1$ has only one root. If $P\langle X \rangle = (X-a)^k Q\langle X \rangle$ with $Q(a) \neq 0$ we say that a is a root of $P\langle X \rangle$ of multiplicity k. For $k = 2$ the root is called a double root, for $k = 1$ the root is simple. It can be shown that for *any* polynomial in $F[X]$ of degree at least one there exists a field $\bar{F} \supset F$ in which the polynomial has a root. The proof cannot be given here but we shall make some remarks concerning this result about field extensions in Chapter 7. Theorem 6.2 has the following very important consequence.

Theorem 6.3   *If F is a field and $P\langle X \rangle \in F[X]$ is of the form*

$$P\langle X \rangle = p_n X^n + \cdots + p_0$$

and has $n + 1$ distinct roots then $p_i = 0$ for $i = 0, 1, \ldots, n$. If F has infinitely many elements and P and Q are two polynomial functions with $P(x) = Q(x)$ for every $x \in F$ then they have the same coefficients.

Proof. $P\langle X \rangle - Q\langle X \rangle$ must be the zero polynomial. $\square$

Exercises

ⓘ **Exercise 6.4.1**   *This exercise uses the notation of Exercise 6.2.11. Let $A \subset S$, $A \neq S$ and assume that A has infinitely many elements. Show that the polynomial $1_{S \setminus A} X \in \mathfrak{F}[X]$ has infinitely many roots. [Hint: $X = 1_B$ with $B \subset A$.]*

Exercise 6.4.2 *Show that if **F** is finite then there exists a polynomial which is not the zero polynomial and for which every element in **F** is a root.*

6.5   The Taylor polynomial

In this section we explore Equation (6.19) further. The coefficients of $Q\langle X \rangle = q_{n-1}X^{n-1} + \cdots + q_0$ and the value $P(a)$ in Equation (6.19) can be easily obtained by comparing the coefficients.

$$p_n = q_{n-1},$$
$$p_{n-1} = -aq_{n-1} + q_{n-2},$$
$$\vdots \qquad\qquad\qquad (6.20)$$
$$p_1 = -aq_1 + q_0$$
$$p_0 = -aq_0 + P(a).$$

One finds q_{n-1} from the first equation. q_{n-2} can be found from the second equation, and by continuing with this process we can find all q_k and also $P(a)$. The whole computation can be arranged in the following array, known as Horner's scheme. Firstly, we write all coefficients of $P\langle X \rangle$ in the first row. It is important to write all coefficients, even the zero ones. Then we leave one row empty and copy p_n in the first column of the third row.

$$\begin{array}{cccccc} p_n & p_{n-1} & p_{n-2} & \cdots p_1 & p_0 \end{array}$$

$$p_n = q_{n-1}$$

With this done we multiply each element of the third row by a, place it in the next column in the row above, add up the elements in that column (by using equations (6.20) in the form $q_{n-2} = p_{n-1} + aq_{n-1}$, etc.) and successively fill in the whole array.

$$\begin{array}{cccccc} p_n & p_{n-1} & p_{n-2} & \cdots & p_1 & p_0 \\ & aq_{n-1} & aq_{n-2} & \cdots & aq_1 & aq_0 \\ \hline q_{n-1} & q_{n-2} & q_{n-3} & \cdots & q_0 & P(a) \end{array} \qquad (6.21)$$

We can divide $Q\langle X \rangle$ by $X - a$, the quotient again by $X - a$ and by continuing this process we form the following scheme.

$$P\langle X \rangle = (X - a)Q_1\langle X \rangle + P(a),$$
$$Q_1\langle X \rangle = (X - a)Q_2\langle X \rangle + Q_1(a),$$
$$\vdots$$
$$Q_{n-1}\langle X \rangle = (X - a)Q_n\langle X \rangle + Q_{n-1}(a),$$
$$Q_n\langle X \rangle = Q_n(a).$$

(6.22)

The coefficients of $Q_k\langle X \rangle$, which we denote by $q_{k,i}$, can be obtained by continuation of the Horner scheme.

p_n	p_{n-1}	p_{n-2}	$\cdots$	p_2	p_1		p_0
	$aq_{1,n-1}$	$aq_{1,n-2}$	$\cdots$	$aq_{1,2}$	$aq_{1,1}$		$aq_{1,0}$
$q_{1,n-1}$	$q_{1,n-2}$	$q_{1,n-3}$	$\cdots$	$q_{1,1}$	$q_{1,0}$		$q_0 = P(a)$
	$aq_{2,n-2}$	$aq_{2,n-3}$	$\cdots$	$aq_{2,1}$	$aq_{2,0}$		
$q_{2,n-2}$	$q_{2,n-3}$	$q_{2,n-4}$	$\cdots$	$q_{2,0}$	$q_1 = Q_1(a)$		
$\cdots$							
$\vdots$							
$q_{n-1,1}$	$q_{n-1,0}$	q_{n-2}					
	$aq_{n,0}$						
$q_{n,0} = q_n$	q_{n-1}						

(6.23)

Finally, by working backwards through the schemes (6.22) and (6.23), we express $P\langle X \rangle$ as a sum of powers of $X - a$.

$$P\langle X \rangle = q_0 + q_1(X - a) + \cdots + q_{n-1}(X - a)^{n-1} + q_n(X - a)^n. \quad (6.24)$$

The right hand side of this equation is called the *Taylor polynomial* or the *Taylor expansion* of $P\langle X \rangle$ around the point $X = a$. In the scheme (6.23) we have $q_{1,n-1} = q_{2,n-2} = \cdots = q_{n-1,1} = q_{n,0} = q_n = p_n$.

To illustrate the procedure we expand $X^4 + 4X^2 + 3X + 2$ in powers of $X - 2$ with the coefficients lying in the field of integers mod 5.

$$
\begin{array}{ccccc}
1 & 0 & 4 & 3 & 2 \\
  & 2 & 4 & 1 & 3 \\
\hline
1 & 2 & 3 & 4 & 0 \\
  &   & 2 & 3 & 2 \\
\hline
1 & 4 & 1 & 1 &   \\
  &   & 2 & 2 &   \\
\hline
1 & 1 & 3 &   &   \\
  &   & 2 &   &   \\
\hline
1 & 3 &   &   &   \\
\end{array}
$$

As a result we have

$$ X^4 + 4X^2 + 3X + 2 = (X - 2) + 3(X - 2)^2 + 3(X - 2)^3 + (X - 2)^4. $$

The coefficients q_k can be determined in other ways than by the scheme (6.23). One can write $X^k = \big((X - a) + a\big)^k$ and use the binomial theorem (Theorem 5.4). Such a procedure might be advantageous if the degree of $P\langle X \rangle$ is very small. Collecting the absolute terms leads to the result $q_0 = P(a)$ which we already know, collecting the coefficients of $X - a$ leads to

$$ q_1 = np_n a^{n-1} + (n - 1)p_{n-1}a^{n-2} + \cdots + 2p_2 a + p_1. $$

We can take some inspiration from this formula and define a derivative of a polynomial (derivatives are treated from a calculus point of view in detail in Chapter 13) by

$$ DP\langle X \rangle = np_n X^{n-1} + (n - 1)p_{n-1}X^{n-2} + \cdots + 2p_2 X + p_1. \qquad (6.25) $$

The formula for q_1 then becomes

$$ q_1 = DP(a). $$

The following formulae are rather important.

$$ D\big(P_1\langle X \rangle + P_2\langle X \rangle\big) = DP_1\langle X \rangle + DP_2\langle X \rangle, \qquad (6.26) $$
$$ D\big(P_1\langle X \rangle P_2\langle X \rangle\big) = P_1\langle X \rangle DP_2\langle X \rangle + P_2\langle X \rangle DP_1\langle X \rangle. \qquad (6.27) $$

Equation (6.26) follows easily from the definition of derivative. To prove (6.27) we write

$$P_1\langle X\rangle = P_1(a) + \big(DP_1(a)\big)(X-A) + (X-a)^2 W_1\langle X-a\rangle,$$
$$P_2\langle X\rangle = P_2(a) + \big(DP_2(a)\big)(X-A) + (X-a)^2 W_2\langle X-a\rangle,$$

where we abbreviated by W_1 and W_2 some polynomials whose form is unimportant for the proof. It follows that

$$P_1\langle X\rangle P_2\langle X\rangle = P_1(a)P_2(a)$$
$$+ \big[P_2(a)DP_1(a) + P_1(a)DP_2(a)\big](X-a) + (X-a)^2 W\langle X-a\rangle,$$

where we again denoted by W some polynomial. From this equation the formula (6.27) follows. Using induction it follows from (6.27) that $D(X-c)^k = k(X-c)^{k-1}$. This can also be verified directly from (6.25). Defining $D^k P\langle X\rangle \overset{\text{def}}{=} D\big(D^{k-1}P\langle X\rangle\big)$ and applying D successively to Equation (6.25) and then substituting $X = a$ we obtain

$$\begin{aligned}
D^2 P(a) &= 2\cdot 1\cdot q_2,\\
D^3 P(a) &= 3\cdot 2\cdot 1\cdot q_3,\\
&\vdots\\
D^n P(a) &= n!\, q_n.
\end{aligned} \tag{6.28}$$

From these equations it is easy to find the coefficients q_k. Some caution is needed: generally speaking, all coefficients of q_k in (6.28) are distinct from zero[6] only if F contains $\mathbb{N}$. This, of course, happens in the most important cases when $F = \mathbb{Q}$ or $\mathbb{R}$. The Taylor formula then takes the form

$$P\langle X\rangle = P(a) + \frac{DP(a)}{1!}(X-a) + \frac{D^2 P(a)}{2!}(X-a)^2 + \cdots + \frac{D^n P(a)}{n!}(X-a)^n. \tag{6.29}$$

The importance of this equation lies in the fact that it gives explicit formulae for the coefficients of powers of $X-a$. In practice, however, the process is sometimes reversed: one finds $D^k P(a)$ as q_k by using the scheme (6.23). If one has a computer and *Maple* available everything becomes easy. The command

```
taylor(P(X),X=a,n)
```

---

[6] For example $n! = 0 \bmod p$ if $n \geq p$.

produces expansion (6.24) with the coefficients q_k evaluated. It does that even if the coefficients of $P\langle X\rangle$ contain some parameters. It is advisable, as we mentioned earlier, to add the command `convert`. Hence the complete command is then

```
taylor(P(X),X=a,n):convert(%,polynom);.
```

Exercises

Exercise 6.5.1   *Use the binomial theorem to find the Taylor expansion of X^3 for $a = -1$.*

Exercise 6.5.2   *Find the Taylor expansion at $a = -2$ for the following polynomials*

(1) $P\langle X\rangle = X^5$;
(2) $P\langle X\rangle = X^5 + 2X^4 + 3X^2 + 2X + 1$.

Decide in each case which formula is more convenient to use, (6.23) or (6.29).

Exercise 6.5.3   *Use* Maple *to find the Taylor expansion of X^2 at $a = 2b/3$.*

6.6   Factorization

In this section we shall assume that $\boldsymbol{F}$ always denotes a field. If not stated otherwise every polynomial is automatically assumed to belong to $\boldsymbol{F}[X]$. In Section 2.5.1 we established some results on divisibility in a Euclidean ring and applied it to polynomials with rational coefficients. Since $\boldsymbol{F}[X]$ is also a Euclidean[7] ring these results remain valid. We recapitulate some of them. The polynomial $P\langle X\rangle$ is a multiple of $H\langle X\rangle$ and $H\langle X\rangle$ is called a *divisor* of $P\langle X\rangle$ if there exists a polynomial $Q\langle X\rangle$ such that

$$P\langle X\rangle = H\langle X\rangle Q\langle X\rangle. \tag{6.30}$$

The word factor is often used instead of divisor. If $H\langle X\rangle$ is a divisor of $P\langle X\rangle$ then we also say that $H\langle X\rangle$ divides $P\langle X\rangle$, or that $P\langle X\rangle$ is divisible by $H\langle X\rangle$. If the polynomial $D\langle X\rangle$ is a divisor of both $P_1\langle X\rangle$ and $P_2\langle X\rangle$ it is called a *common divisor*: if every other common divisor divides $D\langle X\rangle$ then $D\langle X\rangle$ is called the *greatest common divisor* of $P_1\langle X\rangle$ and $P_2\langle X\rangle$. For

---

[7]With $N(r)$ equal to the degree of the polynomial.

any two polynomials $P_1\langle X \rangle \neq 0$ and $P_2\langle X \rangle$ in $F[X]$ the greatest common divisor always exists. It is the polynomial of the smallest degree which is of the form

$$P_1\langle X \rangle H_1\langle X \rangle + P_2\langle X \rangle H_2\langle X \rangle \qquad (6.31)$$

with $H_1\langle X \rangle$ and $H_2\langle X \rangle$ in $F[X]$. The greatest common divisor is uniquely determined in the following sense: if $D_1\langle X \rangle$ and $D_2\langle X \rangle$ are two greatest common divisors of the same two polynomials then there exists an element in $c \in F$ such that $D_1\langle X \rangle = cD_2\langle X \rangle$.

A polynomial $P\langle X \rangle$ of degree $n \geq 1$ is called *reducible* over (in) F if there exists a polynomial $H\langle X \rangle \in F[X]$ of degree greater than zero but less than n such that Equation (6.30) holds. A polynomial $P\langle X \rangle$ of degree $n \geq 1$ is *irreducible* over F if Equation (6.30) implies that $\deg H\langle X \rangle$ is either n or 0. It is important to realize that the concept of being reducible or irreducible depends not only on the polynomial $P\langle X \rangle$ but also on the field in question. For example, $X^2 - 2 = (X - \sqrt{2})(X + \sqrt{2})$ is reducible in $\mathbb{R}[X]$, but it is left as an exercise to prove indirectly that it is irreducible in $\mathbb{Q}[X]$. It follows from the definition that every polynomial of degree one is irreducible over $F[X]$.

Our aim is to show that any polynomial can be decomposed into a product of irreducible polynomials and that such a decomposition is in a certain sense unique. To this purpose we first list some properties of irreducible polynomials.

(a) If $P\langle X \rangle$ is irreducible in $F[X]$ and $0 \neq c \in F$ then $cP\langle X \rangle$ is irreducible.

(b) If $H\langle X \rangle$ is a divisor of an irreducible $P\langle X \rangle$ then either $H\langle X \rangle$ is of zero degree or $H\langle X \rangle = cP\langle X \rangle$ for some nonzero element of F.

(c) If (6.30) holds, and $P\langle X \rangle$ is divisible by some irreducible polynomial $D\langle X \rangle$, then $D\langle X \rangle$ is a divisor of either $H\langle X \rangle$ or $Q\langle X \rangle$.

If $D\langle X \rangle$ does not divide $H\langle X \rangle$ then the greatest common divisor of $D\langle X \rangle$ and $H\langle X \rangle$ is 1 and by (6.31) we have

$$1 = D\langle X \rangle A\langle X \rangle + H\langle X \rangle B\langle X \rangle$$

for some $A\langle X \rangle$ and $B\langle X \rangle$. Multiplying by $Q\langle X \rangle$ gives

$$Q\langle X \rangle = D\langle X \rangle Q\langle X \rangle A\langle X \rangle + \big(H\langle X \rangle Q\langle X \rangle \big) B\langle X \rangle,$$
$$= D\langle X \rangle Q\langle X \rangle A\langle X \rangle + P\langle X \rangle B\langle X \rangle.$$

The right hand side is divisible by $D\langle X \rangle$, and so then is $Q\langle X \rangle$.

(d) If $P\langle X \rangle = Q_1\langle X \rangle Q_2\langle X \rangle \cdots Q_n\langle X \rangle$ and $P\langle X \rangle$ is divisible by an irreducible $F\langle X \rangle$ then $F\langle X \rangle$ is a divisor of one of the $Q_i\langle X \rangle$.

This follows by an easy induction from (c).

(e) Every polynomial of degree at least 1 is divisible by some irreducible polynomial.

The statement is true for polynomials of degree 1. We proceed by complete induction. If the polynomial is not irreducible then it is a product of two polynomials of smaller degree and by the induction hypothesis one of these is divisible by an irreducible polynomial and consequently so is the original polynomial itself.

If $\deg P\langle X \rangle > 0$ then by (e) we have $P\langle X \rangle = P_1\langle X \rangle Q_1\langle X \rangle$ with an irreducible $P_1\langle X \rangle$. If $Q_1\langle X \rangle$ is not of zero degree, then again $Q_1\langle X \rangle$ is divisible by some irreducible polynomial, by continuing with this process we must finally arrive at a polynomial of zero degree and we have

$$P\langle X \rangle = P_1\langle X \rangle P_2\langle X \rangle \cdots P_l\langle X \rangle. \tag{6.32}$$

where all the polynomials $P_i\langle X \rangle$ are irreducible.

Theorem 6.4   *For every polynomial $P\langle X \rangle \in \boldsymbol{F}[X]$ of degree at least one there exist polynomials $P_i\langle X \rangle$, $i = 1, 2, \ldots, l$, irreducible over $\boldsymbol{F}[X]$, such that (6.32) holds. If*

$$P\langle X \rangle = Q_1\langle X \rangle Q_2\langle X \rangle \cdots Q_m\langle X \rangle \tag{6.33}$$

is another representation of $P\langle X \rangle$ with all $Q_i\langle X \rangle$ irreducible over $\boldsymbol{F}[X]$ then $m = l$ and, with appropriate renumbering (if necessary), there are non-zero $c_i \in \boldsymbol{F}$ such that $P_i\langle X \rangle = c_i Q_i\langle X \rangle$ for $i = 1, 2, \ldots, l$.

An informal way of stating the theorem in an easy to remember form is: the decomposition of a polynomial into irreducible factors is unique.

Proof.   The proof is inductive. The theorem is true if $P\langle X \rangle$ is of degree one, because such polynomials are irreducible. Let us now assume that the theorem is true for all polynomials of degree less than n and let $P\langle X \rangle$ be of degree n. Since $P_1\langle X \rangle$ divides $P\langle X \rangle$ it must, by (d), divide one of the $Q_i\langle X \rangle$. Renumbering it $Q_1\langle X \rangle$ (if required, and the others correspondingly), we have $Q_1\langle X \rangle = c_1 P_1\langle X \rangle$ with a non-zero c_1 because $Q_1\langle X \rangle$ is irreducible over $\boldsymbol{F}[X]$ and $P\langle X \rangle$ is of positive degree. Substituting this into Equation (6.32) we obtain

$$P_1\langle X\rangle P_2\langle X\rangle \cdots P_l\langle X\rangle = c_1 P_1\langle X\rangle Q_2\langle X\rangle \cdots Q_m\langle X\rangle.$$

It follows that

$$P_2\langle X\rangle \cdots P_l\langle X\rangle = \big(c_1 Q_2\langle X\rangle\big) \cdots Q_m\langle X\rangle.$$

Since the polynomials in this equation are of degree less than n the theorem can be applied: it follows that $l - 1 = m - 1$, consequently $l = m$ and also $P_2\langle X\rangle = c_2 Q_2\langle X\rangle$, ..., $P_l\langle X\rangle = c_l Q_l\langle X\rangle$. $\qquad\square$

Remark 6.1   The theorem does not assert that the polynomials $P_i\langle X\rangle$ are distinct. If we require that the leading coefficients in all polynomials are 1 and group together identical polynomials, the factorization takes the form

$$P\langle X\rangle = c\big(P_1\langle X\rangle\big)^{n_1}\big(P_2\langle X\rangle\big)^{n_2}\cdots\big(P_s\langle X\rangle\big)^{n_s}. \qquad (6.34)$$

The polynomials $P_i\langle X\rangle$, the natural numbers n_i and c, a non-zero element of F, are uniquely determined,

In Theorem 6.4 we established the existence and uniqueness of the decomposition of a polynomial into irreducible factors. Actual finding of the factors can be quite difficult, even in $\mathbb{Q}[X]$. Merely to decide whether or not a polynomial is irreducible can also be quite difficult. *Maple* can be of great help with these tasks.

The command to use is

`factor(P(X), option);`

A shorter name can be given for $P(X)$ using the assignment operator. If option is left out *Maple* tries to factorize the polynomial in the smallest field which contains the coefficients. For `option` we discuss here three possibilities.

(a) One types in the word real. The polynomial is then factored in $\mathbb{R}[X]$;
(b) A number or a list of numbers or a set of numbers is inserted for `option`. The number or each number in the list or set should be a root of a polynomial with rational coefficients, for example $\sqrt{2}$, $\sqrt[3]{24}$. The factorization is carried out in the smallest field which contains the given number or all the numbers in the list or set, respectively.
(c) Similar to (b) but the numbers are given with `RootOf`. For example, instead of $\sqrt{2}$ as in (b) the option is `RootOf`$(x^2 - 2)$.

The following session illustrates the use of `factor`.

```
>    # to save typing we name the polynomial;
>    p:=x^4+5*x^3+5*x+1:

>    # then we try to factor it;
>    factor(p);
```

$$x^4 + 5\,x^3 + 5\,x + 1$$

Maple returned the polynomial to be factored. This means that the polynomial is irreducible over the rationals. Now we try to factor the polynomial in the smallest field which contains square roots of the first few primes. First we produce a list of these roots, the *Maple* command ithprime produces what it promises, the i-th prime.

```
>    L:=[seq(sqrt(ithprime(i)),i=1..5)];
```

$$L := [\sqrt{2},\ \sqrt{3},\ \sqrt{5},\ \sqrt{7},\ \sqrt{11}]$$

```
>    # Then we try to factor again;
>    factor(p,L);
```

$$\frac{1}{4}\,(2\,x^2 + 5\,x + \sqrt{11}\,\sqrt{3}\,x + 2)\,(2\,x^2 + 5\,x - \sqrt{11}\,\sqrt{3}\,x + 2)$$

We were lucky and found a factorization. A natural question arises as to how many primes should we include in the list. It is hard to guess; it is, however, not advisable to make the list too long. It can take a long time or *Maple* can refuse to do it, because the task is just too big. Finally, we factor the polynomial over the reals.

```
>    factor(p,real);
```

$$(x + 5.179201362)\,(x + .1930799616)\,(x^2 - .3722813232\,x + .9999999999)$$

Two things should be noted. Firstly, the factorization uses floating point approximations and is not exact: it is not an exact factorization in the smallest possible subfield of reals. Secondly, the polynomial

$$2x^2 + 5x + \sqrt{11}\sqrt{3}x + 2$$

can be factorized by solving the quadratic equation and a complete exact factorization in the smallest possible subfield of reals obtained.

```
>   solve(2*x^2+5*x+11^(1/2)*3^(1/2)*x+2);
```

$$-\frac{5}{4} - \frac{1}{4}\sqrt{11}\sqrt{3} + \frac{1}{4}\sqrt{42 + 10\sqrt{11}\sqrt{3}}, \quad -\frac{5}{4} - \frac{1}{4}\sqrt{11}\sqrt{3} - \frac{1}{4}\sqrt{42 + 10\sqrt{11}\sqrt{3}}$$

Denoting the first root as a and the second as b we then have

$$p = (x - a)(x - b)(2x^2 + 5x - \sqrt{11}\sqrt{3}x + 2).$$

The capitalized command **Factor** works similarly: it factorizes a polynomial in the field of polynomials with coefficients integers modulo a prime.

```
>   Factor(x^4+1) mod 101;
```

$$(x^2 + 10)(x^2 + 91)$$

We are unable to prove it here but it is an interesting fact that the polynomial $X^4 + 1$ is reducible in the field of integers mod p for every prime p and is irreducible over $\mathbb{Q}$.

Exercises

(*i*) **Exercise 6.6.1**   Let F be a field and $P_1\langle X\rangle$, $P_2\langle X\rangle$ two polynomials in $F[X]$ of degree n and m respectively. Assume that the greatest common divisor of $P_1\langle X\rangle$ and $P_2\langle X\rangle$ is of the form (6.31). Prove that the polynomials $H_1\langle X\rangle$ and $H_2\langle X\rangle$ can be so chosen that $\deg H_1\langle X\rangle \leq m - 1$ and $\deg H_2\langle X\rangle \leq n - 1$.

Exercise 6.6.2   Find the greatest common divisor of each of the following pairs of polynomials

(1) $x^6 + 3x^5 + 3x^4 + 3x + 4x^2 + 6x + 4$, $x^4 + x^3 - 3x^2 - x + 2$;
(2) $x^6 - 3x^5 + x^3 + 2x^2 - x + 3$, $x^3 - 2x^2 + 3x - 1$;
(3) $x^{100} + x^2 + px + q$, $x^2 + px + q$;

in the field $\mathbb{Q}[x]$ and in the fields of equivalence classes mod 3 and mod 5.

Exercise 6.6.3   *Prove that the polynomial $x^2 + px + q$ is irreducible over the reals if and only if $p^2 - 4q < 0$.*

Exercise 6.6.4   *Show that the polynomial $X^4 + 1$ is irreducible over $\mathbb{Q}$ and is reducible in the smallest field which contains $\sqrt{2}$.*

Exercise 6.6.5   *Find all integers a for which the polynomial $x^3 + ax + 1$ is reducible in $\mathbb{Q}[x]$. [Hint: ± 1 must be the roots.]*

Exercise 6.6.6   *Factorize $X^8 + X^7 + X^6 + X^5 + X^4 + X^3 + X^2 + X + 1$ in $\mathbb{Q}[X]$ and over the field which contains the root of the polynomial $X^2 + X + 1$.*

Chapter 7

Complex Numbers

We introduce complex numbers; that is numbers of the form $a + bi$ where the number i satisfies $i^2 = -1$. Mathematicians were led to complex numbers in their efforts to solve so-called algebraic equations; that is equations of the form $a_n x^n + a_{n-1} x^{n-1} + \cdots + a_0 = 0$, with $a_k \in \mathbb{R}$, $n \in \mathbb{N}$. Our introduction follows the same idea although in a modern mathematical setting. Complex numbers now play important roles in physics, hydrodynamics, electromagnetic theory, electrical engineering as well as pure mathematics.

7.1 Field extensions

During the history of civilisation the concept of a number was unceasingly extended, from integers to rationals, from positive numbers to negative numbers, from rationals to reals, etc. We now embark on an extension of reals to a field in which the equation

$$\xi^2 + 1 = 0 \tag{7.1}$$

has a solution. This will be the field of complex numbers.

Let us consider the following question. Is it possible to extend the field of rationals to a larger field in which the equation $\xi^2 - 2 = 0$ is solvable? The obvious answer is yes: the reals. Is there a smaller field? The answer is again yes: there is a smallest field which contains rationals and $\sqrt{2}$, namely the intersection of all fields which contain $\mathbb{Q}$ and the (real) number $\sqrt{2}$. Can this field be constructed directly without using the existence of $\mathbb{R}$? The answer is contained in Exercises 7.1.1–7.1.3.

The process of extension of a field can be most easily carried out in the case of a finite field. Let us consider the following problem: is it possible to extend $GF(2)$ in such a way that the equation[1]

$$\xi^2 + \xi + 1 = 0 \tag{7.2}$$

has a solution in the extended field. The following Table 7.1 describes addition and multiplication in GF(2).

Table 7.1   Addition and multiplication in $GF(2)$

<table>
<tr><td colspan="3">Add</td><td colspan="3">Multiply</td></tr>
<tr><td>+</td><td>0</td><td>1</td><td>×</td><td>0</td><td>1</td></tr>
<tr><td>0</td><td>0</td><td>1</td><td>0</td><td>0</td><td>0</td></tr>
<tr><td>1</td><td>1</td><td>0</td><td>1</td><td>0</td><td>1</td></tr>
</table>

Denoting by k the solution of the equation we shall try to augment these tables by additional rows and columns in order to obtain addition and multiplication tables for the extended field. Firstly, we add a row and column headed by k. The entry in the second row and third column of the addition table cannot be equal to 0 or to 1 or to k. So it is another 'new' element, let us denote it by $1 + k$ and augment the table by another column and another row. (See Table 7.2.) It is now easy to fill the rest of the addition table: remember $1 + 1 = 0$ in $GF(2)$ and this must also hold in the extended field.

Table 7.2   Addition in $GF(4)$
Add

+	0	1	k	$1 + k$
0	0	1	k	$1 + k$
1	1	0	$1 + k$	k
k	k	$1 + k$	0	1
$1 + k$	$1 + k$	k	1	0

Filling the augmented multiplication table is a bit more involved and we use the Equation (7.2) to do it. For instance, $k^2 = -k - 1 = k + 1$, $(1 + k)k = k^2 + k = -1 = 1$. (See Multiply in Table 7.3.) If the extension existed, addition and multiplication would be described by the augmented

---

[1]It is easily checked the the equation is not solvable in $GF(2)$.

Table 7.3   Multiplication in $GF(4)$
Multiply

×	0	1	k	$1+k$
0	0	0	0	0
1	0	1	k	$1+k$
k	0	k	$1+k$	1
$1+k$	0	$1+k$	1	k

tables. We now reverse the process. For the four elements 0, 1, k, $1+k$ we *define* addition and multiplication by Tables 7.2 and 7.3. The verification of the field axioms for the new field, which is called $GF(4)$, is easy and we leave it to the readers.[2] We succeeded in our task of extending $GF(2)$ in such a way that in the extended field the Equation (7.1) has a solution.

Exercises

(i) **Exercise 7.1.1**   *Prove: The set $\{a + b\sqrt{2};\ a \in \mathbb{Q},\ b \in \mathbb{Q}\} \subset \mathbb{R}$, together with the same addition and multiplication as in $\mathbb{R}$ constitutes a field. This field contains $\mathbb{Q}$ and $\sqrt{2}$ and is the smallest field with this property. This field is often denoted by $\mathbb{Q}(\sqrt{2})$.* [Hint: $(a+b\sqrt{2})(a-b\sqrt{2})(a^2-b^2)^{-1} = 1$.]

(i) **Exercise 7.1.2**   *Let $\mathbf{F}_{\sqrt{2}}$ be the set of ordered pairs of rational numbers with the following definitions of addition and multiplication*

$$(a,b) + (c,d) = (a+b, c+d),$$
$$(a,b)(c,d) = (ac + 2bd, bc + ad).$$

Prove that $\mathbf{F}_{\sqrt{2}}$ is a field.

(i) **Exercise 7.1.3**   *Show that the field $\mathbb{Q}(\sqrt{2})$ from Exercise 7.1.1 is isomorphic to the field $\mathbf{F}_{\sqrt{2}}$ from Exercise 7.1.2.* [Hint: $(a,b) \to a + b\sqrt{2}$]

Exercise 7.1.4   *Write down the addition and multiplication tables for the field of three elements (denoted by $GF(3)$).* [Hint: The third element must be equal to $1+1$. Denote it by 2 (or by $\bar{2}$, if you wish to emphasize that it is not the natural number 2). Then realize that $2+1 = 0$ and $2 \times 2 = 1$ and the rest is easy.]

---

[2]E.g. the inverse for addition for the element $1+k$ is this element itself, the multiplicative inverse to k is $1+k$.

Exercise 7.1.5   *Extend $GF(3)$ to a field of nine elements in which the Equation (7.1) is solvable.* [Hint: Use the same method of extending the addition and multiplication tables which we used for construction of $GF(4)$.]

7.2   Complex numbers

The field of complex numbers $\mathbb{C}$ is the smallest field which contain $\mathbb{R}$ and the root of the Equation (7.1). Before we construct it, we examine its structure, assuming that $\mathbb{C}$ exists. Firstly, if $P\langle X\rangle$ is a polynomial with real coefficients and $\imath$ a solution to the equation $x^2 + 1 = 0$ then all the numbers $P(\imath)$ must be in $\mathbb{C}$. We now show that there are no other elements in $\mathbb{C}$ by showing that these numbers[3] themselves form a field and so they form the smallest field which contains the reals and the solution to Equation (7.1). Only axioms A_3, A_4, M_3 and M_4 from Table 4.1 need verification. If Z is the zero polynomial then axiom A_3 is satisfied with the zero element $Z(i)$. The inverse element to $P(\imath)$ under addition is $-P(\imath)$, hence we have A_4. Verifying axiom M_3 is also easy: if

$$U\langle X\rangle = 0 \cdot X^n + 0 \cdot X^{n-1} + \cdots + 0 \cdot X + 1 = 1,$$

then $U(i)$ is the multiplicative unit. Before we deal with M_4 we prove the following statement:

If $P\langle X\rangle \in \mathbb{R}[X]$ then

$$P(\imath) = 0 \tag{7.3}$$

if and only if $X^2 + 1$ divides $P\langle X\rangle$.

The 'if' part is obvious. Let (7.3) hold. Dividing $P\langle X\rangle$ by $X^2 + 1$ gives

$$P\langle X\rangle = Q\langle X\rangle(X^2 + 1) + bX + a, \tag{7.4}$$

with $a, b \in \mathbb{R}$. Substituting[4] $X = \imath$ gives $a + b\imath = 0$. Now $b = 0$ otherwise $\imath = -a/b \in \mathbb{R}$. It follows that $a = 0$ and consequently $P\langle X\rangle$ is divisible by $X^2 + 1$.

In order to show that $P(\imath) \neq 0$ has a multiplicative inverse we note first that $P\langle X\rangle$ is not divisible by $X^2 + 1$. Since $X^2 + 1$ is irreducible over $\mathbb{R}$, the greatest common divisor of $P\langle X\rangle$ and $X^2 + 1$ is 1 and there exist

---

[3] Addition and multiplication of these elements is understood in the obvious way, namely $P_1(\imath) + P_2(\imath) = (P_1 + P_2)(\imath)$ and $P_1(\imath)P_2(\imath) = (P_1P_2)(\imath)$.
[4] See Substitution rule on page 173

polynomials $H\langle X\rangle, H_1\langle X\rangle$ such that

$$P\langle X\rangle H\langle X\rangle + (X^2 + 1)H_1\langle X\rangle = 1. \tag{7.5}$$

Putting $X = \imath$ shows that $H(\imath)$ is the multiplicative inverse to $P(\imath)$.

The similarity in addition and multiplication of numbers $P(\imath)$ and polynomials with real coefficients leads naturally to the idea of using $\mathbb{R}[X]$ for the construction of the complex number field. There is, however, a crucial difference. Polynomials are equal if and only if they have the same coefficients. As we have just seen $P(\imath) = Q(\imath)$ if and only if $P\langle X\rangle - Q\langle X\rangle$ is divisible by $X^2 + 1$. Motivated by this we introduce an equivalence relation into $\mathbb{R}[X]$ by saying that $P\langle X\rangle$ and $Q\langle X\rangle$ are equivalent if their difference is divisible by $X^2 + 1$. We denote that by $P\langle X\rangle \equiv Q\langle X\rangle \mod (X^2 + 1)$. This relation is indeed an equivalence: it is obviously symmetric and reflexive. The proof that it is transitive is easy and left as an exercise.

Addition and multiplication in $\mathbb{R}[X]$ is consistent with equivalence $\mod (X^2 + 1)$. More precisely, if $P_1\langle X\rangle \equiv P_2\langle X\rangle \mod (X^2 + 1)$ and $Q_1\langle X\rangle \equiv Q_2\langle X\rangle \mod (X^2 + 1)$ then

$$P_1\langle X\rangle + Q_1\langle X\rangle \equiv P_2\langle X\rangle + Q_2\langle X\rangle \mod (X^2 + 1) \tag{7.6}$$
$$P_1\langle X\rangle \cdot Q_1\langle X\rangle \equiv P_2\langle X\rangle \cdot Q_2\langle X\rangle \mod (X^2 + 1) \tag{7.7}$$

We mentioned in Remark 3.1 that elements of an equivalence class are sometimes said to be equal. For example, by the equivalence for common fractions we identify $\frac{a}{b}$ with $\frac{2a}{2b}$. In the same vein we identify here X^2 with -1 since $X^2 - (-1) \equiv 0 \mod (X^2 + 1)$, and more generally, we identify $P\langle X\rangle$ with $Q\langle X\rangle$ if $P\langle X\rangle \equiv Q\langle X\rangle \mod (X^2 + 1)$. If, in addition we replace the summands by equal[5] elements the result will be only seemingly affected: by (7.6), no matter how the result appears, it will always consist of equal elements. Similarly for multiplication. Since $\mathbb{R}[X]$ is a ring, it is clear that with addition, multiplication and identification just explained the resulting algebraic structure is a ring. We denote it by $\mathbb{C}$, call its elements complex numbers and in the course of this section we shall prove that $\mathbb{C}$ is a field. Since we identified X^2 with -1 it follows that the indeterminate X itself is a solution to Equation (7.1) and as such we denote it by $\imath$. Every element of $\mathbb{C}$ is of the form $a + b\imath$ with $a, b \in \mathbb{R}$. Indeed, by (7.4) we have $\alpha = P(\imath) = a + b\imath$. We call a, b the *real part* of α and the *imaginary part* of α, respectively. In symbols, $a = \Re\alpha$, $b = \Im\alpha$. It is worth noting that

---

[5]That is, equivalent $\mod (X^2 + 1)$.

the imaginary part of a complex number is a real number. If $b \neq 0$ then α is called an imaginary number[6]: if $a = 0$ then α is purely imaginary. The number $a - bi$ is called the *complex conjugate* of $a + bi$, and we denote it by $\bar{\alpha}$. It is easy to see that for $\alpha, \beta \in \mathbb{C}$

$$\bar{\alpha} + \bar{\beta} = \overline{\alpha + \beta}, \tag{7.8}$$

$$\bar{\alpha} \cdot \bar{\beta} = \overline{\alpha \cdot \beta}. \tag{7.9}$$

It follows that if $P\langle X \rangle$ is a polynomial with *real* coefficients then $P(\bar{\alpha}) = \overline{(P\alpha)}$. This is very often used.

7.2.1   *Absolute value of a complex number*

The number $\sqrt{a^2 + b^2}$ is called the absolute value of the complex number $\alpha = a + bi$. We denote it by $|\alpha|$. It is always non-negative and if $b = 0$ then α is real and the notation is consistent with that introduced in Section 4.3 for the absolute value of real numbers. Obviously $|\alpha| \geq \Re\alpha$, $|\alpha| \geq \Im\alpha$ and $|\alpha| = |\bar{\alpha}|$.

Theorem 7.1   *If α and β are complex numbers then*

   (i) $|\alpha| = \sqrt{\alpha\bar{\alpha}} = |\bar{\alpha}|$;

   (ii) $|\alpha\beta| = |\alpha||\beta|$;

   (iii) $|\alpha + \beta| \leq |\alpha| + |\beta|$;

   (iv) $\big||\alpha| - |\beta|\big| \leq |\alpha - \beta|$.

Proof.   The first item is obvious. For (ii) we have $|\alpha\beta| = \sqrt{(\alpha\beta)\overline{(\alpha\beta)}} = \sqrt{\alpha\bar{\alpha}}\sqrt{\beta\bar{\beta}} = |\alpha||\beta|$. To prove (iii) we first note that $\alpha\bar{\beta} + \bar{\alpha}\beta$ is real and consequently it is equal to its real part. Then we have

$$\begin{aligned}
|\alpha + \beta|^2 &= (\alpha + \beta)\overline{(\alpha + \beta)} = \alpha\bar{\alpha} + \beta\bar{\beta} + \Re(\alpha\bar{\beta} + \bar{\alpha}\beta) \\
&= |\alpha|^2 + |\beta|^2 + \Re(\alpha\bar{\beta}) + \Re(\bar{\alpha}\beta) \\
&\leq |\alpha|^2 + |\beta|^2 + |\alpha\bar{\beta}| + |\bar{\alpha}\beta| \\
&= |\alpha|^2 + |\beta|^2 + 2|\alpha||\beta| \\
&= (|\alpha| + |\beta|)^2.
\end{aligned}$$

---

[6]This terminology comes from times when mathematicians regarded complex numbers as mysterious. These days an imaginary number is a concrete mathematical object.

In the last step we used (ii) and (i). Now it is sufficient to take square roots.

Since the proof of (iv) is similar to the proof of (ii) of Theorem 4.3 we leave it as an exercise. $\qquad\square$

By (i) we have $\alpha\bar{\alpha} = |\alpha|^2$ and if $\alpha \neq 0$ then $\alpha\dfrac{\bar{\alpha}}{|\alpha|^2} = 1$. The multiplicative inverse of α is $\dfrac{\bar{\alpha}}{|\alpha|^2}$, consequently $\mathbb{C}$ is a field. Let us emphasize

> $\mathbb{C}$ is the set of numbers of the form $a + b\imath$, where a, b are real and $\imath^2 = -1$.
> $\mathbb{C}$ is a field.

Exercises

ⓘ **Exercise 7.2.1**   *Prove: If $n \in \mathbb{N}$ and $z \in \mathbb{C}$ then $|z^n| = |z|^n$. If $z \neq 0$ then this equation also holds for $n \in \mathbb{Z}$.*

ⓘ **Exercise 7.2.2**   *Prove (iv) of Theorem 7.1.* [Hint: Mimic the proof of (ii) of Theorem 4.3.]

ⓘ **Exercise 7.2.3**   *For complex numbers $\alpha_1, \alpha_2, \ldots, \alpha_n$ prove:*

(1) $|\alpha_1\alpha_2 \cdots \alpha_n| = |\alpha_1||\alpha_2| \cdots |\alpha_n|$;
(2) $|\alpha_1 + \alpha_2 + \cdots + \alpha_n| \leq |\alpha_1| + |\alpha_2| + \cdots + |\alpha_n|$.

7.2.2   *Square root of a complex number*

In this subsection we prove that every complex number α has a square root, that is a number whose square is α. Let us recall that if α is real[7] and non-negative then we agreed in Definition 5.2 to denote by $\sqrt{\alpha}$ the *non-negative* number β such that $\beta^2 = \alpha$. If $\alpha < 0$ then we *define* $\sqrt{\alpha} = \imath\sqrt{|\alpha|}$. Now we turn our attention to the case $\alpha = a + \imath b$ with $b \neq 0$. If $\beta^2 = \alpha$ then also $(-\beta)^2 = \alpha$, hence in our search for the square root of α we can look for a β with a non-negative real part. The following calculation yields the real part of β.

---

[7] $\Im\alpha = 0$.

$$\beta^2 = a + \imath b\,; \tag{7.10}$$

$$\beta\bar{\beta} = |\beta|^2 = |\beta^2| = \sqrt{a^2 + b^2}\,; \tag{7.11}$$

$$\overline{\beta^2} = \bar{\beta}^2 = a - \imath b\,; \tag{7.12}$$

$$(\beta + \bar{\beta})^2 = 2(a + \sqrt{a^2 + b^2}) \tag{7.13}$$

$$2\Re\,\beta = \beta + \bar{\beta} = \sqrt{2}\sqrt{a + \sqrt{a^2 + b^2}}. \tag{7.14}$$

To determine the imaginary part of β we note first that $2\Re\,\beta\,\Im\,\beta = \Im\,\beta^2$ and then by using Equation (7.10) we have

$$\Im\,\beta = \frac{b}{2\Re\,\beta}, \tag{7.15}$$

$$\Im\,\beta = \frac{1}{\sqrt{2}}\frac{b}{\sqrt{\sqrt{a^2 + b^2} + a}} = \frac{b}{|b|\sqrt{2}}\sqrt{\sqrt{a^2 + b^2} - a}. \tag{7.16}$$

Consequently

$$\beta = \frac{1}{\sqrt{2}}\sqrt{a + \sqrt{a^2 + b^2}} + \imath\frac{b}{|b|\sqrt{2}}\sqrt{\sqrt{a^2 + b^2} - a}. \tag{7.17}$$

Let us summarize:

Theorem 7.2 (Square root of a complex number)   *For a real number $a \neq 0$ there are two distinct square roots. For $a > 0$ these are $\sqrt{a}$ (Definition 5.2) and $-\sqrt{a}$: for $a < 0$ these are $\imath\sqrt{|a|}$ and $-\imath\sqrt{|a|}$. For an imaginary number $\alpha = a + \imath b$ with $b \neq 0$ there are also two square roots: one is given by Equation (7.17) and the other is $-\beta$.*

Remark 7.1   We deliberately avoided the use of the symbol $\sqrt{\alpha}$ for the square root of the complex number α, although this symbol is commonly used. Readers are warned that the rules which govern operations with square roots of *positive reals* as given in Theorem 5.6 are not valid for complex numbers. Some counter-examples are given in Exercises 7.2.4 and 7.2.5.

Exercises

Exercise 7.2.4   Give an example showing that the equation $\sqrt{z}\sqrt{w} = \sqrt{zw}$ is false if z, w are complex numbers. [Hint: Choose $z = w = -1$.]

Exercise 7.2.5 *Show that the equation* $\sqrt{\dfrac{1}{z}} = \dfrac{1}{\sqrt{z}}$ *is false for* $z = -1$.

7.2.3 *Maple and complex numbers*

Maple uses the letter I for $\imath$. The arithmetic of complex number in *Maple* brings nothing new. Readers must not forget the operator $*$ when entering a complex number, for instance $1 + 3\imath$ *must* be typed as 1+3*I. To obtain $\bar{z}$ enter `conjugate(z)`. For instance, `conjugate((4+6*I)/(I-1));` yields $1 + 5I$. The absolute value of a complex number z is entered into *Maple* as `abs(z)`, for instance `abs(1+I);` gives $\sqrt{2}$. If the square root of a complex number is relatively simple expression then *Maple* provides the exact answer. For instance, `sqrt(2*I);` returns $1 + I$ and `sqrt(24-10*I);` gives $5 - I$. Quite often *Maple* returns the question, e. g. `sqrt(4+I);` gives only $\sqrt{4+I}$. In such a case we have two options. We can use a floating point approximation, `sqrt(4.0+I)` yields $2.015329455 + .2480983934I$. Or if we insist on obtaining an exact result we use the command `evalc(sqrt(4+I))`. The command `evalc` simplifies a complicated complex number to the standard form $a + b\imath$. We can also use Example A.2 in Appendix A which provides a *Maple* program which calculates the exact value of the square root of a complex number according to the formula (7.17).

7.2.4 *Geometric representation of complex numbers. Trigonometric form of a complex number*

A complex number $\alpha = a_1 + \imath a_2$ is represented in the plane with coordinate axis x, y as a segment with the initial point at the origin and end point (a_1, a_2). The length of this segment is $|\alpha|$. The sum $\alpha + \beta$ with $\beta = b_1 + \imath b_2$ is then represented by the diagonal of the parallelogram constructed with sides α and β; see Figure 7.1. The plane in which the complex numbers are depicted is often referred to as the complex plane or the Gaussian plane or the Argand diagram.

A very useful representation of complex numbers is achieved with the help of trigonometric functions. They are rigorously defined and their properties established in Chapter 14. In this and the next subsection we rely on readers' knowledge of trigonometric functions from their previous work. With the notation as in Figure 7.2 we have

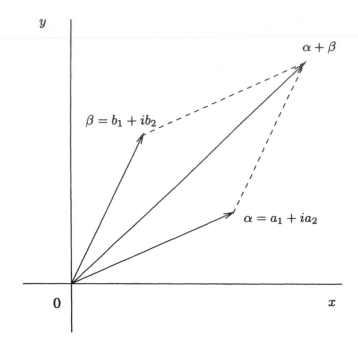

Fig. 7.1   Sum of two complex numbers

$$z = |z|(\cos\phi + i\sin\phi). \tag{7.18}$$

For $z \neq 0$ the angle ϕ is determined up to an integer multiple of 2π: it is uniquely determined by Equation (7.18) if it is additionally required that $-\pi < \phi \leq \pi$ and then it is called the *argument* of z. In *Maple* we have argument(z). The inequalities $0 \leq \phi < 2\pi$ together with Equation (7.18) also uniquely determine ϕ. If $w = |w|(\cos\psi + i\sin\psi)$ then we obtain for the product

$$\begin{aligned}
zw &= |z||w|(\cos\phi + i\sin\phi)(\cos\psi + i\sin\psi), \\
&= |z||w|(\cos\phi\cos\psi - \sin\phi\sin\psi + i\cos\phi\sin\psi + i\sin\phi\cos\psi), \\
&= |z||w|\big(\cos(\phi+\psi) + i\sin(\phi+\psi)\big).
\end{aligned}$$

The absolute value of the product of two complex numbers is the product of absolute values. The sum of the arguments can be used as the angle which

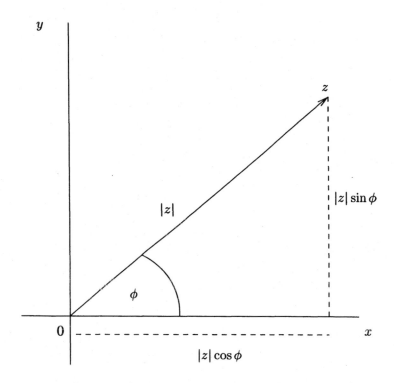

Fig. 7.2   Trigonometric form of a complex number

is subtended by the representation of the product and the real axis.[8] This is illustrated in Figure 7.3, where for sake of simplicity we took $|z| = |w| = 1$. For $z = w$ we have

$$z^2 = |z|^2\big(\cos(2\phi) + \imath\sin(2\phi)\big).$$

and by simple induction, for $n \in \mathbb{N}$,

$$z^n = |z|^n\big(\cos(n\phi) + \imath\sin(n\phi)\big). \tag{7.19}$$

This equation is called Moivre's formula.

---

[8]One has to be careful not to say that that the argument of the product is the sum of the arguments of the factors: it is, however, if the sum lies in the interval $]-\pi, \pi]$.

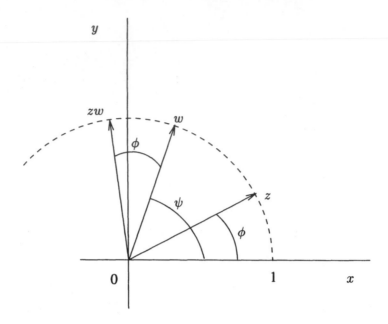

Fig. 7.3   Multiplication of complex numbers

Exercises

Exercise 7.2.6   *Use Equations (7.18), (7.19) and the Binomial Theorem to express* $\sin 5\phi$ *in terms of* $\sin \phi$ *and* $\cos \phi$.

Exercise 7.2.7   *Express* $\sin^4 \phi$ *in terms of* $\cos 2\phi$ *and* $\cos 4\phi$. [Hint: Set $z = \cos \phi + \imath \sin \phi$ and then calculate $(z + \bar{z})^4$ in two different ways. The *Maple* command (combine((sin(x))^4); can also be used.]

7.2.5   *The binomial equation*

The equation of the form

$$x^n = a \qquad\qquad (7.20)$$

with $n \in \mathbb{N}$ is called the binomial equation. In this subsection we assume $a \neq 0$, since the case $a = 0$ is trivial. Let us try *Maple* for $a = 1$ and some n, say $n = 3$ and $n = 10$.

```
>   solve(x^3=1);
```

$$1, -\frac{1}{2} + \frac{1}{2} I \sqrt{3}, -\frac{1}{2} - \frac{1}{2} I \sqrt{3}$$

This is a nice result, but unfortunately for larger n *Maple* provide a complicated answer.

```
>  solve(x^(10)-1=0,x);
```

$$-1, 1, -\sqrt{-\frac{1}{4} + \frac{1}{4}\sqrt{5} + \frac{1}{4} I \sqrt{2}\,\%1}, -\sqrt{-\frac{1}{4} - \frac{1}{4}\sqrt{5} + \frac{1}{4} I \sqrt{2}\,\%2},$$

$$-\sqrt{-\frac{1}{4} - \frac{1}{4}\sqrt{5} - \frac{1}{4} I \sqrt{2}\,\%2}, -\sqrt{-\frac{1}{4} + \frac{1}{4}\sqrt{5} - \frac{1}{4} I \sqrt{2}\,\%1},$$

$$\sqrt{-\frac{1}{4} + \frac{1}{4}\sqrt{5} + \frac{1}{4} I \sqrt{2}\,\%1}, \sqrt{-\frac{1}{4} - \frac{1}{4}\sqrt{5} + \frac{1}{4} I \sqrt{2}\,\%2},$$

$$\sqrt{-\frac{1}{4} - \frac{1}{4}\sqrt{5} - \frac{1}{4} I \sqrt{2}\,\%2}, \sqrt{-\frac{1}{4} + \frac{1}{4}\sqrt{5} - \frac{1}{4} I \sqrt{2}\,\%1}$$

$$\%1 := \sqrt{5 + \sqrt{5}}$$

$$\%2 := \sqrt{5 - \sqrt{5}}$$

The result is complicated but can be simplified by using the command evalc. One way to do it is to to put the solutions in a list first, say, L:=[solve(x^(10)-1,x)] and employ seq(evalc(L[i]),i=1..10). Fortunately there is a simple way of dealing with the binomial equation for any n. The key is the Moivre formula (7.19). We use it for $x = |x|(\cos \phi + \imath \sin \phi)$ and write $a = |a|(\cos \alpha + \imath \sin \alpha)$ with $0 \le \alpha < 2\pi$. Equation (7.20) is satisfied if and only if

$$|x| = \sqrt[n]{|a|} \tag{7.21}$$

$$\cos n\phi = \cos \alpha \tag{7.22}$$

$$\sin n\phi = \sin \alpha. \tag{7.23}$$

The last two equations in turn are satisfied if and only if

$$\phi = \frac{1}{n}(\alpha + 2k\pi), \tag{7.24}$$

for some integer k. Obviously not all these values of ϕ can be distinct: by the uniqueness of the trigonometric representation they will be if

$$0 \le \alpha + 2k\pi < 2n\pi. \tag{7.25}$$

This leads to the condition

$$-\frac{\alpha}{2n} \le k < n - \frac{\alpha}{2n}. \tag{7.26}$$

Consequently $k = 0, 1, 2, \ldots, n - 1$. Now all the numbers

$$x_k = \sqrt[n]{|a|}\left(\cos\frac{\alpha + 2k\pi}{n} + \sin\frac{\alpha + 2k\pi}{n}\right)$$

are distinct and therefore they represent all solutions to Equation (7.20). For $a = 1$ they are denoted ε_k and by Moivre's formula $\varepsilon_k = \varepsilon_1^k$. We shall call ε_1 the nth primitive root of unity. In the Gaussian plane the numbers ε_k are equidistantly placed on the unit circle: ε_{k+1} is obtained by rotating ε_k by an angle $2\pi/n$. The solutions to Equation (7.20) are called nth roots of a and are denoted by $\sqrt[n]{a}$. There are n roots and in order to give a well defined meaning to the symbol[9] $\sqrt[n]{a}$ some choice must be made. In order to avoid confusion or even a mistake the following convention is made: *whenever a choice for the symbol $\sqrt[n]{a}$ is made that choice must be kept consistently during a calculation or a particular investigation.* It is possible to derive some rules for computation with $\sqrt[n]{a}$. One such obvious rule is $\left(\sqrt[n]{a}\right)^n = a$ for any complex a. Some other rules are left for Exercises 7.2.10–7.2.12. One must be aware that if some choice is made for the symbol $\sqrt[n]{a}$ this choice might be inconsistent with previous definitions, in particular with Definition 5.2. This is undesirable and it is best to avoid the use of the symbol $\sqrt[n]{a}$ for a complex a as much as possible. We explained it because this notation is commonly used and because it helps to understand what *Maple* does with nth roots of a. For instance, we agreed on Page 154 that $\sqrt[3]{-8/27}$ would be $-2/3$ but if you ask *Maple* root[3](-1); or root[3](-8/27); the answers are $(-1)^{\frac{1}{3}}$ and $\frac{2}{3}(-1)^{\frac{1}{3}}$; *Maple* simplifies what can be simplified but leaves the choice of $(-1)^{\frac{1}{3}} = \sqrt[3]{-1}$ to the readers. However, if we use the floating point approximations with the root command, *Maple* will made a choice for us! It finds the floating point approximation of the root for which the argument has the smallest absolute value and for which the imaginary part is non-negative. The following *Maple* worksheet illustrates this; *Maple's* choice might surprise you.

```
>    root[3](-1);
```

$$(-1)^{(1/3)}$$

---

[9]For $n = 2$ we write simply $\sqrt{a}$ instead of $\sqrt[2]{a}$.

The choice is up to us.

```
>   root[3](-1.0);
```

$$.5000000001 + .8660254037 \, I$$

The number .8660254037 is a floating point approximation to $\dfrac{\sqrt{3}}{2}$. We did not get the answer -1.000 because this root has the maximal argument and *Maple* chooses the approximation with the smallest argument.

```
>   (-1+I)^3;
```

$$2 + 2 \, I$$

but

```
>   root[3](2.0+2.0*I);
```

$$1.366025404 + .3660254037 \, I$$

This is an example where $\sqrt[n]{a^n} \neq a$, as shown by *Maple*.

Exercises

Exercise 7.2.8   *Show that Equations (7.22) and (7.23) are satisfied if and only if (7.24)i holds. [Hint: multiply Equation (7.22) by $\cos\alpha$ and Equation (7.23) by $\sin\alpha$ and add.]*

Exercise 7.2.9   *Solve Equation (7.20) for*

(1) $a = 1$ and $n = 8$;
(2) $a = -1$ and $n = 4$;
(3) $a = -2 + 2i$ and $n = 3$.

Exercise 7.2.10   *Prove: If $a \neq 0$ and $\sqrt[n]{a}$ is a fixed solution to Equation (7.20) then*

$$\left\{ \sqrt[n]{a}, \; \varepsilon_1 \sqrt[n]{a}, \; \varepsilon_1^2 \sqrt[n]{a}, \ldots, \varepsilon_1^{n-1} \sqrt[n]{a} \right\}$$

is the set of all solutions to Equation (7.20), provided ε_1 denotes the nth primitive root of unity. [Hint: All the numbers in the set are solutions to Equation (7.20) and are distinct.]

Exercise 7.2.11   *Prove: If a, b are complex numbers, $n \in \mathbb{N}$ and the choice of value for each of $\sqrt[n]{a}$, $\sqrt[n]{b}$ and $\sqrt[n]{ab}$ is fixed and if ε_1 denotes the nth primitive root of unity then there exists a $k \in \mathbb{N}$ such that $\sqrt[n]{a}\,\sqrt[n]{b} =$*

$\varepsilon_1^k \sqrt[n]{ab}$. [Hint: Both numbers $\sqrt[n]{a}\sqrt[n]{b}$ and $\sqrt[n]{ab}$ are solutions to the equation $x^n - ab = 0$. Use the previous exercise.]

(*i*) **Exercise 7.2.12**   *Prove: If a, b are complex numbers, $n \in \mathbb{N}$ and the choice of value for each of $\sqrt[n]{a}$, $\sqrt[n]{b}$ is fixed then it is possible to define $\sqrt[n]{ab}$ in such a way that $\sqrt[n]{a}\sqrt[n]{b} = \sqrt[n]{ab}$. Consider the case of $a = b = \imath$. [Hint: $x = \sqrt[n]{a}\sqrt[n]{b}$ satisfies Equation (7.20).]*

Chapter 8

Solving Equations

In this chapter we discuss existence and uniqueness of solutions
to various equations and show how to use *Maple* to find solutions.
We deal mainly with polynomial equations in one unknown and
add only some basic facts about systems of linear equations.

8.1 General remarks

Given a function f, solving the equation[1]

$$f(x) = 0 \tag{8.1}$$

means finding all x in dom f which satisfy Equation (8.1). Any x which
satisfies Equation (8.1) is called a solution of Equation (8.1). In this chapter
we assume that dom f is always part of $\mathbb{R}$ or $\mathbb{C}$ and it will be clear from the
context which set is meant. Solving an equation like (8.1) usually consists
of a chain of implications, starting with the equation itself and ending with
an equation (or equations) of the form $x = a$. For instance:

$$\frac{1}{x+1} + \frac{1}{x+2} = 0 \Rightarrow x + 2 + x + 1 = 0,$$
$$\Rightarrow 2x + 3 = 0, \tag{8.2}$$
$$\Rightarrow x = -\frac{3}{2}.$$

For greater clarity we printed the implication signs, but implications are
always understood automatically. It is a good habit always to check the solution by substituting the found value back into the original equation. This

---

[1]A more complicated equation of the form $F(x) = G(x)$ can be reduced to the form
given by taking $f = F - G$.

is not a mere verification of the calculations, it has a far more fundamental reason. Solving an equation starts with an assumption that the equation *is* satisfied for some x, in other words, it is assumed that a solution exists. If it does not then the chain of implication can lead to a wrong result.[2] Consider the following example[3]

$$x - 1 = \sqrt{1 - 2x},$$
$$x^2 - 2x + 1 = 1 - 2x,$$
$$x^2 = 0, \tag{8.3}$$
$$x = 0.$$

Substituting 0 in the original equation gives $-1 = 1$. The conclusion is clear and easy: the original equation had no solution.

It is preferable to use equivalences rather than implications. It is then possible to conclude that the value found in the last equation is indeed a solution. The verification is not necessary from the logical point of view but it is still useful for checking the calculations.

It is not always possible to employ equivalences and this was the case in Equations (8.3): the second equation does not imply the first. If we do not use equivalences then the found value(s) need not be a solution and we must perform the check. Let us illustrate it by one more example.

$$x - 1 = \sqrt{5 - 2x},$$
$$x^2 - 2x + 1 = 5 - 2x,$$
$$x^2 = 4, \tag{8.4}$$
$$x = 2 \text{ or } x = -2.$$

Back substitution shows that 2 is a solution and -2 is not.

Exercises

Exercise 8.1.1 *Solve the following equations*

(1) $\dfrac{x}{x - 1} + \dfrac{x}{x + 1} = \dfrac{4x - 2}{x^2 - 1}$;

(2) $\dfrac{x}{x - 1} + \dfrac{x}{x + 1} = \dfrac{9x - 7}{x^2 - 1}$.

---

[2] Remember, a false proposition can lead to anything.

[3] In this book unless otherwise stated the symbol $\sqrt{y}$ means the square root of a non-negative number y: if y contains an unknown or unknowns then the search is for quantities which make y non-negative.

Exercise 8.1.2   *Solve the equation* $x - 1 = \sqrt{a - 2x}$. [Hint: Mimic Equations (8.3). $(\sqrt{a - 1} - 1)^2 = a - 2\sqrt{a - 1}$.]

8.2   *Maple* commands `solve` and `fsolve`

Maple can efficiently solve practically any equation. There are two commands: `solve` provides an exact solution and `fsolve` gives a numerical solution. Although generally speaking it is better to obtain an exact solution, for some equations this might take too long and for some equations it might even be impossible. Moreover, sometimes `fsolve` gives a more useful result than `solve`. For example, it is usually preferable to use `fsolve` on an equation like $x^2 + x - 1 = 0$ rather than solve it exactly. An exact solution can be converted to a floating point result by the command `evalf`. The following *Maple* session illustrates this.[4]

```
>   fsolve(x^2+x-1=0);
```
$$-1.618033989, .6180339887$$
```
>   solve(x^2+x-1=0);
```
$$-\frac{1}{2} + \frac{1}{2}\sqrt{5}, \ -\frac{1}{2} - \frac{1}{2}\sqrt{5}$$
```
>   evalf(%);
```
$$.6180339800, \quad 1.618033980$$

Both commands have some options and sometimes it is important to employ these options. The structure for `solve` is

$$\text{solve(equation,unknown),}$$

or if there are several equations and several unknowns

$$\text{solve(equations, unknowns).}$$

The structure for `fsolve` is

$$\text{fsolve(equation, unknown, range).}$$

---

[4]Note that the *Maple* floatign point results differ in low decimal places. This is a natural result of computational inaccuracies.

The options **unknown** or **unknowns** identify the unknown or unknowns for which the equation should be solved. If there is only one unknown in the equation, this option can be omitted. However, if there are several unknowns or equations they should be entered as sets, for instance with { and } around them. The range could be given as a..b, where a, b are real numbers, or could be -infinity, infinity. If the option **range** is given as above, *Maple* will find a solution[5] although there might be many solutions in that range. The range can also be given by the word *complex*. In this case, **fsolve** searches for a complex solution. If the equation is a polynomial equation then **fsolve** will find all real solutions, or all complex solutions if the option **complex** is given. There is additional useful and important information on **solve** and **fsolve** in **Help**.

Example 8.1   Now we use *Maple* for solving (8.3) and (8.4).

>   Eq:=x-1=sqrt(1-2*x);

$$Eq := x - 1 = \sqrt{1 - 2\,x}$$

>   solve(Eq,x);

Maple correctly did not return any result: there is no result to be obtained. As we know, the equation has no solution.

>   eqn:=x-1=sqrt(5-2*x);

$$eqn := x - 1 = \sqrt{5 - 2\,x}$$

>   solve(eqn,x);

$$2$$

This is very pleasing: *Maple* returned the correct answer and omitted the spurious 'solution' x=-2. *Maple* can also check the solution.

>   x:=2;eqn;

$$x := 2$$

$$1 = 1$$

Maple has its own command **eval** for checking.

>   eval(eqn,x=2);

---

[5]If there is one.

$$1 = 1$$

Example 8.2 An agent offers an annuity of 12 years duration with yearly payments of $10 000 for $103 000. We wish to compare this offer with other investments and therefore want to know the interest rate. In order to find it we use fsolve and annuity.[6]

```
> with(finance);
```

```
> fsolve(annuity(10000,x,12)=103000);
```

$$-1.781236788$$

An interest rate of -178% is clearly wrong. We shall look for an explanation shortly, but first we try to employ options with fsolve. We know that the rate of interest must be positive and less then 100%, hence we try

---

```
> fsolve(annuity(10000,x,12)=103000,x,0..1);
```

$$.02432202661$$

Why was the first result wrong? The explanation lies in the formula for annuity.

```
> annuity(10000,x,12);
```

$$10000\,\frac{1}{x} - \frac{10000}{x\,(1+x)^{12}}$$

So the equation to be solved reads $10(1+x)^{12} - 10 = 103x(1+x)^{12}$. This is a polynomial equation and fsolve will provide all real solutions.

```
> fsolve(10*(1+x)^(12)-10-103*x*(1+x)^(12));
```

$$-1.781236788,\ 0.,\ .02432202661$$

---

[6]The finance package was introduced in Example 1.4.

The discrepancy has disappeared. The annuity problem leads to a polynomial equation with three real solutions. The first attempt picked up the negative root of the polynomial equation, and this root was not the solution of the original problem. The second attempt was successful. The moral of this story is that we should always use the options with fsolve, provided we are certain in what range the desired solution lies.

Example 8.3   This example was used by mathematicians to tease lay people. It is not difficult but it requires a numerical solution.

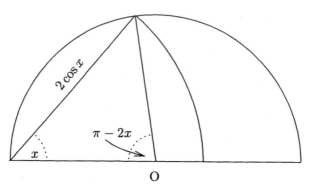

Fig. 8.1   Grazing goat

A goat is tied up at the circumference of a circular meadow of radius one. How long should the rope be which allows the goat to graze an area of half the circle? Figure 8.1 shows half of the circular yard, with the goat tied at the angle x. Using the notation from this figure we immediately recognize that the angle $x = \pi/4$ produces rope that is too long and that the angle $x = \pi/3$ produces rope which is too short. The length of the required rope is $2\cos x$ and the area the goat can graze is $4x\cos^2 x + \pi - 2x - \sin(\pi - 2x)$

```
>    fsolve(x*2*(cos(x))^2+Pi/2-x-(1/2)*sin(2*x) =
>    Pi/4,x,Pi/4..Pi/3);
```

$$.9528478647$$

```
>    X:=(%);
```

$$X := .9528478647$$

```
>    2*cos(X);
```

$$1.158728473$$

The value for the length of the rope looks about right: however to be sure
we found the right result, we calculate the area which the goat cannot graze
and compare it with the area of the half-circle, which is $\pi/2$.

```
>  Pi-4*X*(cos(X))^2-2*((Pi-2*X)*(1/2)-(1/2)*sin(Pi-2*X));
```

$$1.570796326$$

```
>  evalf(Pi/2);
```

$$1.570796327$$

This shows that we obtained the correct answer.

8.3 Algebraic equations

An equation of the form

$$P(x) = 0, \qquad (8.5)$$

where P is a polynomial, is called algebraic. The degree of P and the
coefficients of P are also the degree of the equation and the coefficients of
the equation, respectively. We shall consider only equations with complex
coefficients.[7] A zero of P is obviously a solution to (8.5) and is also called
a root of Equation (8.5). From ancient times mathematicians tried to solve
algebraic equations and it was only the German mathematician K. F. Gauss
at the end of the eighteenth century who proved that algebraic equations
always have solutions.

Theorem 8.1 (The Fundamental Theorem of Algebra) *An
algebraic equation with complex coefficients and of degree at least one
always has a solution within the field of complex numbers.*

We shall not prove the theorem here, since it requires too much prepara-
tion. A proof accessible with our level of knowledge can be found in Kurosh
(1980, pp. 142–151). The name is slightly misleading: a better name would
be "The Fundamental Theorem of algebra of complex numbers", but this
name is still not perfect. There were many proofs given, however all of them
use the concept of continuity, which will be considered in Chapter 12, and
of completeness of reals. These parts of mathematics are not considered to
belong to algebra.

---

[7]This, of course, includes also the real and the rational coefficients.

An important consequence of the Fundamental Theorem of Algebra is that the only irreducible polynomials in $\mathbb{C}[X]$ are linear. Indeed, if $P\langle X\rangle \in \mathbb{C}[X]$ is of degree at least two then it has a root, say α, and is therefore divisible by $X - \alpha$. Hence it is not irreducible. Writing $P(x)$ as a product of its irreducible, that is linear, factors we have

$$P(x) \equiv x^n + a_1 x^{n-1} + \ldots + a_n = (x - x_1)(x - x_2)\cdots(x - x_n). \quad (8.6)$$

Not all x_i need be distinct, however it follows that an algebraic equation of nth degree has n complex roots, provided each root is counted as many times as its multiplicity. Multiplying through on the right hand side of Equation (8.6) and comparing coefficients leads to the so-called Vieta's formulae[8]

$$a_1 = -(x_1 + x_2 + \cdots + x_n),$$
$$a_2 = x_1 x_2 + x_1 x_3 + \cdots + x_1 x_n + x_2 x_3 + \cdots + x_{n-1} x_n,$$
$$\cdots \quad (8.7)$$
$$a_{n-1} = (-1)^{n-1}(x_1 x_2 \cdots x_{n-1} + x_1 x_2 \cdots x_{n-2} x_n + \cdots x_2 x_3 \cdots x_n)$$
$$a_n = (-1)^n x_1 \cdot x_2 \cdots x_{n-1} \cdot x_n.$$

The polynomials on the right hand sides of Equations (8.7) are called elementary symmetric polynomials. The so-called fundamental theorem on symmetric polynomials states that any symmetric polynomial, that is a polynomial which is not changed by a permutation of indeterminates, can by expressed as a polynomial in elementary symmetric polynomials. We do not have much use for this otherwise important theorem, so we do not prove it. However, we illustrate it by a simple example. If $W = (x_1 - x_2)^2$ then it is easy to see that $W = a_1{}^2 - 2a_2$.

Rational roots

It is clear (for instance from the Vieta formula for a_n) that if the equation $P(x) = 0$ has an integer root x_i then x_i divides a_n. The next example shows how to use this for finding an integer root.

Example 8.4   First we define the polynomial

```
>   f:=y->y^6 + 2*y^5 + 435*y^4 + 11025*y^3
>             - 37125*y^2 - 253125*y + 759375;
```

---

[8]Readers are probably familiar with these for $n = 2$, when $P(x) = x^2 + px + q$ and $p = -(x_1 + x_2)$, $q = x_1 x_2$.

$$f := y \to y^6 + 2\,y^5 + 435\,y^4 + 11025\,y^3 - 37125\,y^2 - 253125\,y + 759375$$

Next we use the *Maple* command divisors to find the positive divisors of the absolute term. For that we need the number theory package. We must not use letter D to denote the divisors, since it is protected. *Maple* provides the set of divisors, which we put into a list we call podi (for positive divisors).

> `with(numtheory,divisors);`

> `podi:=convert(divisors(759375),list);`

$podi := [1, 3, 5, 9, 15, 25, 27, 45, 75, 81, 125, 135, 225, 243, 375, 405,$
$625, 675, 1125, 1215, 1875, 2025, 3125, 3375, 5625, 6075, 9375, 10125,$
$16875, 28125, 30375, 50625, 84375, 151875, 253125, 759375]$

The next step is to create a list of all divisors.

> `di:=[op(podi),op(-podi)];`

This gives a long list di, from which we select those x for which $f(x) = 0$. The procedure for selecting elements from a list according to a given condition was discussed in Subsection 3.3.3 on page 80.

> `good:=x->is(f(x)=0);`

$$good := x \to \text{is}(\text{f}(x) = 0)$$

> `select(good,di);`

$$[3, -5]$$

Since there are very many very large entries in di to do this work with pen and paper would be almost forbidding, but with *Maple* it was easy.

Rational roots can be found similarly. Let

$$Q(x) = a_0 x^n + a_1 x^{n-1} + \cdots + a_n. \tag{8.8}$$

If p and q are relatively prime integers and p/q is a zero of Q then p divides a_n and q divides a_0. Indeed, we have

$$a_n q^n = -p(a_{n-1} + \cdots + a_0 p^{n-1}).$$

Since p and q are relatively prime, p divides a_n. It follows similarly that q divides a_0. Therefore the rational zeros of Q can be found among fractions of the form

$$\frac{\text{divisor of } a_n}{\text{divisor of } a_0}. \tag{8.9}$$

(See also Exercises 8.3.2–8.3.5.) If $Q(x) = 4x^2 + 4x - 3$ we have to check 12 possibilities to find that $-3/2$ and $1/2$ are the rational roots. Obviously, for a polynomial of higher degree with large coefficients the computer is needed. The next example shows the use of *Maple* for finding rational roots. The method we employ is not recommended when using mere pen and paper, but with *Maple* it is easier to implement. We make a substitution $x = y/a_0$ and seek the *integer* roots of $a_0^{n-1}Q(y)$. For justification see Exercise 8.3.1.

Example 8.5   Let us now find the rational roots of $Q(x) = 6x^3 - 17x^2 + 22x - 35$.

```
>   Q:=x->6*x^3-17*x^2+22*x-35;
```
$$Q := x \rightarrow 6x^3 - 17x^2 + 22x - 35$$
```
>   x:=y/6:36*Q(x);
```
$$y^3 - 17y^2 + 132y - 1260$$
```
>   P:=y->y^3-17*y^2+132*y-1260:
>   with(numtheory,divisors):
>   L:=convert(divisors(1260),list):div:=[op(L),op(-L)]:
>   good:=t->is(P(t)=0):
>   select(good,div);
```
$$[14]$$

P has a root 14, consequently, the rational root of Q is $7/3$.

Maple has a ready-made command for finding rational roots of polynomials. It is **roots**, and *Maple* answers with a list of pairs [*root, multiplicity*]. If there are no rational roots *Maple* returns an empty list. We now compare **roots** with our previous results.

```
>   roots(f(x));
```
$$[[3, 1], [-5, 1]]$$
```
>   roots(P(t));
```
$$[[\frac{7}{3}, 1]]$$

Multiple roots

It is important that solving of an algebraic equation can always be reduced to solving another algebraic equation with simple roots. Let k be a positive integer and

$$P(x) = (x - a)^k Q(x), \qquad (8.10)$$

with $Q(a) \neq 0$. Then by Equation (6.27)

$$DP(x) = k(x - a)^{k-1} Q(x) + (x - a)^k DQ(x) = (x - a)^{k-1} F(x), \quad (8.11)$$

with $F(a) = kQ(a) \neq 0$. Consequently a is a root of $DP(x)$ of multiplicity $k - 1$. A root of P of multiplicity k is a root of multiplicity $k - 1$ of the greatest common divisor of P and DP. Thus $P(x)$ divided by the greatest common divisor will have only simple roots.

```
> P:=x->x^7-x^5+x^6-x^4-x^3-x^2+x+1;
```

$$P := x \rightarrow x^7 - x^5 + x^6 - x^4 - x^3 - x^2 + x + 1$$

To investigate the multiplicity of roots of P, we look at[9]

```
> gcd(P(x),D(P)(x));
```

$$x^3 + x^2 - x - 1$$

This polynomial is of degree three, so P must have three multiple roots. To find whether or not some have multiplicity greater than two, we again look for the greatest common divisor of this polynomial and its derivative

```
> gcd(%,3*x^2+2*x-1),
```

$$x + 1$$

-1 is a triple root of P and it is easy to conclude that 1 is a double root.

We could also find by inspection that 1 was a root of P, and use the Taylor polynomial to determine its multiplicity.

```
> taylor(P(x),x=1,8):convert(%,polynom);
```

$$16(x - 1)^2 + 40(x - 1)^3 + 44(x - 1)^4 + 26(x - 1)^5 + 8(x - 1)^6 + (x - 1)^7$$

```
> quo(%,(x-1)^2,x);
```

$$x^5 + 3x^4 + 4x^3 + 4x^2 + 3x + 1$$

-1 is now an obvious root

```
> taylor(%,x=-1,6):convert(%,polynom);
```

---

[9]The command `D(P)(x)` in *Maple* produces the derivative of `P(x)`.

$$2(x+1)^3 - 2(x+1)^4 + (x+1)^5$$

As before we obtained that -1 is a triple root.

Real roots

In this subsection we consider only polynomials with real coefficients. If α is an imaginary root of P then so is $\bar{\alpha}$. Indeed

$$
\begin{aligned}
P(\bar{\alpha}) &= a_0(\bar{\alpha})^n + a_1(\bar{\alpha})^{n-1} + \cdots + a_n \\
&= a_0\overline{(\alpha)^n} + a_1\overline{(\alpha)^{n-1}} + \cdots + a_n \\
&= \overline{a_0\alpha^n} + \overline{a_1\alpha^{n-1}} + \cdots + \overline{a_n} \\
&= \overline{a_0\alpha^n + a_1\alpha^{n-1} + \cdots + a_n} = \overline{P(\alpha)} = 0,
\end{aligned}
$$

since $\bar{a}_i = a_i$ because the coefficients are real. This leads to

---

Theorem 8.2   *If α is a root of multiplicity k of an algebraic equation with real coefficients then so is $\bar{\alpha}$. The number of imaginary roots of an equation with real coefficients is even.*

Proof.   We have just seen that the theorem is true for $k = 1$. We proceed by complete induction. If P has an imaginary root α then $P(x) = (x^2 - (\alpha + \bar{\alpha})x + \alpha\bar{\alpha})Q(x)$, with Q having α as a root of multiplicity smaller than that of P. By the induction hypothesis the multiplicity of $\bar{\alpha}$ for Q is the same as that of α and the assertion follows.   □

Mathematicians throughout the ages found many theorems which help to determine the number of real roots of a given polynomial and isolate them in separate intervals. The best known are the theorems of Budan-Fourier, Sturm and the so called rule of Descartes. The interested reader is referred to Kurosh (1980). *Maple* has a tool for an application of the Sturm Theorem. We shall not discuss these matters here, but mention a *Maple* command `realroot`. It responds with a list of intervals each containing just one root, selects in some reasonable way the length of these intervals and sometimes returns an interval $[a, a]$ which indicates that a is a root. Multiplicity of roots is not taken into account.

```
>   realroot(x^6+x^5-x^2+1);
```

$$[[-1, -1], [-2, -1]]$$

This means the polynomial has a root -1 and a real root in the interval $[-2,-1]$.

If there are no real roots the command returns an empty list.

```
>  realroot(x^4+1);
```

$$[]$$

The concept of an algebraic solution

By an algebraic solution of an algebraic equation we understand formulae for the roots which use only the coefficients of the equation and only operations of addition, subtraction, multiplication, division and extraction of integer roots. This is a useful description but it is not precise enough.

We shall say that we found an algebraic solution to the equation

$$a_0 x^n + a_1 x^{n-1} + \cdots + a_n = 0, \qquad (AE)$$

with complex coefficients[10] or that the equation is solvable in radicals if we define a finite chain of *binomial* equations

$$y_1^{m_1} = A_1, \qquad (A1)$$

$$y_2^{m_2} = A_2, \qquad (A2)$$

$$\cdots$$

$$y_k^{m_k} = A_k, \qquad (Ak)$$

with the following properties:

(1) the numbers m_i, $i = 1, .., k$ are positive integers;
(2) A_1 is an element of a field F which is an extension of $\mathbb{Q}$ by the coefficients of (AE);
(3) A_i for $i = 2, \ldots, k$ is an element of the smallest field which contains F and the roots of previous binomial equations;
(4) all the roots of Equation (AE) lie in the smallest field which contains F and roots of the binomial equations (A1)–(Ak).

Example 8.6 Consider the equation $(x^3 - a)^2 + 2 = 0$. We see that the

---

[10]The concept of an algebraic solution to an equations with coefficients in an arbitrary field can be defined rather similarly but we shall not do it here.

required chain of binomial equations is

$$y_1^2 = -2,$$
$$y_2^3 = a + i\sqrt{2},$$
$$y_2^3 = a - i\sqrt{2}.$$

It might be plausible to think that every algebraic equation is solvable in radicals: the equation was formed from the roots of the equation using addition, subtraction, multiplication, division and taking powers so it should be possible by using only addition, subtraction, multiplication, division and by extracting roots from complex numbers to untangle the equation and get back to the roots of the equation. Mathematicians for a long time held this false belief. It was the Italian mathematician Paulo Ruffini who lived in the second half of the eighteenth and first half of the nineteenth century who proved that there are equations of the fifth degree which cannot be solved algebraically. This was followed by important works of N.H. Abel and E. Galois. Today we have a theory, called the Galois theory (interested readers are referred for example to Hadlock (1978): however we add a warning that at this stage they might not be sufficiently equipped to handle it) by which it is possible to decide whether or not a given algebraic equation can be solved algebraically. The impossibility of solving an equation in radicals should not be confused with the existence of the solution. By the Fundamental Theorem of Algebra the solution always exists no matter whether or not a *solution in radicals* can be found. Solving an equation algebraically means finding the chain of Equations (A1)–(Ak); that is, finding the roots of the equation by using *very special equations*, namely the binomial equations. From this point of view it would be surprising if every algebraic equation was solvable in radicals. Obviously it is easy to be wise after the event. Equations of degree up to four are always solvable in radicals, other equations might or might not be algebraically solvable. *Maple* is fantastic in this regard: if there is a solution in radicals, *Maple* will find it.[11]

Quadratic equations

In this section we show that the 'usual' formula for the quadratic equation with which the readers are probably familiar from school is valid for a quadratic equation with complex coefficients. Rewriting the equation $x^2 +$

---

[11]Within reasonable limits of your computer capabilities.

$px + q = 0$ in the form $(x + \frac{p}{2})^2 = \frac{1}{4}D$ with $D = p^2 - 4q$ we see immediately that if $D = 0$ the equation has a double root $\frac{-p}{2}$ and if $D > 0$ then the roots are

$$\frac{-p}{2} + \frac{1}{2}\sqrt{D} \text{ and } \frac{-p}{2} - \frac{1}{2}\sqrt{D}. \tag{8.12}$$

$\sqrt{D}$ is understood in the sense of Definition 5.2, the square root of a positive real number. However, the Formulae (8.12) also hold if D is negative. The symbol $\sqrt{D}$ then means $\imath\sqrt{|D|}$. Finally, if D is imaginary, we know that there are two complex numbers which squared equal to D. We denote one of them[12] $\sqrt{D}$ and the formulae (8.12) still hold. It is customary to choose $\sqrt{D}$ as β in Equation (7.17). For example, the roots of the equation $x^2 + 2\imath x + 3$ are $\imath$ and $-3\imath$, the roots of the equation $x^2 - 2x + 1 - 2\imath$ are $2 + \imath$ and $-\imath$. It is useful to know by heart the Formulae (8.12) for the solution but *Maple* provides the solution instantly, also for an equation of the form $ax^2 + bx + c = 0$. However, for an equation with numerical coefficients the formulae for the solution should *not* be used but the command `solve` applied directly to the given equation.

Cubic equations

Scipione del Ferro, who lived between 1496 and 1526, was the first to solve the general cubic equation. H. Cardano published formulae for the solution in his book *Ars Magna* in 1545 and the formulae for the solution are now often referred to as the Cardano formulae. Solving the cubic equation was undoubtedly a great intellectual achievement but the Cardano formulae themselves have little practical value. Often they give the solution in a rather complicated form and in case of an irreducible equation with real coefficients and three real roots the formulae involve cube roots of complex numbers and an effort to simplify them leads back to solving the equation. For this reason this case is called *casus irreducibilis* (Latin for "the irreducible case").

Let us consider a cubic equation in more detail. An equation

$$x^n + a_1 x^{n-1} + \cdots + a_n = 0 \tag{8.13}$$

---

[12]It does not matter which.

can be rewritten in the form

$$\left(x + \frac{a_1}{n}\right)^n + \text{ terms of degree less than } n \; = 0.$$

After the substitution $y = x + a_1/n$, Equation (8.13) takes the form

$$y^n + c_{n-2}y^{n-2} + \cdots + c_n = 0.$$

In particular, for cubic equations, only the cubic

$$x^3 + px + q = 0 \tag{8.14}$$

needs to be solved.

```
>   ceq:=x^3+p*x+q=0;
```

$$ceq := x^3 + px + q = 0$$

The *Maple* command `solve` can now provide the Cardano formulae, not in the original form, but *Maple*'s form has an advantage. The Cardano formulae require the (possibly imaginary) cube root of $12p^3 + 81q^2$; the correct choice of root has to be made. In the *Maple* formulae, this root can be chosen arbitrarily out of three possible values. However the same value must be kept throughout the formulae. The result is valid only for $p \neq 0$, but this is no obstacle since for $p = 0$ the equation is binomial.

```
>   Sol:=[solve(ceq,x)];
```

$$Sol := [\frac{1}{6}\%1^{(1/3)} - \frac{2\,p}{\%1^{(1/3)}},$$

$$-\frac{1}{12}\%1^{(1/3)} + \frac{p}{\%1^{(1/3)}} + \frac{1}{2}I\sqrt{3}\,(\frac{1}{6}\%1^{(1/3)} + \frac{2\,p}{\%1^{(1/3)}}),$$

$$-\frac{1}{12}\%1^{(1/3)} + \frac{p}{\%1^{(1/3)}} - \frac{1}{2}I\sqrt{3}\,(\frac{1}{6}\%1^{(1/3)} + \frac{2\,p}{\%1^{(1/3)}})]$$

$$\%1 := -108\,q + 12\,\sqrt{12\,p^3 + 81\,q^2}$$

For $p = 3$ and $q = -4$ the equation has an obvious root 1. But if we substitute p and q into the previous *Maple* result, we get:

```
>   p:=3;q:=-4;Sol[1];
```

$$p := 3$$

$$q := -4$$

$$\frac{1}{6}(432 + 12\sqrt{1620})^{(1/3)} - \frac{6}{(432 + 12\sqrt{1620})^{(1/3)}}$$

```
>  simplify(%);
```

$$\frac{(2+\sqrt{5})^{(2/3)}-1}{(2+\sqrt{5})^{(1/3)}}$$

```
>  rationalize(%);
```

$$((2+\sqrt{5})^{(2/3)}-1)\,(2+\sqrt{5})^{(2/3)}\,(-2+\sqrt{5})$$

```
>  simplify(%);
```

$$((2+\sqrt{5})^{(2/3)}-1)\,(2+\sqrt{5})^{(2/3)}\,(-2+\sqrt{5})$$

We still cannot obtain 1. Of course, our approach was clumsy. Applying solve directly to the equation with numeric coefficients immediately gives the correct answer.

```
>  solve(x^3+3*x-4=0,x);
```

$$1,\ -\frac{1}{2}+\frac{1}{2}\,I\,\sqrt{15},\ -\frac{1}{2}-\frac{1}{2}\,I\,\sqrt{15}$$

If x_1, x_2, x_3 are roots of the equation then $(x_1-x_2)^2(x_1-x_3)^2(x_2-x_3)^2$ is called the discriminant of the equation. By the fundamental theorem on symmetric polynomials (see page 214) the discriminant can be expressed in terms of the coefficients. This is a laborious task and we let *Maple* do it for us. We use the command simplify in a new way, specifying eqs as a *list* of equations, which *Maple* should use in expressing SP in a new form, thus

$$\text{simplify(PS,eqs)};$$

```
>  eqns:=[x[1]+x[2]+x[3]=0,x[1]*x[2]+x[1]*x[3]+x[2]*x[3]
>  =p,-x[1]*x[2]*x[ 3]=q];
```

$$eqns := [x_1+x_2+x_3=0,\ x_1\,x_2+x_1\,x_3+x_2\,x_3=p,\ -x_1\,x_2\,x_3=q]$$

```
>  disc:=(x[1]-x[2])^2*(x[1]-x[3])^2*(x[2]-x[3])^2;
```

$$disc := (x_1-x_2)^2\,(x_1-x_3)^2\,(x_2-x_3)^2$$

```
>  simplify(disc,eqns);
```

$$-4\,p^3-27\,q^2$$

If p, q are real then the equation has three real roots if and only if the discriminant is positive, however the Cardano formulae then contains complex numbers (in a rather complicated way). But solve might also give the roots in a complex form. Consider the equation

---

> Eq:=x^3-3*x+1=0;

$$Eq := x^3 - 3\,x + 1 = 0$$

Looking at the graph or evaluating the discriminant $-4p^3 - 27q^2 = 81$ shows that this equation has three real roots. However

---

> solve(Eq,x);

$$\frac{1}{2}\,\%1 + \frac{2}{(-4 + 4\,I\,\sqrt{3})^{(1/3)}},$$

$$-\frac{1}{4}\,\%1 - \frac{1}{(-4 + 4\,I\,\sqrt{3})^{(1/3)}} + \frac{1}{2}\,I\,\sqrt{3}\,(\frac{1}{2}\,\%1 - \frac{2}{(-4 + 4\,I\,\sqrt{3})^{(1/3)}}),$$

$$-\frac{1}{4}\,\%1 - \frac{1}{(-4 + 4\,I\,\sqrt{3})^{(1/3)}} - \frac{1}{2}\,I\,\sqrt{3}\,(\frac{1}{2}\,\%1 - \frac{2}{(-4 + 4\,I\,\sqrt{3})^{(1/3)}})$$

$$\%1 := (-4 + 4\,I\,\sqrt{3})^{(1/3)}$$

One way to proceed is to use fsolve.

> fsolve(Eq,x);

$$-1.879385242, .3472963553, 1.532088886$$

If we want an exact solution we use the command evalc similarly as we did in Subsection 7.2.5 with the binomial equation.

> L:=[solve(Eq)]:seq(simplify(evalc(L[i])),i=1..3);

$$2\cos(\frac{2}{9}\,\pi), \ -\cos(\frac{2}{9}\,\pi) - \sqrt{3}\sin(\frac{2}{9}\,\pi), \ -\cos(\frac{2}{9}\,\pi) + \sqrt{3}\sin(\frac{2}{9}\,\pi)$$

For an irreducible cubic equation it is almost invariably better to use fsolve rather than solve, not only in the case of three real roots. On the other hand if the equation is reducible in the field implied by the coefficients, solve will find all roots in a neat form even if all roots are real.

> solve(8*x^3-6*x+sqrt(2));

$$-\frac{1}{4}\,\sqrt{2} + \frac{1}{4}\,\sqrt{6}, \ -\frac{1}{4}\,\sqrt{2} - \frac{1}{4}\,\sqrt{6}, \ \frac{1}{2}\,\sqrt{2}$$

The case of an irreducible cubic equation with three real roots is called *casus irreducibilis*. The Cardano formulae (or solve) express the roots in a complicated form involving imaginary numbers and it can be shown that no amount of algebraic manipulation can simplify it to the standard form $a + bi$. It is therefore very pleasing that with *Maple* you can obtain satisfactory result by using `evalc` together with `solve`. However *Maple* does not always do it 'algebraically': note that in the previous example trigonometric functions were used.

8.3.1 *Equations of higher orders and* `fsolve`

Although equations of fourth order can be solved algebraically and equations of order five or higher, generally speaking, cannot, we recommend, unless the equation has a strikingly simple form, to use `fsolve` rather than `solve`. For comparison consider the next equation which can be solved algebraically (by `solve`).

```
> solve(x^5+x+1);
```

$$-\frac{1}{2} + \frac{1}{2} I \sqrt{3}, \ -\frac{1}{2} - \frac{1}{2} I \sqrt{3}, \ -\frac{1}{6}\%1 - \frac{2}{3}\frac{1}{(100 + 12\sqrt{69})^{(1/3)}} + \frac{1}{3},$$

$$\frac{1}{12}\%1 + \frac{\frac{1}{3}}{(100 + 12\sqrt{69})^{(1/3)}} + \frac{1}{3} + \frac{1}{2}I\sqrt{3}\left(-\frac{1}{6}\%1 + \frac{\frac{2}{3}}{(100 + 12\sqrt{69})^{(1/3)}}\right),$$

$$\frac{1}{12}\%1 + \frac{\frac{1}{3}}{(100 + 12\sqrt{69})^{(1/3)}} + \frac{1}{3} - \frac{1}{2}I\sqrt{3}\left(-\frac{1}{6}\%1 + \frac{\frac{2}{3}}{(100 + 12\sqrt{69})^{(1/3)}}\right)$$

$$\%1 := (100 + 12\sqrt{69})^{(1/3)}$$

```
> fsolve(x^5+x+1);
```

$$-.7548776662$$

```
> fsolve(x^5+x+1,x, complex);
```

$$-.7548776662,$$
$$-.5000000000 - .8660254038\ I, \ -.5000000000 + .8660254038\ I,$$
$$.8774388331 - .7448617666\ I, \ .8774388331 + .7448617666\ I$$

Although we recommend using `fsolve` for solving equations of degree higher than three, even with the aid of a computer, numerical calculations can have problems of their own. The next example illustrates this.

```
>   Digits:=5;
```

$$Digits := 5$$

```
>   badpol:=expand((x-1.0001)^2*(x-1.0002)^2*(x-1.0003)^2);
```

$$badpol := x^6 - 6.0012\,x^5 + 15.006\,x^4 - 20.012\,x^3$$
$$+15.012\,x^2 - 6.0060\,x + 1.0012$$

```
>   fsolve(badpol);
```

$$1., 1., 1., 1., 1., 1.0012$$

The results says that the polynomial has a root of multiplicity 5 (which it does not) and the error for the last root is fairly large. This was caused by the number of digits used and rounding of all numbers to 5 digits. However similar phenomena can occur with any number of digits. It shows that the analysis of rounding errors can be important, but a lot of deep mathematics is needed for that. To see the reason why we obtained such an unsatisfactory result we graph the polynomial.

```
>   plot(badpol,x=0..2, scaling=CONSTRAINED);
```

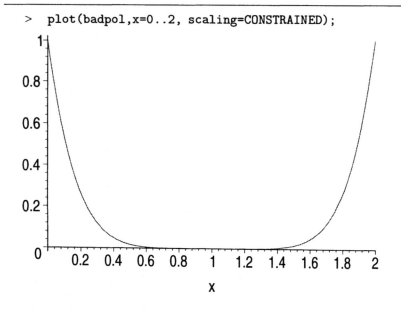

We see that the graph is almost indistinguishable from the x axis between 0.6 and 1.4 and therefore to locate the zeros is difficult. By increasing the number of digits we obtain a perfect result.

```
>  restart;
>  Digits:=30;
```

$$Digits := 30$$

```
>  badpol:=expand((x-1.0001)^2*(x-1.0002)^2*(x-1.0003)^2);
```

$$badpol := x^6 - 6.0012\,x^5 + 15.00600058\,x^4$$
$$- 20.012002320144\,x^3 + 15.0120034804320193\,x^2$$
$$- 6.006002320432038860132\,x$$
$$+ 1.0012005801440193013200036$$

```
>  fsolve(badpol);
```

1.000100000000000000000000000000, 1.000100000000000000000000000000,
1.000200000000000000000000000000, 1.000200000000000000000000000000,
1.000300000000000000000000000000, 1.000300000000000000000000000000

It should be realized that increasing the number of digits may not always be a viable option and, when the numbers come from experimental data, could be meaningless.

Exercises

ⓘ **Exercise 8.3.1** *Prove: A rational root of a polynomial with integer coefficients and leading coefficient 1 is an integer.* [Hint: q divides 1.]

Exercise 8.3.2 *Test the numbers of the form (8.9) to find the rational roots of $2x^6 + x^5 - 9x^4 - 6x^3 - 5x^2 - 7x + 6$ and then compare your results with that obtained by the Maple command* roots. [Hint: Make a list L of divisors of 2 and a list K of positive divisors of 6. Then [seq(op(L)/K[i],i=1..nops(K))] is the list to be tested.]

Exercise 8.3.3 *Let S be a polynomial with integer coefficients, α a rational root of S and $S(x) = (x - \alpha)H(x)$. Show that H has integer coefficients.* [Hint: Compare coefficients!]

ⓘ **Exercise 8.3.4** *Use notation as in the previous exercise with $\alpha = p/q$ and p, q relatively prime. Prove that $p-q$ and $p+q$ divide $S(1)$ and $S(-1)$,* respectively. [Hint: $qS(1) = (q - p)H(1)$. Similarly for $p + q$.]

Exercise 8.3.5   *Use the previous exercise to find the rational roots of* $S(x) = 6x^5 + 11x^4 - x^3 + 5x - 6$.

Exercise 8.3.6   *Find the multiple roots of* $P(x) = x^9 + 2x^7 + 2x^5 + x^4 + 2x^3 + 2x^2 + 1$.

ⓘ **Exercise 8.3.7**   *Use* `fsolve` *on the equations* $\prod_{j=1}^{20}(x - j) + 10/11 = 0$, $\prod_{j=1}^{20}(x - j) + 1.1 = 0$. *Increase the number of Digits to 30 and repeat.*

8.4   Linear equations in several unknowns

Linear equations in several unknowns are important because many problems in Science and engineering lead ultimately to solving large systems of linear equations. For instance, in shipbuilding, equations with over 40000 unknowns were encountered. The branch of mathematics which deals with linear equations is linear algebra and *Maple* has two useful packages, `linalg` and **Linear Algebra**. This section is not meant even as a modest introduction to linear algebra or to these packages. It is a simple guide on how to solve small systems with exact coefficients.

We start with a simple example which exhibits the behavior of linear systems. Consider the system

$$x + y = 0, \tag{8.15}$$
$$x + cy = b. \tag{8.16}$$

Subtracting the first equation from the second gives $(c - 1)y = b$ and if $c \neq 1$, y is uniquely determined as $y = b/(c - 1)$ and x is then uniquely determined from the first equation $x = -b/(1 - c)$. It is easy to check that the found values are indeed a solution. So we have

The first case: *The system has a unique solution.*

If $c = 1$ and $b \neq 0$ then the second equations contradicts the first. So we have

The second case: *The system has no solution.*

And if $c = 1$ and $b = 0$ then the second equation is superfluous: we can choose y arbitrarily and $x = -y$ is a solution. So we have

The third case: *The system has infinitely many solutions and some of the unknowns can be chosen arbitrarily.*

These three cases are typical for linear systems no matter how large the number of unknowns. Linear sytems have another feature: the solution always lies in the same field as the coefficients. We now look at some examples. The first one would require a lot of work with pen and paper, but using *Maple* the only work needed is to type the equations correctly.

```
>   restart;
>   eqs:={x+2*y+3*z+4*u+5*v+6*w=1,
>   2*x+3*y+4*z+5*u+6*v+w=-2,
>   3*x+4*y+5*z+6*u+v+2*w=3,
>   4*x+5*y+6*z+u+2*v+3*w=-4,
>   5*x+6*y+z+2*u+3*v+4*w=5,
>   6*x+y+2*z+3*u+4*v+5*w=-6};
```

$eqs :=$
$$\{x + 2y + 3z + 4u + 5v + 6w = 1, 2x + 3y + 4z + 5u + 6v + w = -2,$$
$$3x + 4y + 5z + 6u + v + 2w = 3, 4x + 5y + 6z + u + 2v + 3w = -4,$$
$$5x + 6y + z + 2u + 3v + 4w = 5, 6x + y + 2z + 3u + 4v + 5w = -6\}$$

```
>   solve(eqs,{x,y,z,u,v,w});
```

$$\{w = \frac{10}{21}, z = \frac{-32}{21}, u = \frac{8}{7}, v = \frac{-6}{7}, y = \frac{38}{21}, x = \frac{-25}{21}\}$$

Returning to our introductory example

```
>   eq1:={x+y=0,x+c*y=b};
```

$$eq1 := \{x + y = 0, x + cy = b\}$$

```
>   solve(eq1,{x,y});
```

$$\{y = \frac{b}{-1+c}, x = -\frac{b}{-1+c}\}$$

Maple returned the answer only for the 'general' case. This shows that even in simple examples when symbolic solution is obtained, caution is needed.

```
>   restart;
>   equations:={x+y+z=1,x+y+a*z=1,x+b*y+z=1};
```

$$equations := \{x + y + z = 1, x + by + z = 1, x + y + az = 1\}$$

```
>   solve(equations,{x,y,z});
```

$$\{x = 1, \, z = 0, \, y = 0\}$$

Maple again provided the solution in the general case. Unfortunately, in this example, it did not give any hint as to which values of a, b are exceptional. However, subtracting the first equation from the second and from the third leads to $(b-1)y = 0$, $(a-1)z = 0$. If $b = 1$ and $a \neq 0$[13] then $z = 0$, $x = -y$, and y can be chosen arbitrarily. This illustrates two facts: *Maple* did not provide a complete solutions for all possible values of a and b and in the third case *some* unknowns can be chosen but the system might determine which unknowns cannot be chosen. In this example z cannot be chosen but x or y can. Finally, if $a = b = 1$ any two unknowns can be chosen, say x, y and the third one is then uniquely determined, $z = 1-x-y$.

Exercises

Exercise 8.4.1   *Use* Maple *to solve the systems*

(1) $x - z = 0$, $y - u = 0$, $-x + z - v = 0$, $-y + u - w = 0$, $-z + v = 0$, $-u + w = 0$;

(2) $x - z + v = 0$, $y - u + w = 0$, $x - y + v - w = 0$, $y - z + w = 0$, $x - u + v = 0$;

(3) $x + y + z = 1$, $ax + by + cz = d$, $a^2x + b^2y + c^2z = d^2$.

---

[13]The case $a = 1$ and $b \neq 0$ is similar.

Chapter 9

Sets Revisited

In this chapter we introduce the concept of equivalence for sets and study countable sets. We also discuss briefly the axiom of choice.

9.1 Equivalent sets

Two finite sets A and B have the same number of elements if there exists a bijection of A onto B. We can count the visitors in a sold-out theatre by counting the seats. The concept of equivalence for sets is an extension of the concept of two sets having the same number of elements, generalised to infinite sets.

> **Definition 9.1** A set A is said to be equivalent to a set B if there exists a bijection of A onto B; we then write $A \sim B$.

Example 9.1

(a) The set $\mathbb{N}$ and the set of even positive integers are equivalent. Indeed, $n \mapsto 2n$ is a bijection of $\mathbb{N}$ onto $\{2, 4, 6, \ldots\}$.

(b) $\mathbb{Z} \sim \mathbb{N}$. Let $f : n \mapsto 2n$ for $n > 0$, and $f : n \mapsto -2n + 1$ for $n \leq 0$. Then f is obviously onto and it is easy to check that it is one-to-one. Hence it is a bijection of $\mathbb{Z}$ onto $\mathbb{N}$.

(c) $]0, 1[\sim]0, \infty[$. The required bijection is $x \mapsto \dfrac{1}{x} - 1$.

We observe that the relation $A \sim B$ is reflexive, symmetric and transitive, and thus it is an equivalence relation. Reflexivity follows from the use of the function id_A, which provides a bijection of A onto itself. If the function f is a bijection of A onto B then it has an inverse and f_{-1} is a bijection of B onto A, so $A \sim B \Rightarrow B \sim A$. Finally, if f and g are bijections of B onto C and of A onto B, respectively, then $f \circ g$ is a bijection of A onto C, so $(A \sim B,\ B \sim C) \Rightarrow A \sim C$. In view of the symmetry property we may put $A \sim B$ into words as 'the sets A and B are equivalent'.

Displaying the first few terms of a sequence often gives a good insight into the behaviour of the sequence. For example, for the sequence from Example 9.1 (b) we have

$$0,\ 1,\ -1,\ 2,\ -2,\ 3,\ -3,\ \ldots \tag{9.1}$$

From this it is clear that each $n \in \mathbb{Z}$ appears in a uniquely determined place in (9.1), which means that this sequence is a bijection of $\mathbb{N}$ onto $\mathbb{Z}$.

A set A is said to be *enumerable* if it is equivalent to $\mathbb{N}$, that is, if there exists a bijection of $\mathbb{N}$ onto A; this bijection is called an *enumeration* of A. The set of even positive integers and $\mathbb{Z}$ are both enumerable by Example 9.1. A set is called *countable* if it is finite or enumerable. Note that 'countable' in mathematics differs from the meaning this word may have in other contexts.

Warning: Some authors use the words 'enumerable' and 'countable' with slightly different meanings. Also the word 'denumerable' is sometimes used instead of 'countable'.

Theorem 9.1   *The range of a sequence is countable.*

Let the sequence be $n \mapsto u_n$. We list the members

$$u_1,\ u_2,\ u_3,\ \ldots \tag{9.2}$$

and cross out all repetitions when 'moving along the sequence (9.2) from left to right'. We are then left with either a finite sequence or we obtain an enumeration of $\operatorname{rg} u$. The formal proof is as follows.

Proof.   The theorem is obvious if $\operatorname{rg} u$ is finite. Define $n_1 = 1$, and if $n_1, n_2, \ldots, n_k$ have been defined let n_{k+1} be the smallest integer n such that $u_n \notin \{u_{n_1}, u_{n_2}, \ldots, u_{n_k}\}$. Then the sequence $k \mapsto u_{n_k}$ is an enumeration of $\operatorname{rg} u$. □

It is worth noting that in the course of this proof we have tacitly used the recursion theorem (Theorem 5.13) and the well ordering principle (Theorem 4.10).

> **Theorem 9.2** *A non-empty countable set is the range of some sequence.*

Proof. If A is enumerable then A is the range of its enumeration. If A is finite then $A = \{a_1, a_2, \ldots, a_N\}$ for some N. In this case define $a_n = a_N$ for $n > N$. Then $n \mapsto a_n$ is a sequence and A is its range. $\qquad\square$

> **Theorem 9.3** *If A is countable and $f : A \mapsto B$ is onto, then B is countable.*

Proof. If A is finite then so is B. If A is enumerable let u be the enumeration of A. The function $f \circ u$ is a sequence and its range is B. Hence B is countable by Theorem 9.1. $\qquad\square$

> **Theorem 9.4** *If $B \subset A$ and A is countable then so is B.*

This is rather intuitive: a subset of a countable set is countable.

Proof. If $B = \emptyset$ or $B = A$ there is nothing to prove. Otherwise choose some element $b \in B$. Now consider the identity function on B, id_B, and let h be the extension of this to A such that $h(x) = b$ for $x \in A \setminus B$. Clearly $h : A \mapsto B$ is onto, and hence B is countable by Theorem 9.3. $\qquad\sqcap$

> **Theorem 9.5** $\mathbb{N} \times \mathbb{N}$ *is countable.*

We can arrange the elements of $\mathbb{N} \times \mathbb{N}$ into an infinite table.

$$
\begin{array}{cccccc}
(1,1) & (1,2)\to(1,3) & (1,4)\to(1,5) & \cdots \\
(2,1) & (2,2) & (2,3) & (2,4) & (2,5) & \cdots \\
(3,1) & (3,2) & (3,3) & (3,4) & (3,5) & \cdots \\
(4,1) & (4,2) & (4,3) & (4,4) & (4,5) & \cdots \\
\vdots & \vdots & \vdots & \vdots & \vdots & \ddots
\end{array}
$$

By following the arrows we can arrange the elements of $\mathbb{N} \times \mathbb{N}$ into a sequence

$$(1, 1), \ (2, 1), \ (1, 2), \ (1, 3), \ (2, 2), \ (3, 1), \ (4, 1), \ \ldots$$

and this sequence enumerates $\mathbb{N} \times \mathbb{N}$. For a reader who feels uncomfortable with this proof, because we have not explicitly defined the enumeration, we give another proof. However, a formal proof based on the above ideas can also be given.

Proof. Let $u : (n, k) \mapsto 2^{n-1}(2k-1)$. By Example 5.7 and Exercise 5.9.6 u is one-to-one and onto $\mathbb{N}$. Consequently u_{-1} is an enumeration of $\mathbb{N} \times \mathbb{N}$. $\square$

Theorem 9.6   Let A_k be countable for every $k \in \mathbb{N}$. Then $\displaystyle\bigcup_{k=1}^{\infty} A_k$ is countable.

A simple memory aid for Theorem 9.6 is: a countable union of countable sets is countable.

Proof. Without loss of generality we can assume that each A_k is non-empty. By Theorem 9.2 each A_k is the range of some sequence, say

$$a_k : \ m \mapsto a_k(m).$$

Let w be an enumeration of $\mathbb{N} \times \mathbb{N}$, $w : n \mapsto (m, k)$. Then $n \mapsto a_k(m)$ maps $\mathbb{N}$ onto $\displaystyle\bigcup_{k=1}^{\infty} A_k$, and by Theorem 9.1 this set is countable. $\square$

Theorem 9.7   *The set $\mathbb{Q}$ is countable.*

Proof. For $k \in \mathbb{N}$ define $A_k = \left\{ \dfrac{m}{k}; \ m \in \mathbb{Z} \right\}$. Since $m \mapsto \dfrac{m}{k}$ maps $\mathbb{Z}$ onto A_k, the set A_k is countable by Theorem 9.3. Since $\mathbb{Q} = \displaystyle\bigcup_{k=1}^{\infty} A_k$, we see that $\mathbb{Q}$ is countable by Theorem 9.5. $\square$

Theorem 9.8 $]0, 1[$ *is not countable.*

The method of proof used here is known as *Cantor's diagonal process* (or Cantor's diagonalization process), and is often used on other occasions (see, for example, Exercise 9.1.6). The proof uses decimal fractions for real numbers, as expounded in Example 5.8.

Proof. Assume $]0,\ 1[$ is countable, then $]0,\ 1[$ is a range of some sequence $a:\ n \mapsto a_n$. With each a_n we associate its decimal fraction $0.a_1^n a_2^n a_3^n \ldots$ (here n is an upper suffix, not the index of some power). We now define a sequence b_n by $b_n = 5$ if $a_n^n \neq 5$ and $b_n = 1$ if $a_n^n = 5$. This process is illustrated in the following diagram where $a_1^1 \neq b_1$, $a_2^2 \neq b_2$, $a_3^3 \neq b_3$, $\ldots$, $a_i^i \neq b_i$, $\ldots$, are marked, showing the reason for the adjective "diagonal".

$$a_1 = 0.\cancel{a_1^1}a_2^1 a_3^1 a_4^1 a_5^1 \ldots a_i^1 \ldots$$
$$a_2 = 0.a_1^2 \cancel{a_2^2} a_3^2 a_4^2 a_5^2 \ldots a_i^2 \ldots$$
$$a_3 = 0.a_1^3 a_2^3 \cancel{a_3^3} a_4^3 a_5^3 \ldots a_i^3 \ldots$$
$$a_4 = 0.a_1^4 a_2^4 a_3^4 \cancel{a_4^4} a_5^4 \ldots a_i^4 \ldots$$

$$\vdots$$

$$a_i = 0.a_1^i a_2^i a_3^i a_4^i a_5^i \ldots \cancel{a_i^i} \ldots$$

$$\vdots$$

and

$$b = 0.b_1 b_2 b_3 b_4 b_5 \ldots b_i \ldots$$

Let $B = \sup\left\{ B_n;\ B_n = \dfrac{b_1}{10} + \dfrac{b_2}{10^2} + \dfrac{b_3}{10^3} + \cdots + \dfrac{b_n}{10^n} \right\}$. For $m > n$ we have

$$B_m = B_n + \frac{b_{n+1}}{10^{n+1}} + \frac{b_{n+2}}{10^{n+2}} + \cdots + \frac{b_m}{10^m},$$
$$< B_n + \frac{5}{10^{n+1}} \frac{1}{1 - \frac{1}{10}},$$
$$= B_n + \frac{5}{9} \frac{1}{10^n},$$
$$< B_n + \frac{1}{10^n}.$$

It follows from these inequalities that $B \leq B_n + \dfrac{5}{9} \dfrac{1}{10^n} < B_n + \dfrac{1}{10^n}$. From the definition of B it is obvious that $B_n \leq B$, and so the decimal fraction associated with B is $0.b_1 b_2 b_3 \ldots$; for the definition of decimal fraction see Example 5.8. It is also easy to see that $0 < B < 1$. The decimal fraction for B differs in at least one place from the decimal fraction for any a_n. Consequently $B \neq a_n$ for every $n \in \mathbb{N}$, contradicting the assumption that the range of a was $]0, 1[$. $\qquad\square$

An obvious consequence of Theorem 9.8 is that $\mathbb{R}$ is not countable. Any interval $]a, b[$ is not countable either because $b : x \mapsto \dfrac{x - a}{b - a}$ is a bijection of $]a, b[$ onto $]0, 1[$. Also the set of irrationals, that is, $\mathbb{R} \setminus \mathbb{Q}$ is not countable: if it were then $\mathbb{R}$ would be countable as the union of $\mathbb{R} \setminus \mathbb{Q}$ and $\mathbb{Q}$.

Theorem 9.8 shows that there is at least one infinite set which is 'larger' than the set of natural numbers.

The next theorem describes enumerable sets as the smallest infinite sets, or, in other words, that the natural numbers are equivalent to any of the smallest infinite sets.

Theorem 9.9 *Every infinite set contains an enumerable part.*

Proof. [**First attempt**] Let A be infinite, and choose $a_1 \in A$. The set $A \setminus \{a_1\}$ is non-empty (it is even infinite) and we can choose $a_2 \in A \setminus \{a_1\}$. Continuing with this process indefinitely we select

$$a_{n+1} \in A \setminus \{a_1, a_2, a_3, \ldots, a_n\}.$$

Then $B = \{a_1, a_2, a_3, \ldots\} \subset A$ and $n \mapsto a_n$ is an enumeration of B. $\quad\square$

The phrase "continuing with this process indefinitely" in the above 'proof' looks suspect. Can the proof be made impeccable by replacing the offending phrase by a reference to one of the recursion theorems? The trouble is that we have no function g as in Theorem 5.2 or g_n as in Theorem 5.13 at our disposal, and since the theorem we are trying to prove refers to "every infinite set" there seems little hope of devising such functions.

Arguments such as those used in the above 'proof' had been freely used in mathematics until it was realised that their acceptance hinges on the validity of the following statement—which has come to be known as the axiom of choice.

Axiom of Choice *If* $\mathbf{S}$ *is a family of non-empty sets then there exists a function G such that $G(S) \in S$ for every $S \in \mathbf{S}$.*

The great German mathematician David Hilbert compared the family **S** and the function G to a group of societies each selecting its president. From this point of view the axiom of choice is very plausible. However, the difficulty with its acceptance in mathematics lies in its great generality: it asserts the existence of G, for *any* infinite family of sets **S**, and, furthermore, each of the sets S in the family may also be infinite.

We now return to the proof of Theorem 9.9: this time, using the axiom of choice, we can provide an impeccable proof.

Proof. From the axiom of choice there exists a function G which associates with every non-empty subset $B \subset A$ an element $G(B) \in B$. We now define

$$A_0 = A,$$
$$a_1 = G(A_0),$$
$$A_1 = A_0 \setminus \{a_1\} = A_0 \setminus \{G(A_0)\},$$

and inductively, for $n \in \mathbb{N}$,

$$a_{n+1} = G(A_n),$$
$$A_{n+1} = A_n \setminus \{G(A_n)\}.$$

The sequences $n \mapsto A_n$ and $n \mapsto a_n$ are uniquely determined by the recursion theorem. To see this we define f, S and g in Theorem 5.13 as follows:

$$S = \{x, X\} \quad \text{where } x \in A \text{ and } X \subset A,$$
$$g(x, X) = (G(x), X \setminus \{G(x)\}),$$
$$f(1) = (G(A), A - G(A)),$$
$$f(n + 1) = g(f(n)).$$

Now a_n becomes the first element in the couple $f(n)$ and $n \mapsto a_n$ is an enumeration of a part of A. $\qquad \square$

Exercises

① **Exercise 9.1.1**   *Define a bijection for each of the pairs of sets given below.*

(1)	$\mathbb{N}$, $\mathbb{N}_0$	(2)	$[0, 1]$, $[3, 7]$;
(3)	$(0, \infty)$, $\mathbb{R}$;	(4)	$[a, b]$, $[d, c]$;
(5)	$[0, 1]$, $]0, 1[$;	(6)	$\mathbb{Z} \times \mathbb{Z}$, $\mathbb{Z}$;
(7)	$\mathbb{Z} \times \mathbb{Z}$, $\mathbb{N}$;	(8)	$\mathbb{Q} \times \mathbb{Q}$, $\mathbb{N}$.

(i) **Exercise 9.1.2**   Prove that if A is not countable and B is countable then $A \setminus B$ is not countable.

Exercise 9.1.3   Show that the following sets are countable:

(1) $\{a + b\sqrt{3};\ a \in \mathbb{Q},\ b \in \mathbb{Q}\}$;

(2) $\{x :\ x^2 + px + q = 0;\ p \in \mathbb{Q},\ q \in \mathbb{Q}\}$.

(i) **Exercise 9.1.4**   If $n \in \mathbb{N}$ and A_1, A_2, A_3, $\ldots$, A_n are countable, then the set $A_1 \times A_2 \times A_3 \times \cdots \times A_n$ is also countable. Prove this.

(i) **Exercise 9.1.5**   If L is countable and for every $l \in L$ the set A_l is countable, then so is $\bigcup_{l \in L} A_l$. Prove this slight generalisation of Theorem 9.5.

(i) **Exercise 9.1.6**   Use the diagonalization process from Theorem 9.8 to prove the following theorem. Let S be the set of all sequences having $\{0, 1\}$ as their ranges. Then S is not countable.

(i) **Exercise 9.1.7**   Let S be the set of real numbers in $]0, 1[$ such that their decimal fractions do not contain the digit 7. Show that S is not countable. [Hint: Most of the work has already been done in the proof of Theorem 9.8].

(i) **Exercise 9.1.8**   Prove, without using the axiom of choice, that the set $B = \{x;\ x$ irrational, $0 < x < 1\}$ has an enumerable part.

[Hint: consider $\left\{x;\ x = \dfrac{1}{n\sqrt{2}},\ n \in \mathbb{N}\right\}$.]

(i) **Exercise 9.1.9**   The range of the sequence

$$1, \frac{1}{2}, \frac{2}{2}, \frac{1}{3}, \frac{2}{3}, \frac{3}{3}, \frac{1}{4}, \frac{2}{4}, \frac{3}{4}, \frac{4}{4}, \frac{1}{5}, \frac{2}{5}, \frac{3}{5}, \frac{4}{5}, \cdots$$

contains all rationals between 0 and 1. Use this sequence to define a bijection $\mathbb{Q}$ onto $\mathbb{N}$.

Chapter 10

Limits of Sequences

In the early sections of this chapter we introduce the idea of the limit of a sequence and prove basic theorems on limits. The concept of a limit is central to subsequent chapters of this book. The later sections are devoted to the general principle of convergence and more advanced concepts of limits superior and limits inferior of a sequence.

10.1 The concept of a limit

If we observe the behaviour of several sequences, for example

$$n \mapsto \frac{1}{n+1}; \ \frac{1}{2}, \frac{1}{3}, \frac{1}{4}, \frac{1}{5}, \ldots \tag{10.1}$$

$$n \mapsto \frac{(-1)^{n+1}}{n}; \ 1, \frac{-1}{2}, \frac{1}{3}, \frac{-1}{4}, \frac{1}{5}, \ldots \tag{10.2}$$

$$n \mapsto \frac{n}{n+1}; \ \frac{1}{2}, \frac{2}{3}, \frac{3}{4}, \frac{4}{5}, \frac{5}{6}, \ldots \tag{10.3}$$

$$n \mapsto \frac{1+(-1)^{n+1}}{2^n}; \ 1, 0, \frac{1}{4}, 0, \frac{1}{16}, 0, \ldots \tag{10.4}$$

we see that as n becomes larger the terms of each sequence approach a certain number: for Sequences (10.1), (10.2) and (10.4) it is 0, for (10.3) it is 1. In Sequences (10.1), (10.2), (10.3) and (10.4) there is a clear pattern in this approach. In the next example the terms of the sequence approach a certain number quite irregularly. The following table is a record of an experiment: casting of two dice and noting as a success whenever the sum is 7. The columns headed n in Table 10.1 indicate the number of throws, and the other columns gives the values of s/n where s is the number of successes in n trials.

Table 10.1   Results from throwing two dice

n	s/n	n	s/n
1000000	.166553	1000016	.166553335
1000001	.166552833	1000017	.166553169
1000002	.166552667	1000018	.166553002
1000003	.1665525	1000019	.166552835
1000004	.166552334	1000020	.166552669
1000005	.166552167	1000021	.166552502
1000006	.166553001	1000022	.166552336
1000007	.166553834	1000023	.166552169
1000008	.166553668	1000024	.166552003
1000009	.166553501	1000025	.166551836
1000010	.166553334	1000026	.16655167
1000011	.166553168	1000027	.166551503
1000012	.166553001	1000028	.166551337
1000013	.166552835	1000029	.16655117
1000014	.166552668	1000030	.166552003
1000015	.166553502		

There is no discernible pattern this time but (assuming the dice are un-biased) the terms of the sequence of values of s/n approach the number .16666... = 1/6. What is characteristic is that after one million terms, all subsequent terms of the sequence are approximately equal to .1666... within three decimal places. If we required higher accuracy we would have to keep casting the dice longer but the same phenomenon would occur; after casting the dice sufficiently often the terms of the sequence will be approximately equal to 1/6 within the required accuracy. This leads us to:

Tentative temporary definition. The sequence $n \mapsto x_n$ has a *limit* l if, for any prescribed accuracy, there exists N such that for $n > N$, the terms x_n are approximately equal to l.

In the example of casting two dice for accuracy to three decimal places the number N was one million. In Sequences (10.1) and (10.2) the limit is clearly zero and for accuracy of k decimal places the number N would be

$2 \cdot 10^k$. For Sequence (10.3) the limit is again zero but it is not so obvious what N should be. Since $\dfrac{1 + (-1)^{n+1}}{2n} \le \dfrac{1}{2^{n-1}} < \dfrac{1}{n}$ for $n > 3$ it is clear that $\dfrac{1 + (-1)^n}{2n}$ will be approximately equal to zero to k decimal places if $n > 2 \cdot 10^k$. Hence $2 \cdot 10^k$ will do for N. Perhaps this N is unnecessarily large but that is of no importance; the tentative definition requires existence of some number N (with the required property) and soon as some N is found, the job is done.

It is undesirable and often would be inconvenient to use a decimal fraction to measure accuracy. It is simpler to use a positive number, say ε, and say that x_n is approximately equal to l within ε if

$$|x_n - l| < \varepsilon$$

Definition 10.1 (Limit of a sequence)   A sequence $n \mapsto x_n \in \mathbb{C}$ is said *to have a limit* l if for every positive ϵ there exists a real number N such that

$$|x_n - l| < \varepsilon$$

whenever $n > N$. If the sequence $n \mapsto x_n$ has a limit l we write

$$\lim_{n \to \infty} = l \quad \text{or} \quad x_n \to l.$$

A sequence which has a limit is called *convergent*. A sequence which is not convergent is called *divergent*. Definition 10.1 is often referred to as the ε–N definition of the limit.

The phrases "the sequence converges to l" or the "sequence approaches l" are used with the same meaning as "the sequence has a limit l".

The arrow in the notation $x_n \to l$ indicating that l is the limit of the sequence x must not be confused with the arrow indicating the functional dependence as in $n \mapsto x_n$. Hence $n \mapsto l$ defines a sequence with every term equal to l, whereas $x_n \to l$ states that l is the limit of the sequence $n \mapsto x_n$.

Example 10.1   The sequence $n \mapsto x_n = c$ has a limit c. Since $|x_n - c| = 0$ for every n we can choose any number for N, for instance $N = 1$, and

$$0 = |x_n - l| < \varepsilon$$

whenever $n > N$.

Example 10.2　　The sequence $n \mapsto \dfrac{1}{n+1}$ is convergent to zero, that is $\dfrac{1}{n+1} \to 0$. Let $\varepsilon > 0$. The set $\mathbb{N}$ is not bounded from above by Theorem 4.8 and consequently there exists $N \in \mathbb{N}$ such that $1/\varepsilon < N$. If $n > N$ then $0 < \dfrac{1}{n+1} < \dfrac{1}{N} < \varepsilon$. This proves that $\lim\limits_{n \to \infty} \dfrac{1}{n+1} = 0$.

Theorem 10.1　　*Every sequence has at most one limit.*

Proof.　　Assume, contrary to what we want to prove, that the sequence x has two distinct limits l, L, with $|L - l| > 0$. Choose $\varepsilon = \frac{1}{2}|L - l|$. Then by the definition of the limit there exist numbers N_1, N_2 such that $|x_n - L| < \varepsilon$ whenever $n > N_1$ and $|x_n - l| < \varepsilon$ whenever $n > N_2$. Let $N = \operatorname{Max}(N_1, N_2)$. If $n > N$ then

$$|L - l| = |L - x_n + x_n - l| \le |x_n - L| + |x_n - l| < 2\varepsilon = |L - l|.$$

This is manifestly wrong and we reached a contradiction.　　　　□

We now make some important comments concerning the definition of a limit. (Definition 10.1

(A) The number N from the definition depends on ϵ: this is sometimes emphasised by writing $N(\epsilon)$. Generally, the smaller the value of ϵ, the larger the value of N.

(B) If $n \mapsto x_n$ has a limit l then there are many N such that $|x_n - l| < \epsilon$ for $n > N$. As soon as we have a suitable N, then, for example, $N + 1$, $N + 13$ and $2N$ are also suitable: any number greater than N is suitable.

(C) The meaning of Definition 10.1 is not changed if, instead of permitting N to be any real number we restrict it to $N \in \mathbb{N}$.

(D) The symbol $\lim\limits_{n \to \infty} x_n$ is to be understood as an indivisible quantity. It simply denotes the limit of the sequence x and *does not introduce the symbol* ∞ *in itself*. In particular ∞ does not denote a number.

(E) The definition makes it possible to check whether or not a number l is the limit of a specified sequence. It provides no assistance with finding the value of l. If we can guess what the limit should be then we can use the definition to *prove* that our guess was correct, or show that our guess was incorrect. Obviously this might be difficult or not feasible. It is, however, possible to prove powerful theorems concerning limits from the definition, and then to use these theorems to find limits. Some such theorems are given in the next section.

(F) Consider now a real sequence x, that is, a sequence whose terms are real numbers x_n. Since the inequality $|x_n - l| < \varepsilon$ is equivalent by Theorem 4.4 to

$$l - \varepsilon < x_n < l + \varepsilon$$

all terms of the sequence with an index greater than N lie in the interval $]l - \varepsilon, l + \varepsilon[$. This means that outside this interval can lie only terms of the sequence with $n \le N$, which are finitely many. If the terms x_n are graphed as points on the number line then they cluster near l in the interval $]l - \varepsilon, l + \varepsilon[$ (see Figure 10.1).

Fig. 10.1   Limit of a real sequence

(G) For a complex sequence x, that is, a sequence $x_n \in \mathbb{C}$ for all n, the inequality $|x_n - l| < \varepsilon$ means that the distance between x_n and l is less than ε. Geometrically this means that, in the complex plane, x_n lies in the circle centered at l and of radius ε. If the sequence has a limit l the points x_n in the complex plane cluster near l in the circle centered at l and of radius ε, see Figure 10.2, where the terms of the sequence starting with x_5 lie within the circle.

(H) It goes without saying that the letter n in the symbol $\lim_{n \to \infty} x_n = l$ can be replaced by any other letter without altering the meaning. Thus $\lim_{n \to \infty} x_n = \lim_{k \to \infty} x_k = \lim_{m \to \infty} x_m$.

(I) Sometimes $\lim_{n \to \infty} x_n$ is abbreviated to $\lim x_n$. We shall not do that in this book. If several letters are involved in the formula for x_n, confusion can arise. It is not clear what is the meaning of $\lim \left(\dfrac{n}{n+2} + \dfrac{2}{k} \right)$. It is left as an exercise to show that

$$\lim_{n \to \infty} \left(\frac{n}{n+2} + \frac{2}{k} \right) = 1 + \frac{2}{k} \quad \text{and} \quad \lim_{k \to \infty} \left(\frac{n}{n+2} + \frac{2}{k} \right) = \frac{n}{n+2}.$$

(J) The notation $x_n \to l$ has the same disadvantage as abbreviating

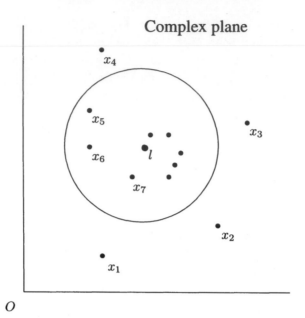

Fig. 10.2   Limit of a complex sequence

$\lim\limits_{n\to\infty} x_n$ to $\lim x_n$. However, if no confusion can arise, as for example in $1/n \to 0$, we shall use this notation for its convenience.

(K) It is easy to see by comparing the definitions that $x_n \to l$ if and only if $|x_n - l| \to 0$. In particular, for $l = 0$, we have that $x_n \to 0$ if and only if $|x_n| \to 0$.

Let a be a sequence of complex numbers and $b_n = a_{n+1}$. Listing a few members of both sequences

$$a_1, a_2, a_3, a_4, \ldots \tag{10.5}$$

and

$$b_1 = a_2, a_3, a_4, \ldots \tag{10.6}$$

shows that the sequence b is obtained from a by omitting the first term and renumbering.

> **Theorem 10.2** *With the notation above* $\lim\limits_{n\to\infty} a_n$ *exists if and only if* $\lim\limits_{n\to\infty} b_n$ *exists and then both limits are equal.*

Proof. Since the wording of the theorem includes 'if and only if' the proof comes in two parts.

I. If $\lim\limits_{n\to\infty} a_n = l$ then for every $\epsilon > 0$ there exists N such that $|a_n - l| < \epsilon$ for $n > N$. Since $b_n = a_{n+1}$, it follows that $|b_n - l| < \epsilon$ for $n > N$.

II. If $\lim\limits_{n\to\infty} b_n = l$ then for every $\epsilon > 0$ there exists N_1 such that $|b_n - l| < \epsilon$ for $n > N_1$. This means that $|a_n - l| < \epsilon$ for $n > N_1 + 1$. $\qquad\square$

The sequence b is obtained from a by omitting the first term. It is clear, by using Theorem 10.2 repeatedly, that omission or change of finite number of terms affects neither the existence nor the value of the limit. Because of this we extend convergence, divergence and limits to functions defined not on all of $\mathbb{N}$ but only on $\mathbb{N} \setminus A$, where A is a finite set and we also call such functions sequences. For instance, if $a_n = 1/(n-5)$ we would say that $\lim\limits_{n\to\infty} 1/(n-5) = 0$ although the function a is not defined for $n = 5$.

Example 10.3 The Sequence (10.3) is convergent to 1, that is, $\lim\limits_{n\to\infty} \dfrac{n}{n+1} = 1$. Consider that $\left| \dfrac{n}{n+1} - 1 \right| = \dfrac{1}{n+1}$. Since $\lim\limits_{n\to\infty} \dfrac{1}{n+1} = 0$, by Example 10.2, for every $\epsilon > 0$ there exists N such that $0 < \dfrac{1}{n+1} < \epsilon$ for $n > N$. Consequently

$$\left| \frac{n}{n+1} - 1 \right| < \epsilon$$

for $n > N$.

In Example 10.3 we used our knowledge of one sequence to establish the limit of another sequence. The same idea is contained in the next theorem.

> **Theorem 10.3** *Assume that $x_n \in \mathbb{R}$ and $y_n \in \mathbb{R}$ for all $n \in \mathbb{N}$. If $x_n \to l$, $y_n \to l$ and there is an N_1 such that*
>
> $$x_n \leq z_n \leq y_n \tag{10.7}$$
>
> *for $n > N_1$ then z_n is convergent to l.*

Proof. For every $\varepsilon > 0$ there exist N_2 and N_3 such that

$$l - \epsilon < x_n < l + \epsilon \tag{10.8}$$

for $n > N_2$ and

$$l - \epsilon < y_n < l + \epsilon \qquad (10.9)$$

for $n > N_3$. Let $N = \text{Max}(N_1, N_2, N_3)$. If $n > N$ then all Inequalities (10.7), (10.8) and (10.9) hold. Consequently

$$l - \epsilon < x_n \leq z_n \leq y_n < l + \epsilon$$

and this implies $|z_n - l| < \epsilon$. $\square$

We shall call Theorem 10.3 the *Squeeze Principle*. This is quite intuitive: z_n is squeezed between x_n and y_n, and as x_n and y_n approach l, so does z_n.

Example 10.4

$$\lim_{n \to \infty} \left(\frac{1}{\sqrt{n^2 + 1}} + \frac{1}{\sqrt{n^2 + 2}} = \cdots + \frac{1}{\sqrt{n^2 + n}} \right) = 1.$$

To prove it we use the Squeeze principle but estimate $\displaystyle\sum_{k=1}^{n} \frac{1}{\sqrt{n^2 + k}}$ first. We have

$$\sum_{k=1}^{n} \frac{1}{\sqrt{n^2 + k}} \leq n \frac{1}{\sqrt{n^2 + 1}} < 1$$

and

$$\sum_{k=1}^{n} \frac{1}{\sqrt{n^2 + k}} \geq n \frac{1}{\sqrt{n^2 + n}} > \frac{n}{\sqrt{n^2 + 2n + 1}} = \frac{n}{n + 1}.$$

Now we set

$$z_n = \sum_{k=1}^{n} \frac{1}{\sqrt{n^2 + k}}, \quad x_n = \frac{n}{n + 1}, \quad y_n = 1.$$

Application of the Squeeze Principle completes the proof.

Example 10.5   Since $\displaystyle\lim_{n \to \infty} \frac{1}{n} = 0$ it follows from the Squeeze Principle that for every $k \in \mathbb{N}$, $\displaystyle\lim_{n \to \infty} \frac{1}{n^k} = 0$.

The Squeeze Principle dealt with *real* sequences. It has, however, an important consequence for complex sequences.

> **Corollary 10.3.1** *Let z be a complex sequence and y a real sequence. If $|z_n - l| < y_n$ for $n \in \mathbb{N}$ and $\lim\limits_{n \to \infty} y_n = 0$ then $\lim\limits_{n \to \infty} z_n = l$.*

Proof. Since $0 \le |z_n - l| < y_n$ the Squeeze Principle implies that $\lim\limits_{n \to \infty} |z_n - l| = 0$ and by (K) $\lim\limits_{n \to \infty} z_n = l$. $\square$

Example 10.6 Let $z_{2k-1} = 1 - \dfrac{1}{2^k}$, $z_{2k} = \dfrac{k+1}{k}$ for $k \in \mathbb{N}$. The first few terms of the sequence are

$$\frac{1}{2}, \ 2, \ \frac{3}{4}, \ \frac{3}{2}, \ \frac{7}{8}, \ \frac{4}{3}, \ \frac{15}{16}, \ \frac{5}{4}, \cdots$$

It is not difficult to see that $z_k \to 1$. We prove it. Since $|z_{2k-1} - 1| = \dfrac{1}{2^k}$, $|z_{2k} - 1| = \dfrac{1}{k}$ we can, in applying the corollary, choose $y_k = \dfrac{2}{k}$ (since by Exercise 5.1.11 we have $2^k \ge k$) and then $|z_k - 1| \le y_k$ for every $k \in \mathbb{N}$. Obviously [1] $y_k \to 0$ and it follows from Corollary 10.3.1 that $z_k \to 1$.

> **Theorem 10.4** *Assume that z is a complex sequence and l a complex number with $x_n = \Re z_n$, $y_n = \Im z_n$, $p = \Re l$ and $q = \Im l$. Then*
>
> $$\lim_{n \to \infty} z_n = l \tag{10.10}$$
>
> *if and only if*
>
> $$\lim_{n \to \infty} x_n = p \tag{10.11}$$
>
> $$\lim_{n \to \infty} y_n = q. \tag{10.12}$$

Proof.
I. Assume (10.10). Since

$$|x_n - p| \le |z_n - l| \tag{10.13}$$

and the right hand side of (10.13) approaches zero by (K) we can apply Corollary 10.3.1 again and we have Equation (10.11). The proof of Equation (10.12) is similar.

II. For $\varepsilon > 0$ there are numbers N_1, N_2 such that $|x_n - p| < \varepsilon/2$ and $y_n - q| < \varepsilon/2$ for $n > N_1$ and $n > N_2$, respectively. Hence

$$|z_n - l| \le |x_n - p| + |y_n - q| < \varepsilon.$$

---

[1] A formal proof is similar to the one given in Example 10.2.

This proves (10.10). □

Not every sequence has a limit. For instance the sequence $n \mapsto (-1)^n$ is divergent. Assume, for an indirect proof, that $(-1)^n \to l$. Then for $\varepsilon = 1$ there exists a number N such that $|(-1)^n - l| < 1$ for all $n > N$. Choose a natural number k with $k > N$. Then

$$|(-1)^k - l| < 1 \quad \text{and} \quad |(-1)^{k+1} - l| < 1.$$

Consequently

$$2 = |(-1)^k - l + l - (-1)^{k+1}| \le |(-1)^k - l| + |(-1)^{k+1} - l| < 2.$$

This contradiction completes the proof that the sequence has no limit.

Exercises

Exercise 10.1.1   *Prove that*

(1) $\lim\limits_{n \to \infty} \dfrac{c}{n} = 0$ *for every* $c \in \mathbb{C}$;

(2) $\lim\limits_{n \to \infty} \dfrac{1}{\sqrt{n}} = 0$;

(3) $\lim\limits_{n \to \infty} \dfrac{n^2 + 2}{n^3 + n} = 0$.

[Hint: For (1) mimic the proof of Example 10.2. Use that the set $\{\sqrt{n}; n \in \mathbb{N}\}$ is not bounded for (2). For (3) show that $\dfrac{n^2 + 2}{n^3 + n} < \dfrac{2}{n}$ and apply the Squeeze principle.]

Exercise 10.1.2   *Let* $a_{2n-1} = \dfrac{4n + 1}{2n}$ *and* $a_{2n} = \dfrac{4n + 1}{2n + 1}$. *Write down the first eight members of the sequence* a. *Find* N_1 *such that* $|a_n - 2| < 10^3$ *for* $n > N_1$. *For* $\varepsilon > 0$ *find* N *such that* $|a_n - 2| < \varepsilon$ *for* $n > N$. [Hint: Simplify $|a_n - 2|$.]

Exercise 10.1.3   *Let* $a_1 = 100$ *and* $a_{n+1} = (a_1 - n)a_n$ *for* $n \in \mathbb{N}$. *Write down the first five terms of the sequence. Find the limit.*

Exercise 10.1.4   *Show that the sequence with terms*

$$1, 2, 3, 1, 2, 3, 1, 2, 3, \ldots$$

has no limit.

ⓘ **Exercise 10.1.5**   Prove that the meaning of Definition 10.1 remains unchanged if

  (1) the inequality $n > N$ is replaced by $n \geq N$;
  (2) the inequality $|x_n - l| < \varepsilon$ is replaced by $|x_n - l| \leq \varepsilon$;
  (3) for some positive ε_0 the requirement that ε is positive is replaced by the requirement that $0 < \varepsilon \leq \varepsilon_0$.

❗ **Exercise 10.1.6**   Prove: If $l \in \mathbb{R}$ and for every $k \in \mathbb{N}$ there exists N such that $\lfloor 10^k x_n \rfloor = \lfloor 10^k l \rfloor$ for $n > N$ then $x_n \to l$. Is the converse true?

ⓘ **Exercise 10.1.7**   Let $a \in \mathbb{C}$ have the property that for every $\varepsilon > 0$ the inequality $|a| < \varepsilon$ is true. Prove that $a = 0$. [Hint: For an indirect proof choose $2\varepsilon = |a|$.]

ⓘ **Exercise 10.1.8**   Let x be a sequence with the property that for every $\varepsilon > 0$ there exists a number N such that $|x_n - l| < \varepsilon$, for $n > N$ and some $l \in \mathbb{R}$. Prove that $x_n = l$ for $n > N$. This shows the importance of the order in which ε and N appear in Definition 10.1. [Hint: Use the previous exercise.]

10.2 Basic theorems

> **Definition 10.2**   A function f is said to be *bounded, bounded below, bounded above on a set* S if the set $f(s)$ is *bounded, bounded below, bounded above*, respectively. The function f is said to be *bounded* if it is bounded on dom f.

This means that a complex sequence x is bounded if and only if there exists a number K such that

$$|x_n| < K \quad \text{for all} \quad n \in \mathbb{N}.$$

> **Theorem 10.5**   *Every convergent sequence is bounded.*

Proof.   Let $x_n \to l$. For $\epsilon = 1$ there exists N such that $|x_n - l| < 1$, for $n > N$. It follows that

$$|x_n| \leq |x_n - l| + |l| < 1 + |l| \tag{10.14}$$

for $n > N$. Let $S_1 = \{x;\ n > N\}$ and $S_2 = \{x;\ n \leq N\}$. The set S_1 is bounded by (10.14) and S_2 is bounded because it is finite. Consequently, $\mathrm{rg}\, x = S_1 \cup S_2$ is bounded. $\qquad\qquad\square$

Theorem 10.6   Let $x_n \in \mathbb{C}$, $y_n \in \mathbb{C}$ and $x_n \to l$, $y_n \to m$. Then

   (i) $|x_n| \to |l|$;

   (ii) $x_n + y_n \to l + m$;

   (iii) $x_n y_n \to lm$:

   (iv) if $m \neq 0$ then $\dfrac{x_n}{y_n} \to \dfrac{l}{m}$.

Remark 10.1   An important special case of (iii) is $y_n = c$ for every $n \in \mathbb{N}$. This yields $c x_n \to cl$. By taking $c = -1$ we obtain $-x_n \to -l$ and then it follows from (ii) that $x_n - y_n \to l - m$. By taking $y_n = x_n$ it follows from (iii) that $x_n^2 \to l^2$ and by simple induction that $x_n^k \to l^k$ for $k \in \mathbb{N}$. For $x_n = 1$ it follows from (iv) that $1/y_n \to 1/m$. It follows similarly from (iii) and (iv) that $y_n^{-k} \to m^{-k}$, provided $m \neq 0$.

Proof.

   (i) by (K) of Section 10.1 $|x_n - l| \to 0$ and since $||x_n| - |l|| \leq |x_n - l|$ it follows from Corollary 10.3.1 that $|x_n| \to |l|$.

  (ii) For $\eta > 0$ there exists N_1 and N_2 such that

$$|x_n - l| < \eta \qquad\qquad (10.15)$$

and

$$|y_n - m| < \eta \qquad\qquad (10.16)$$

for $n > N_1$ and $n > N_2$, respectively. Consequently, for $n > N = \mathrm{Max}(N_1, N_2)$

$$|(x_n + y_n) - (l + m)| \leq |x_n - l| + |y_n - m| < 2\eta. \qquad (10.17)$$

Now take $\varepsilon > 0$, choose $\eta = \dfrac{\varepsilon}{2}$ and for this particular η find N_1 and N_2 as above. Then for $n > N$ we have from (10.17) that

$$|x_n + y_n - (l + m)| < \varepsilon.$$

This proves (ii).

$\qquad\qquad\square$

Remark 10.2   We now pause and reflect on the proof completed thus far. Starting with (10.15) and (10.16) we finish with (10.17). This inequality is almost the one required by the definition of $\lim_{n\to\infty}(x_n+y_n)=l+m$, except that we have there 2η instead of η. The fact that we used η instead of ε is irrelevant, it was an arbitrary positive number. Only a simple adjustment was then needed to remove the factor 2. In a situation like this when we succeed in showing that there is N such that

$$|X_n - L| < M\varepsilon$$

for all $n > N$ and some fixed M (independent of n) we shall regard the proof of $X_n \to L$ as completed.

We now return to the rest of the proof of Theorem 10.6.

Proof.

(iii) Since $x_n \to l$ the sequence x is bounded by Theorem 10.5, that is there is a K such that $|x_n| \le K$ for all $n \in \mathbb{N}$. Now, we have obviously that

$$x_n y_n - lm = x_n(y_n - m) + m(x_n - l)$$

and it follows that

$$|x_n y_n - lm| \le |x_n||y_n - m| + |m||x_n - l| \le K|y_n - m| + |m||x_n - l|.$$

For $\varepsilon > 0$ we can find N_1 and N_2 such that (10.15) and (10.16) hold for $n > N_1$ and $n > N_2$, respectively. If $n > N = \text{Max}(N_1, N_2)$ then

$$|x_n y_n - lm| < (K + |m|)\varepsilon.$$

In view of Remark 10.2 this proves (iii).

(iv) It suffices to prove that $\dfrac{1}{y_n} \to \dfrac{1}{m}$ because as soon as this is established (iv) follows by applying (iii) to y_n replaced by $\dfrac{1}{y_n}$. Let us choose $\varepsilon = \dfrac{m}{2}$ then there exists N such that (10.16) holds for $n > N$. Consequently

$$|y_n| = |m - (m - y_n)| \ge |m| - |y_n - m| > \frac{m}{2} > 0,$$

for $n > N$. In particular $y_n \ne 0$. It follows that

$$\frac{1}{|y_n|} < \frac{2}{m},$$

for $n > N$. Now we have

$$\left| \frac{1}{y_n} - \frac{1}{m} \right| = \frac{|y_n - m|}{|y_n||m|} < \frac{2}{m^2} |y_n - m|.$$

By Remark 10.1 and (K) of Section 10.1

$$\frac{2}{m^2} |y_n - m| \to 0$$

and by Corollary 10.3.1

$$\frac{1}{y_n} \to \frac{1}{m}.$$ □

Theorem 10.6 is important in itself for the further development of the theory but it is also a powerful tool for calculating limits.

Example 10.7

$$\lim_{n\to\infty} \frac{n^2 + 4n - 7}{2n^2 - n + 1} = \lim_{n\to\infty} \frac{1 + \frac{4}{n} - \frac{7}{n^2}}{2 - \frac{1}{n} + \frac{1}{n^2}} = \frac{\lim_{n\to\infty} \left(1 + \frac{4}{n} - \frac{7}{n^2}\right)}{\lim_{n\to\infty} \left(2 - \frac{1}{n} + \frac{1}{n^2}\right)} = \frac{1}{2}.$$

Example 10.8   We wish to find

$$\lim_{n\to\infty} \frac{1 - 2 + 3 - 4 + \cdots - 2n}{\sqrt[3]{n^3 + 2n}}.$$

First we realize that $1 - 2 + 3 - 4 + \cdots - 2n = -n$, then we estimate

$$-1 < \frac{1 - 2 + 3 - \cdots - 2n}{\sqrt[3]{n^3 + 2n}} < -\frac{n}{n+1}. \tag{10.18}$$

Now

$$\lim_{n\to\infty} \frac{n}{n+1} = \lim_{n\to\infty} \frac{1}{1 + \frac{1}{n}} = 1.$$

Using the Squeeze principle and Equation (10.18) above yields

$$\lim_{n\to\infty} \frac{1 - 2 + 3 - 4 + \cdots - 2n}{\sqrt[3]{n^3 + 2n}} = -1.$$

Theorem 10.7   If $x_n \to l$ and $y_n \to m$ with $x_n, y_n \in \mathbb{R}$ and $l < m$ then there exists N such that

$$x_n < y_n,$$

for all $n > N$.

Proof. Let $\epsilon = \dfrac{m-l}{2}$, then there exists N_1 and N_2 such that

$$l - \varepsilon = \frac{l-m}{2} < x_n < l + \varepsilon = \frac{l+m}{2},$$

and

$$m - \varepsilon = \frac{l+m}{2} < y_n < m + \varepsilon,$$

for $n > N_1$ and $n > N_2$, respectively. For $n > N = \text{Max}(N_1, N_2)$ we obtain from these inequalities

$$x_n < \frac{l+m}{2} < y_n.$$

$\square$

A memory aid for Theorem 10.7 is: Inequality between limits implies a similar inequality ultimately for the terms of the sequences.

Corollary 10.7.1 *If $y_m \to m > 0$ then there exists N such that $y_n > 0$ for $n > N$.*

Proof. It suffices to take $x_n = l = 0$ for every n in Theorem 10.7. $\sqcap$

Corollary 10.7.2 *If $x_n \to l < m$ then there exists N such that $x_n < m$ for $n > N$.*

Proof. It suffices to take $y_n = m$ for every n in Theorem 10.7. $\square$

Theorem 10.8 *If $x_n \to l$, $y_n \to m$ with $x_n, y_n \in \mathbb{R}$ and if $x_n \leq y_n$ for $n \in \mathbb{N}$ then $l < m$.*

Proof. If $m < l$ then by Theorem 10.7 (with the roles of x_n and y_n interchanged) $y_n < x_n$ for some $n \in \mathbb{N}$, contrary to our assumption. $\square$

Taking $y_n = m$ yields

Corollary 10.8.1 *If $x_n \to l$ and $x_n \leq m$ then $l \leq m$.*

The next two examples illustrate the role of strict and non-strict inequalities in Theorems 10.7 and 10.8.

Example 10.9 If $x_n = 0$ and $y_n = -1/n$ then $l \leq m$ since $l = m = 0$. However it is not true that $x_n \leq y_n$ for sufficiently large n; we even have $y_n < x_n$ for all $n \in \mathbb{N}$.

Example 10.10   If $x_n = -1/n$ and $y_n = 1/n$ then $x_n < y_n$ for all $n \in \mathbb{N}$. However the inequality $l < m$ is false since $l = m = 0$.

Exercises

Exercise 10.2.1   *Find the following limits*

(1) $\displaystyle\lim_{n\to\infty} \frac{(n+2)^2}{3n^2}$;

(2) $\displaystyle\lim_{n\to\infty} \frac{1000n^3 - 2n + 1}{\frac{1}{10}n^5 - 100n^3 + 17}$;

(3) $\displaystyle\lim_{n\to\infty} \frac{(3n+2)^2 - (n+1)^3}{(3n+2)^3 + (n+1)^3}$

(4) $\displaystyle\lim_{n\to\infty} \left(\frac{1+2+3+\cdots+n}{n+2} - \frac{n}{2} \right)$.

(i) **Exercise 10.2.2**   *Prove: For every $c \in \mathbb{R}$ there are sequences $x_n \to c$, $y_n \to c$ with $x_n \in \mathbb{Q}$, $y_n \notin \mathbb{Q}$. [Hint: Consider $\lfloor nc \rfloor/n$ and $\lfloor nc\sqrt{2} \rfloor/(n\sqrt{2})$.]*

Exercise 10.2.3   *Prove that*

$$\lim_{n\to\infty} \frac{\lfloor x \rfloor + \lfloor 2x \rfloor + \cdots \lfloor nx \rfloor}{n^2} = \frac{x}{2}.$$

[Hint: Use the inequalities $\lfloor kx \rfloor \le kx$, $kx - 1 < \lfloor kx \rfloor$ and the Squeeze principle.]

(i) **Exercise 10.2.4**   *If $x_n \to l \ne 0$ then $\dfrac{x_{n+1}}{x_n} \to 1$. Prove this. Also give examples of convergent sequences $n \mapsto x_n$ such that*

(1) $\dfrac{x_{n+1}}{x_n}$ *diverges;* (ii) $\dfrac{x_{n+1}}{x_n} \to c \ne 1$.

Exercise 10.2.5   *Give examples of divergent sequences x and y such that (i) $x + y$ is convergent; and (ii) xy is convergent.*

Exercise 10.2.6   *For $A > 0$, $B > 0$ and $q \in \mathbb{N}$ prove*

$$\left| \sqrt[q]{A} - \sqrt[q]{B} \right| \le \frac{|A - B|}{\displaystyle\sum_{k=0}^{q-1} \left(\sqrt[q]{A} \right)^{q-1-k} \left(\sqrt[q]{B} \right)^k} \le \frac{|A - B|}{\sqrt[q]{B^{q-1}}}.$$

(i) **Exercise 10.2.7**   *Using the previous exercise and the Squeeze principle show that*

$$x_n \to l > 0 \Rightarrow \sqrt[q]{x_n} \to \sqrt[q]{l},$$

for $q \in \mathbb{N}$.

(i) **Exercise 10.2.8**   *Use the previous exercise to prove that $x_n^r \to l^r$ if $x_n \to l$ and $r \in \mathbb{Q}$.*

(!) **Exercise 10.2.9**   *Find the following limits*

(1) $\lim\limits_{n \to \infty} \left(\sqrt{n^2 + 1} - \sqrt{n^2 - 1} \right);$

(2) $\lim\limits_{n \to \infty} \sqrt[3]{\dfrac{n^2 + 2n + 7}{8n^2 - 1}};$

(3) $\lim\limits_{n \to \infty} \left(\sqrt{n^2 + 2n + 3} - \sqrt{n^2 + n - 7} \right);$

(4) $n^{-1} \left(\sqrt{4n^2 + n - 1} - \sqrt[3]{n^3 + 5} \right);$

(5) $\lim\limits_{n \to \infty} \sqrt[5]{n^3} \left(\sqrt[5]{n^2 + 2n + 3} - \sqrt[5]{n^2 + n - 9} \right);$

(6) $\lim\limits_{n \to \infty} \left(\sqrt{n^2 + 6n} - \sqrt[3]{n^3 + 3n^2} \right);$

(7) $\lim\limits_{n \to \infty} \left(\sqrt{n^2 + n - 1} - \sqrt[3]{n^3 + 5} \right);$

(8) $\lim\limits_{n \to \infty} n \left(\sqrt{n^2 + 2n} - \sqrt[3]{n^3 + 3n^2} \right);$

(9) $\lim\limits_{n \to \infty} \left(n + \sqrt[3]{1 - n^3} \right).$

[Hint: If you are finding some exercises difficult, we will show how to use *Maple* to solve them in the next Section.]

(i) **Exercise 10.2.10**   *Prove: If x, y are real sequences with $x_n \to l$ and $y_n \to m$ then $\mathrm{Max}(x_n, y_n) \to \mathrm{Max}(l, m)$ and $\mathrm{Min}(x_n, y_n) \to \mathrm{Min}(l, m)$.* [Hint: use either Definition 10.1 or the identity $\mathrm{Max}(a, b) = \frac{1}{2}(a + b + |a - b|)$.]

10.3   Limits of sequences in *Maple*

Maple can find limits of sequences quickly and efficiently even if the limit is hard to find otherwise. Simply, instead of writing $\lim\limits_{n \to \infty} f(n)$ we issue the command `limit(f(n),n=infinity)`. On most occasions this would be sufficient but the command is incomplete: we have to tell *Maple* that n (or m or whatever notation we use) is a positive integer. This can be done either with **assume** or **assuming**. We illustrate both possibilities. We wish to calculate $\lim\limits_{n\to\infty} \dfrac{(n+2)^2}{3n^2}$ and $\lim\limits_{n\to\infty} (\sqrt{n^2+2n+3} - \sqrt{n^2+n-1})$.

```
> assume(n::posint);
> limit(((n+2)^2)/(3*n^2),n=infinity);
```
$$\frac{1}{3}$$
```
> limit(sqrt(n^2+2*n+3)-sqrt(n^2+n-1),n=infinity);
```
$$\frac{1}{2}$$
```
> n:='n';
```
$$n := n$$

It is a good habit to clear the assigned variable if it is not needed any more. For calculating just one limit it is more convenient to use **assuming**. In the next example we need a cubic root. The command to use for odd roots is **surd**, since using fractional exponents can bring in complex roots which are undesirable. Now we calculate the limit from Exercise 10.2.9 part (ix).

```
> limit(m+surd(1-m^3,3),m=infinity) assuming m::posint;
```
$$0$$
```
> limit((k^2)*(k+surd(1-k^3,3)),k=infinity) assuming
k::posint;
```
$$\frac{1}{3}$$

Neglecting the **assuming** part of the command can lead to incorrect results. For instance $\lim\limits_{n\to\infty} \sin(n\pi) = 0$ but

```
> limit(sin(n*Pi),n=infinity);
```
$$-1..1$$

This result does not look right. We have to accept that we ask a wrong question and got a wrong answer.

Exercises

Exercise 10.3.1   Use Maple *to evaluate the limits in Exercise 10.2.9.*

(i) **Exercise 10.3.2**   *What happens if you use* Maple *to solve Exercise 10.2.3.* [Hint: *Maple* cannot solve this problem directly.]

Exercise 10.3.3   Let $x_n = \dfrac{3n}{2n+1}$ and $y_n = \sqrt[3]{n^3 + 6n^2 + 8n} - \sqrt[3]{n^3 + n^2 - 9}$. *Find* $\lim\limits_{n\to\infty} \text{Max}(x_n, y_n)$.

10.4   Monotonic sequences

We shall assume in this section that all functions (and sequences) are real valued and we shall not repeat this assumption all the time.

Definition 10.3 (Increasing and decreasing functions)   A function f is said to be *increasing on a set* S if for every $x_1,\ x_2 \in S$

$$x_1 \leq x_2 \Rightarrow f(x_1) \leq f(x_2). \tag{10.19}$$

f is said to be *increasing* if it is increasing on dom f. By replacing the implication (10.19) by

$$x_1 < x_2 \Rightarrow f(x_1) < f(x_2)$$

we obtain the definition of a *strictly increasing* f (on S), and by reversing the inequality $f(x_1) \leq f(x_2)$ or $f(x_1) < f(x_2)$ we obtain the definition of a *decreasing* f or a *strictly decreasing* f, respectively.

For example the function $x \mapsto x^2$ is strictly increasing on the interval $]0, \infty[$, the function $x \mapsto x^3$ is strictly increasing and $1_{\mathbb{P}}$, that is, the characteristic function of the interval $]0, \infty[$, is increasing but not strictly increasing.

Sequences are functions defined on $\mathbb{N}$. Consequently Definition 10.3 applies to sequences as well. Clearly a *sequence* x is *increasing* if and only if $x_n \leq x_{n+1}$ for every $n \in \mathbb{N}$. It should be clear what is meant by a *strictly increasing, decreasing* or *strictly decreasing* sequence.

A function which is either increasing or decreasing is called monotonic. Monotonic sequences play an important role in the theory of limits.

Theorem 10.9   *Every bounded monotonic sequence is convergent.*

Since a strictly increasing sequence is increasing, a strictly decreasing sequence is decreasing, and $-x$ is increasing if x is decreasing, it is sufficient to prove

---

Theorem 10.10   *If $n \mapsto x_n$ is increasing and bounded above then it is convergent and*

$$\lim_{n \to \infty} x_n = \sup\{x_1,\ x_2,\ x_3, \ldots\}. \qquad (10.20)$$

Proof.   Denote by l the supremum on the right hand side of (10.20). Let $\epsilon > 0$. Since $l - \epsilon < l$ there exists, by definition of the least upper bound, a number N such that $l - \epsilon < x_N \leq l$. Since $x_n \geq x_N$ for $n > N$ we have

$$l - \varepsilon < x_n < l + \varepsilon$$

for $n > N$.   □

Remark 10.3   If $n \mapsto x_n$ is decreasing and bounded below then

$$\lim_{n \to \infty} x_n = \inf\{x_1,\ x_2,\ x_3, \ldots\}.$$

Remark 10.4   For an increasing sequence $n \mapsto x_n$ with limit l we shall write $x_n \uparrow l$, and similarly if x_n decreases with limit l we shall write $x_n \downarrow l$.

Although Theorems 10.9 and 10.10 sound theoretical their application often enables us to find limits.

Example 10.11   We show for $b \in \mathbb{C}$ that $b^n \to 0$ if $|b| < 1$. We define $x_n = |b|^n$. Obviously $x_{n+1} = |b|x_n$, hence the sequence $n \mapsto x_n$ is decreasing, it is also bounded below by zero so it has a limit, say l. Passing to the limit in the previous equation gives $l = |b|l$. Consequently $l = 0$ and $|b|^n \downarrow 0$. By (K) of Section 10.1 it follows that $b^n \to 0$.

Remark 10.5   If $|c| > 1$ then $n \mapsto c^n$ cannot be bounded. If it were then $|c^n| \uparrow l$ and the same argument as in Example 10.11 would show that $c^n \to 0$, which is clearly impossible.

Example 10.12   We show that

$$\sqrt[n]{a} \to 1$$

for $a > 0$. The case $a = 1$ is obvious.

Consider first $a > 1$. The sequence $n \to \sqrt[n]{a}$ is decreasing and bounded from below by 1. Hence it is convergent, $\sqrt[n]{a} \downarrow c$. Then $1 \le c \le \sqrt[n]{a}$ and consequently $1 \le c^n \le a$. This means that the sequence $n \to c^n$ is bounded and therefore $|c| \le 1$. Hence we have $|c| \le 1 \le c \le \sqrt[n]{a}$, so clearly $c = 1$.

If $0 < a < 1$ then $1/a = b > 1$. By Equation (5.33) and by what we have just proved $\sqrt[n]{b} \to 1$. By Remark 10.1 we have $\sqrt[n]{a} = \dfrac{1}{\sqrt[n]{b}} \to 1$, proving the result.

The use of the existence assertion from Theorem 10.9 or Theorem 10.10 in Examples 10.11 and 10.12 is essential. Consider the sequence $n \mapsto (-1)^n$ and denote its limit l. Since $(-1)^{n+1} = -(-1)^n$ we obtain by passage to the limit that $l = -l$, that is $l = 0$. This conclusion is false since the sequence in question has no limit.

Example 10.13 The sequence $n \mapsto \left(1 + \dfrac{1}{n}\right)^n$ is convergent and its limit is denoted by e. This number plays an important part in mathematical analysis. We first prove that the sequence is bounded. By the binomial theorem

$$\left(1 + \frac{1}{n}\right)^n = \sum_{k=0}^{n} \binom{n}{k} \frac{1}{n^k}$$

$$= 1 + n\frac{1}{n} + \frac{1}{2!}\left(1 - \frac{1}{n}\right) + \frac{1}{3!}\left(1 - \frac{1}{n}\right)\left(1 - \frac{2}{n}\right) + \cdots$$

$$+ \frac{1}{n!}\left(1 - \frac{1}{n}\right)\left(1 - \frac{2}{n}\right) \cdots \left(1 - \frac{n-1}{n}\right)$$

$$< 1 + 1 + \frac{1}{2!} + \frac{1}{3!} + \cdots + \frac{1}{n!}. \quad (10.21)$$

For $n > 2$ we have $n! > 2^{n-1}$ and consequently

$$\left(1 + \frac{1}{n}\right)^n < 1 + 1 + \frac{1}{2!} + \cdots + \frac{1}{n!}$$

$$< 1 + 1 + \frac{1}{2} + \frac{1}{2^2} + \cdots + \frac{1}{2^{n-1}} < 3 \quad (10.22)$$

We next prove that the sequence is increasing. In the arithmetic-geometric mean inequality (Theorem 5.10)

$$\sqrt[n+1]{a_1 a_2 \ldots a_{n+1}} \le \frac{a_1 + a_2 + \cdots + a_{n+1}}{n+1} \quad (10.23)$$

set $a_1 = a_2 = \cdots = a_n = 1 + \dfrac{1}{n}$, $a_{n+1} = 1$ and obtain

$$\left(1 + \frac{1}{n}\right)^n \le \left(\frac{n+2}{n+1}\right)^{n+1} = \left(1 + \frac{1}{n+1}\right)^{n+1}. \tag{10.24}$$

Example 10.14   Let $y_n = 1 + 1 + \dfrac{1}{2!} + \dfrac{1}{3!} + \cdots \dfrac{1}{n!}$. The sequence y_n is obviously increasing and by (10.22) it is also bounded. Thus it has a limit, $y_n \uparrow E$, say. We are going to show that $E = e$. Firstly, we have from (10.21), on using Theorem 10.8, that $e \le E$. let us now estimate

$$\left(1 - \frac{1}{n}\right)\left(1 - \frac{2}{n}\right)\cdots\left(1 - \frac{r-1}{n}\right)$$

from below. Using Bernoulli's inequality (Theorem 5.9) we have

$$\ge \left(1 - \frac{1}{n}\right)\left(1 - \frac{2}{n}\right)\cdots\left(1 - \frac{r-1}{n}\right) \ge$$

$$\left(1 - \frac{r-1}{n}\right)^{r-1} \ge 1 - \frac{(r-1)^2}{n} \ge 1 - \frac{r(r-1)}{n}$$

for $r, n \in \mathbb{N}$, $1 < r \le n$. Hence

$$\left(1 + \frac{1}{n}\right)^n \ge 2 + \sum_{r-2}^{n}\left(1 - \frac{r(r-1)}{n}\right)\frac{1}{r!} \ge y_n - \frac{1}{n}\sum_{r=2}^{n}\frac{1}{(r-2)!} \ge y_n - \frac{3}{n},$$

since $\displaystyle\sum_{r-2}^{n}\frac{1}{(r-2)!} < 3$ by (10.22). Passing to the limit in the last inequality and using Theorem 10.8 again we have $e \ge E$. Consequently $e = E$ and hence

$$\lim_{n\to\infty}\left(1 + 1 + \frac{1}{2!} + \frac{1}{3!} + \cdots + \frac{1}{n!}\right) = e.$$

This example enables us to calculate e with high accuracy. Firstly we have $y_n < e$. Secondly, for $m > n$, we have

$$y_m = y_n + \frac{1}{(n+1)!} + \frac{1}{(n+2)!} + \cdots + \frac{1}{m!}$$

$$\le y_n + \frac{1}{n!}\left(\frac{1}{n+1} + \frac{1}{(n+1)^2} + \cdots + \frac{1}{(n+1)^{m-n}}\right)$$

$$\le y_n + \frac{1}{n!}\frac{1}{n+1}\frac{1}{1 - \frac{1}{n+1}} = y_n + \frac{1}{n!n}.$$

This shows that $y_n + \dfrac{1}{n!n}$ is an upper bound for y_m with $m > n$ and therefore $e \le y_n + \dfrac{1}{n!n}$. Combining these results yields

$$1 + 1 + \frac{1}{2!} + \frac{1}{3!} + \cdots + \frac{1}{n!} \le e \le 1 + 1 + \frac{1}{2!} + \frac{1}{3!} + \cdots + \frac{1}{n!} + \frac{1}{n!n}. \quad (10.25)$$

The number $1/(n!n)$ decreases rapidly, for instance $\dfrac{1}{10!10} < \dfrac{3}{10^8}$, hence $1 + 1 + \dfrac{1}{2!} + \cdots + \dfrac{1}{10!}$ approximates e with an error less than 10^{-7}. The calculation will show that $e \doteq 2.7182818$. With the aid of computers e has been evaluated to billions of decimal places. With the commands `Digits:=200` and `exp(1)` you obtain e accurately for 200 decimal digits.

We now prove that e is irrational. It follows from Inequalities (10.25) that $0 < e - y_n \le 1/(n!n)$. If e were rational then $e = m/n$ for some $n, m \in \mathbb{N}$ and then

$$0 < \left(\frac{m}{n} - y_n\right) \cdot n! \le \frac{1}{n}.$$

The middle term in this inequality is an integer and this is a contradiction.

The French mathematician C. Hermite proved more, namely that e cannot be a root a of an algebraic equation with rational coefficients. This fact is expressed by saying that e is not algebraic or that e is transcendental.[2] The proof is out of our reach at this stage and the interested reader can find the proof in Spivak (1967, Chapter 20).

Exercises

Exercise 10.4.1   *Prove that the sequence*

$$n \mapsto \frac{1}{n+1} + \frac{1}{n+2} + \cdots + \frac{1}{2n}$$

has a limit. [Hint: Show that the sequence is increasing and bounded above by 1.]

---

[2] A number is called algebraic if it is a root of an algebraic equation with rational coefficients. By multiplying the relevant equation by a sufficiently large integer it will be then a root of an equation with integer coefficients. Trivially every rational number a/b is algebraic as a root of the equation $bx = a$. All concrete irrational numbers encountered so far in this book were algebraic and it is not easy to produce a transcendental number.

Exercise 10.4.2   *Prove that* $\lim\limits_{n\to\infty} \dfrac{c^n}{n!} = 0$ *for any* $c \in \mathbb{C}$. [Hint: Prove $|c|^n/(n!) \downarrow 0$ by an argument similar to that of Example 10.11 and then use (K) of Section 10.1.]

Exercise 10.4.3   *Find the following limits*

(1) $\lim\limits_{n\to\infty} \left(1 + \dfrac{1}{n}\right)^{n+5}$;

(2) $\lim\limits_{n\to\infty} \left(1 + \dfrac{1}{n}\right)^{2n}$;

(3) $\lim\limits_{n\to\infty} \left(1 - \dfrac{1}{n}\right)^{n}$;

(4) $\lim\limits_{n\to\infty} \left(1 - \dfrac{1}{n^2}\right)^{n}$;

(5) $\lim\limits_{n\to\infty} \left(1 + \dfrac{1}{n^2}\right)^{n}$.

[Hint:

(1) $\left(1 + \dfrac{1}{n}\right)^{n+5} = \left(1 + \dfrac{1}{n}\right)^{n}\left(1 + \dfrac{1}{n}\right)^{5}$;

(2) $\left(1 + \dfrac{1}{n}\right)^{2n} = \left[\left(1 + \dfrac{1}{n}\right)^{n}\right]^{2}$;

(3) $\left(1 - \dfrac{1}{n}\right)^{n} = \left(1 + \dfrac{1}{n-1}\right)^{-n}$;

(4) $\left(1 - \dfrac{1}{n^2}\right) = \left(1 + \dfrac{1}{n}\right)\left(1 - \dfrac{1}{n}\right)$;

(5) Show that $\left(1 + \dfrac{1}{n^2}\right)^{n^2} \le e$, then $\left(1 + \dfrac{1}{n^2}\right)^{n} \le \sqrt[n]{e}$.

Use the Squeeze principle and Example 10.12.]

Exercise 10.4.4   *Find the following limits*

(1) $\lim\limits_{n\to\infty} \dfrac{2^{n+1} + 3^n}{2^n + 3^{n+1}}$.

(2) $\lim\limits_{n\to\infty} \dfrac{5^n + 3^n}{5^n + 1}$.

(3) $\lim\limits_{n\to\infty} \dfrac{2^n + 3^n}{3^n + 5^n}$.

[Hint: Divide by the dominating term.]

Exercise 10.4.5   *Find the following limits*

(1) $\lim\limits_{n\to\infty} \sqrt[2n]{a}$ for $a > 0$;

(2) $\lim\limits_{n\to\infty} \sqrt[n]{3 + (-1)^n}$;

(3) $\lim\limits_{n\to\infty} \sqrt[n]{2^n + 5^n + 13^n}$;

(4) $\lim\limits_{n\to\infty} \sqrt[n]{n + 6^n}$

[Hint: For (2)–(4) use the Squeeze principle.]

Ⓘ **Exercise 10.4.6**   Let $a > 0$, $x_1 = 1$, $x_{n+1} = \dfrac{1}{2}\left(x_n + \dfrac{a}{x_n}\right)$. Prove that $\lim\limits_{n\to\infty} x_n$ exists and equals $\sqrt{a}$. [Hint: $x_{n+1} \geq \sqrt{a}$ by the arithmetic-geometric mean inequality. Show that the sequence is decreasing after the second term, then pass to the limit in the defining equation.]

Ⓘ **Exercise 10.4.7**   Prove that $na^n \to 0$ for $|a| < 1$. [Hint: Use the method of Example 10.11. Let $x_n = n|a|^n$. Then $x_{n+1}/x_n = (n+1)|a|/n$ and consequently the sequence $n \mapsto x_n$ is decreasing.]

Ⓘ **Exercise 10.4.8**   Let $p \in \mathbb{N}$. Prove that $n^p a^n \to 0$ for $|a| < 1$. [Hint: use a similar procedure to that used in the previous exercise.]

Exercise 10.4.9   Prove that if $x_n > 0$ for $n \in \mathbb{N}$ and $\dfrac{x_{n+1}}{x_n} \to l < 1$ then $x_n \to 0$. [Hint: $x_{n+1} < x_n$ for large n, if $x_m \to L$ then $L = lL$.]

10.5 Infinite limits

In this section we shall consider only *real* sequences. Consider the first few terms of the following sequences.

$$n \mapsto a_n : \ 1, 2, 4, 8, 16, \ldots ; \tag{10.26}$$

$$n \mapsto b_n : \ 1, 4, \frac{3}{2}, 16, \frac{5}{2}, 36, \frac{7}{2}, \ldots ; \tag{10.27}$$

$$n \mapsto c_n : \ 1, 2, 1, 2, 1, 2, \ldots ; \tag{10.28}$$

$$n \mapsto d_n : \ 1, 2, 1, 3, 1, 4, 1, \ldots ; \tag{10.29}$$

None of these sequences has a limit but (10.26) and (10.27) behave quite differently compared with (10.28) or (10.29). It is natural to say that the terms of sequences (10.26) and (10.27) increase without limit. We would not say that about (10.29) although this sequence, too, is obviously unbounded.

Definition 10.4 (Positive infinite limits)   A sequence x is said to have the *infinite limit* $+\infty$ (or just ∞) if for every K there exists N such that

$$x_n > K \tag{10.30}$$

for all $n > N$. If x has the infinite limit ∞ we write $\lim\limits_{n \to \infty} x_n = \infty$ or $x_n \to \infty$.

The limit in the sense of Definition 10.1 is sometimes called finite, particularly if one wants to emphasize that the limit is not infinite.

Sequence (10.26) has the infinite limit ∞ since $a_n = 2^{n-1} \geq n$. For a given K it is sufficient to set $N = K$.

Sequence (10.27) also has limit ∞, for we have $b_n \geq n/2$ and it is sufficient to choose $N = 2K$.

Sequence (10.28) is bounded and therefore cannot have an infinite limit. We know that it has no finite limit either.

Sequence (10.29) is unbounded but has no limit either because all odd terms are equal 1.

Definition 10.5 (Negative infinite limits)   A sequence x is said to have the *infinite limit* $-\infty$ (or just $-\infty$) if for every k there exists N such that

$$x_n < k \tag{10.31}$$

for all $n > N$. If x has the infinite limit $-\infty$ we write $\lim\limits_{n \to \infty} x_n = -\infty$ or $x_n \to -\infty$.

Instead of saying that a sequence has limit ∞ (or $-\infty$) we may occasionally say that the sequence *diverges to* ∞ (or $-\infty$).

Obviously $\lim\limits_{n \to \infty} (-n) = -\infty$ whereas the sequence $n \mapsto (-1)^n n$ has no limit, finite or infinite.

We emphasize again that we have not introduced any infinite numbers. The statement $\lim\limits_{n \to \infty} x_n = \infty$ or $-\infty$ refers to the behaviour of the sequence x and the symbol ∞ itself has no meaning of its own. Perhaps we should add that in mathematics infinite numbers are indeed introduced and used advantageously, but very little, if anything could be gained by introducing them here and now.

There is some similarity between the definitions of finite and infinite limits. Some theorems concerning finite limits can be extended to infinite limits but not all can and care is needed. For instance, changing finitely many terms of the sequence affects neither the existence nor the value of the limit. Also a sequence can have at most one limit, finite or infinite. The next theorem is the analog of the Squeeze principle.

Theorem 10.11 If $x_n \geq y_n$ for $n \in N$ and $y_n \to \infty \, (x_n \to -\infty)$ then $x_n \to \infty \, (y_n \to -\infty)$ also.

Proof. Let K be given. There exists N such that $y_n > K$ for $n > N$. It follows from the assumption that $x_n > K$ for $n > N$, which proves that $x_n \to \infty$. The proof for $x_n \to -\infty$ is similar. □

Theorem 10.12 If $x_n \to \infty \, (-\infty)$ and y_n is bounded, then $x_n + y_n \to \infty \, (-\infty)$.

Proof. There exists M such that $|y_n| < M$ for all $n \subset \mathbb{N}$. Since $x_n \to \infty$ for an arbitrary K there exists N such that $x_n > K + M$ for $n > N$. Consequently $x_n + y_n > K + M - |y_n| > K$ for $n > N$. The proof for $x_n \to -\infty$ is similar. □

If $n \to y_n$ is not bounded (for instance if $y_n \to -\infty$) then the conclusion of Theorem 10.12 need not hold. Let $x_n = 2n$, $y_n = a - 2n$, $z_n = -3n$, $u_n = -n$, $w_n = -(2 + (-1)^n)n$. Then $x_n \to \infty$ and all the other sequences diverge to $-\infty$, while

$$x_n + y_n \to a, \tag{10.32}$$

$$x_n + z_n \to -\infty, \tag{10.33}$$

$$x_n + u_n \to +\infty \tag{10.34}$$

$$x_n + w_n \quad \text{has no limit, finite or infinite.} \tag{10.35}$$

Theorem 10.13 Every monotonic sequence has a limit (finite or infinite).

Proof. It suffices to consider an increasing sequence $n \mapsto x_n$. If it is bounded then it has a finite limit by Theorem 10.10. If it is not bounded then for every K there exists $N \in \mathbb{N}$ such that $x_N > K$. Since $n \to x_n$ is increasing it follows that $x_n > K$ for $n > N$. □

Exercises

ⓘ **Exercise 10.5.1**   *Prove that the meaning of Definition 10.4 remains unchanged if*

(1) *the inequality $n > N$ is replaced by $n \geq N$;*
(2) *it is additionally required that $N \in \mathbb{N}$;*
(3) *it is additionally required that $K \in \mathbb{N}$;*
(4) *the inequality $x_n > K$ is replaced by $x_n \geq K$;*
(5) *it is additionally required that for some fixed number a (for example, $a = 0$) $K > a$.*

Some of the following exercises are stated as theorems. The task is to prove them.

ⓘ **Exercise 10.5.2**   *If $x_n \to \infty$ and $y_n \to c$ and $c > 0$ then $x_n y_n \to \infty$. If $c < 0$ then $x_n y_n \to -\infty$.*

Exercise 10.5.3   *Give examples of x_n and y_n such that $x_n \to \infty$, $y_n \to 0$ and*

(1) *$x_n y_n \to \infty$;*
(2) *for a given $a \in \mathbb{R}$ it is true that $x_n y_n \to a$.*
(3) *the sequence $n \to x_n y_n$ has no limit, finite or infinite.*

ⓘ **Exercise 10.5.4**   *Prove: if $x_n \to \infty$ or $x_n \to -\infty$ then $|x_n| \to \infty$.*

ⓘ **Exercise 10.5.5**   *Prove: if $|x_n| \to \infty$ then $\dfrac{1}{x_n} \to 0$.*

ⓘ **Exercise 10.5.6**   *Prove: if $a > 1$ then $a^n \to \infty$.*

Exercise 10.5.7   *Find the following limits*

(1) $\lim\limits_{n \to \infty} \dfrac{n^2 + 3n - 10000}{n + \sqrt{n}}$;

(2) $\lim\limits_{n \to \infty} \dfrac{1 - n^3}{n^2 + 10n - 1}$;

(3) *for $\alpha \geq 0$ evaluate $\lim\limits_{n \to \infty} n^\alpha \left(\sqrt{n^2 + 1} - 1 \right)$.*

Exercise 10.5.8   *If $x_n > 0$ and $\sqrt[n]{x_n} \to l > 1$ then $x_n \to \infty$.*

Exercise 10.5.9   *If $x_n > 0$ and $\dfrac{x_{n+1}}{x_n} \to l > 1$ then $x_n \to \infty$.*

10.6   Subsequences

If $k \mapsto n_k$ is a strictly increasing sequence of natural numbers, that is $n_k \in \mathbb{N}$ and $n_k < n_{k+1}$, then the sequence $k \mapsto x_{n_k}$ is called a *subsequence* of the sequence $n \mapsto x_n$. If we list the terms of the sequence

$$x_1, \ x_2, \ x_3, \ \ldots \tag{10.36}$$

we obtain a record of the terms of a subsequence by omitting some members in (10.36). For instance

$$x_2, \ x_5, \ x_9, \ x_{14}, \ x_{20}, \ \ldots$$

are terms of a subsequence of (10.36).

It is intuitive that if x_n approaches a number then so does x_{n_k}. This leads to

> **Theorem 10.14**   *If a complex sequence x has a limit l then any subsequence $k \mapsto x_{n_k}$ also has the limit l.*

Proof.   For $\epsilon > 0$ there exists N such that $|x_n - l| < \varepsilon$ for $n > N$. If $k > N$ then $n_k \geq k > N$ and therefore

$$|x_{n_k} - l| < \varepsilon.$$

This proves that $\lim\limits_{n \to \infty} x_{n_k} = l$. $\square$

One can prove similarly that if a sequence diverges to ∞ (or to $-\infty$) the so does every subsequence.

Theorem 10.14 can often be used conveniently for proving that a given sequence is not convergent by finding two subsequences with distinct limits. For instance the sequence $n \mapsto 1 + (-1)^n$ has two subsequences $k \mapsto 1 + (-1)^{2k} \to 2$ and $k \mapsto 1 + (-1)^{2k+1} \to 0$ and is therefore divergent.

Example 10.15

$$\sqrt[n]{n} \to 1.$$

Since for $n \geq 3$

$$\left(1 + \frac{1}{n}\right)^n \leq 3 \leq n,$$

$$(n+1)^n \leq n^{n+1},$$

$$(n+1)^{1/(n+1)} \leq n^{1/n},$$

the sequence is decreasing after the third term. It is also obviously bounded below, hence convergent. Let the limit be α. The subsequence $k \mapsto \sqrt[2k]{2k}$ has by Theorem 10.14 the same limit. By Remark 10.1, Exercise 10.2.7 and Example 10.12 we have $\sqrt[k]{2k} = \left(\sqrt[2k]{2k} \right)^2 \to \alpha^2$ and also $\sqrt[k]{2k} = \sqrt[k]{2}\sqrt[k]{k} \to \alpha$, so we have $\alpha = \alpha^2$. Moreover $\alpha \geq 1$ since $\sqrt[n]{n} \geq 1$. Consequently $\alpha = 1$.

Exercises

ⓘ **Exercise 10.6.1** *Use the method of Example 10.15 to show that*

(1) $a^n \to 0$ for $0 < a < 1$;

(2) $\sqrt[n]{b} \to 1$ for $0 < b$;

(3) $\displaystyle\lim_{n \to \infty} \left(1 + \frac{3}{n} \right)^n = e^3$;

(4) $\displaystyle\lim_{n \to \infty} \left(1 + \frac{3}{5n} \right)^n = e^{\frac{3}{5}}$;

❗ **Exercise 10.6.2** *Find* $\displaystyle\lim_{n \to \infty} \left(\sqrt[n]{n} - 1 \right)^n$. [Hint: Use Example 10.15 and the Squeeze principle.]

❗ **Exercise 10.6.3** *Prove that* $\displaystyle\lim_{n \to \infty} n \left(\sqrt[n]{n} - 1 \right) = \infty$. [Hint: For $K \in \mathbb{N}$ we have $\left(1 + \dfrac{K}{n} \right)^n \to e^K < n$ for large n. It follows then $K < n \left(\sqrt[n]{n} - 1 \right)$.]

10.7 Existence theorems

Monotonic sequences[3] have the nice property that they always have limits. With a sequence $n \mapsto x_n \in \mathbb{R}$ bounded below we can associate an increasing sequence $n \mapsto \alpha_n$ as follows:

$$\alpha_n = \inf\{x_n, x_{n+1}, x_{n+2}, \ldots\}. \tag{10.37}$$

Similarly, if the sequence is bounded above, there is a decreasing sequence

$$\beta_n = \sup\{x_n, x_{n+1}, x_{n+2}, \ldots\}. \tag{10.38}$$

Obviously

$$\alpha_n \leq x_n \leq \beta_n \tag{10.39}$$

---

[3]Monotonic sequences are automatically assumed to be real.

for every $n \in \mathbb{N}$. We define: $\lim_{n\to\infty} \alpha_n$ as $\liminf_{n\to\infty} x_n$ and $\lim_{n\to\infty} \beta_n$ as $\limsup_{n\to\infty} x_n$. These are called respectively the *limit inferior* and *limit superior* of the sequence $n \mapsto x_n$. Clearly

$$\liminf_{n\to\infty} x_n \leq \limsup_{n\to\infty} x_n.$$

We say that $\limsup_{n\to\infty} x_n = \infty$ if $x_n \in \mathbb{R}$ and the sequence is not bounded above. We say $\limsup_{n\to\infty} x_n = -\infty$ if $\lim_{n\to\infty} x_n = -\infty$. We employ similar conventions for the limit inferior.

Example 10.16   Let $x_{3n-2} = 1$, $x_{3n-1} = 0$, $x_{3n} = -1$ for $n \in \mathbb{N}$. We have that

$$\{x_n, x_{n+1}, x_{n+2}, \ldots\} = \{1, 0 - 1\}$$

for every n and consequently $\alpha = -1$, $\beta = 1$, hence

$$\liminf_{n\to\infty} x_n = -1 \quad \text{and} \quad \limsup_{n\to\infty} x_n = 1$$

Example 10.17   If $x_n \uparrow l$ then clearly $\alpha_n = x_n$, $\beta_n = l$, $\limsup_{n\to\infty} x_n = \liminf_{n\to\infty} x_n = l$. Similarly if $x_n \downarrow l$ then $\alpha_n = l$, $\beta_n = x_n$ and again $\limsup_{n\to\infty} x_n = \liminf_{n\to\infty} x_n = l$.

Example 10.18   Let $x_n = (-1)^{n+1} \dfrac{n+2}{4\lfloor \frac{n}{4} \rfloor + 1}$. The following Table 10.2 displays x_n, α_n and β_n for the first few n. It is left as an exercise to show that $\limsup_{n\to\infty} x_n = 1$, $\liminf_{n\to\infty} x_n = -1$.

Table 10.2   Example of α_n and β_n

n	1	2	3	4	5	6	7	8	9	10	11	12
x_n	3	-4	5	$-\frac{6}{5}$	$\frac{7}{5}$	$-\frac{8}{5}$	$\frac{9}{5}$	$-\frac{10}{9}$	$\frac{11}{9}$	$-\frac{12}{9}$	$\frac{13}{9}$	$-\frac{14}{13}$
α_n	-4	-4	$-\frac{8}{5}$	$-\frac{8}{5}$	$-\frac{8}{5}$	$-\frac{8}{5}$	$-\frac{12}{9}$	$-\frac{12}{9}$	$-\frac{12}{9}$	$-\frac{12}{9}$	$-\frac{16}{13}$	$-\frac{16}{13}$
β_n	5	5	5	$\frac{9}{5}$	$\frac{9}{5}$	$\frac{9}{5}$	$\frac{9}{5}$	$\frac{13}{9}$	$\frac{13}{9}$	$\frac{13}{9}$	$\frac{13}{9}$	$\frac{17}{13}$

Theorem 10.15   *A real bounded sequence* $n \mapsto x_n$ *has a limit if and only if*

$$\limsup_{n \to \infty} x_n = \liminf_{n \to \infty} x_n$$

and then

$$\limsup_{n \to \infty} x_n = \liminf_{n \to \infty} x_n = \lim_{n \to \infty} x_n .$$

Proof.   Again we separate the proof into two parts.

I. If $\lim_{n \to \infty} x_n = l$ then for every $\epsilon > 0$ there exists N such that

$$l - \frac{\varepsilon}{2} < x_n < l + \frac{\varepsilon}{2}$$

for $n > N$. Consequently

$$l - \varepsilon < l - \frac{\varepsilon}{2} \le \alpha_n \le x_n \le \beta_n \le l + \frac{\varepsilon}{2} < l + \varepsilon.$$

This proves that

$$\lim_{n \to \infty} \alpha_n = \lim_{n \to \infty} \beta_n = l$$

II. If $\limsup_{n \to \infty} x_n = \liminf_{n \to \infty} x_n = l$ then $\lim_{n \to \infty} x_n = l$ by the Squeeze Principle since $\alpha_n \le x_n \le \beta_n$. $\square$

Remark 10.6   Theorem 10.15 can be extended to any real sequence. In other words, the Theorem remains valid if the word bounded is omitted. For instance, $\liminf_{n \to \infty} x_n = \infty$ if and only if $\lim_{n \to \infty} x_n = \infty$.

A *sequence* is said to be a *Cauchy sequence* or simply *Cauchy* if for every $\varepsilon > 0$ there exists N such that

$$|x_n - x_m| < \varepsilon \qquad (10.40)$$

for all $n > N$ and $m > N$. The limit of a sequence can be described informally as a number to which members of the sequence are ultimately approximately equal within a prescribed accuracy. A Cauchy sequence has the property that ultimately the members of the sequence are approximately equal to each other within any prescribed accuracy. It is fairly obvious that a convergent sequence is Cauchy. Indeed, if $x_n \to l$ then for every ε there exists N such that

$$|x_k - l| < \frac{\varepsilon}{2} \quad \text{for} \quad k > N.$$

Now we have

$$|x_n - x_m| \le |x_n - l| + |x_m - l| < \frac{\varepsilon}{2} + \frac{\varepsilon}{2} = \varepsilon,$$

for $n > N$ and $m > N$. The converse is also true, and is proved below, yielding

---

Theorem 10.16 (Bolzano–Cauchy) *A sequence with complex terms is convergent if and only if it is Cauchy.*

Proof. We need to prove the if part. Let $n \mapsto x_n$ be a Cauchy sequence. We first assume that x_n are real. There exists a natural N such that

$$|x_n - x_m| < 1,$$

for n and m greater than N. Choose a particular $m > n$ (for instance $m = N + 1$) then

$$|x_n| \le |x_m| + |x_n - x_m| \le |x_m| + 1,$$

for $n > N$. This means that all terms of the sequence with $n > N$ are bounded by the (fixed) number $|x_m| + 1$. Since the finite set of those x_n with $n \le N$ is also bounded, say $|x_n| < K$ for $n \le N$, we have

$$|x_n| < K + |x_m| + 1$$

for all n and hence the sequence is bounded. This allows us to make use of Theorem 10.15. Assume for an indirect proof that

$$\underline{l} = \liminf_{x_n \to \infty} x_n < \limsup_{x_n \to \infty} x_n = \bar{l}.$$

Set $\varepsilon = (\bar{l} - \underline{l})/2$. Since x_n is Cauchy there exists N_1 such that

$$x_n < x_m + \epsilon \tag{10.41}$$

for $n > N_1$ and $m > N_1$. In (10.41) we keep m fixed and run n through $k, k+1, \ldots$. We obtain

$$\beta_k = \sup \{x_k, x_{k+1}, x_{k+2}, \ldots\} \le x_m + \varepsilon.$$

By the definition of limit superior and by Corollary 10.8.1 on preservation of inequalities it follows

$$\bar{l} - \varepsilon \leq x_m.$$

$\bar{l} - \varepsilon$ is a lower bound for x_m for $m > N_1$ and therefore

$$\bar{l} - \varepsilon \leq \inf \{x_m, x_{m+1}, x_{m+2}\}.$$

Taking the limit as $m \to \infty$ gives $\bar{l} - \varepsilon \leq \underline{l}$. Since $\bar{l} - \varepsilon = (\underline{l} + \bar{l})/2$ it follows

$$\underline{l} < \frac{\underline{l} + \bar{l}}{2} \leq \underline{l}.$$

This contradiction completes the proof in case of a real sequence.

For a Cauchy sequence with complex terms x_n we define $u_n = \Re x_n$ and $v_n = \Im x_n$. Clearly

$$|u_n - u_m| \leq |x_n - x_m| \quad \text{and} \quad |v_n - v_m| \leq |x_n - x_m|.$$

Consequently, the sequences $n \to u_n$ and $n \to v_n$ are Cauchy and by what we have already proved for real sequences they converge. By Theorem 10.4 it follows that the sequence x converges. $\square$

The Bolzano–Cauchy Theorem is often referred to as the general convergence principle or the Cauchy convergence principle.

Example 10.19   Let $x_1 = 1$, $x_2 = 2$, $x_{n+2} = \dfrac{x_n + x_{n+1}}{2}$. If we knew that the sequence were convergent with limit l we could find l easily. Indeed

$$x_{n+2} + \frac{1}{2}x_{n+1} = x_{n+1} + \frac{1}{2}x_n = \cdots = x_2 + \frac{1}{2}x_1 = \frac{5}{2},$$

and therefore $l + \dfrac{1}{2}l = \dfrac{5}{2}$, $l = \dfrac{5}{3}$. We prove the convergence by showing that x is Cauchy. Firstly we have

$$x_{n+2} - \frac{1}{2}x_{n+1} = -\frac{1}{2}(x_{n+1} - x_n) = \frac{1}{2^2}(x_n - x_{n-1}) = \cdots$$

$$= \frac{(-1)^n}{2^n}(x_2 - x_1) = \frac{(-1)^n}{2^n}. \quad (10.42)$$

Obviously x_{n+2} lies between x_{n+1} and x_n and x_{n+3} remains there by lying between x_{n+2} and x_{n+1}. It is now clear by induction that for $m \geq n + 2$ the number x_m lies between x_{n+1} and x_n. By Equation (10.42) we have

$$|x_n - x_m| \leq |x_{n+1} - x_n| = \frac{1}{2^{n-1}},$$

for $m > n$. For every $\varepsilon > 0$ there exists N such that $2^{-N+1} < \varepsilon$ since $2^{-n+1} \to 0$. Consequently for $n > N$ and $m > N$ we have

$$|x_n - x_m| < \varepsilon.$$

The next theorem is very plausible and is often used. It says that shrinking closed intervals have a point in common.

Theorem 10.17 (Nested Intervals Lemma) *Let $[a_n, b_n]$ be a sequence of closed bounded intervals such that $b_n - a_n \to 0$ and $[a_{n+1}, b_{n+1}] \subset [a_n, b_n]$. Then there exists a unique c such that $c \in [a_n, b_n]$ for all $n \in \mathbb{N}$.*

Remark 10.7   The conclusion of the Theorem can be restated as

$$\bigcap_{n=1}^{\infty} [a_n, b_n] = \{c\}$$

Proof.   The sequence $n \to a_n$ is increasing and bounded from above by b_1 and therefore has a limit, say c and then $a_n \leq c$ for all natural n. The sequence $n \to b_n$ is decreasing and has the same limit, since $b_n = a_n + (b_n - a_n)$ and $b_n - a_n \to 0$. Since $b_{n+1} \leq b_n$, it follows that $c \leq b_n$ for all natural n. We have now proved that $a_n \leq c \leq b_n$ for all $n \in \mathbb{N}$.

If also $\gamma \subset [a_n, b_n]$ for $n \in \mathbb{N}$ then $|\gamma - c| \leq b_n - a_n$ for all natural n. Sending $n \to \infty$ gives $|\gamma - c| \leq 0$, that is $\gamma = c$. $\square$

The proof of the next theorem contains a typical application of the Nested Intervals Lemma.

Theorem 10.18 (Bolzano–Weierstrass)   *Every bounded sequence contains a convergent subsequence.*

Proof.   We assume first that the sequence x is real. Let $A \leq x_n < B$ for all $n \in \mathbb{N}$. Set $a_1 = A, b_1 = B$ and $n_1 = 1$. One of the intervals $\left[A, \dfrac{A+B}{2}\right]$ and $\left[\dfrac{A+B}{2}, B\right]$ contains infinitely many terms of the sequence, we denote this interval by $[a_2, b_2]$ and choose $n_2 > n_1$ such that $x_{n_2} \in [a_2, b_2]$. Continuing with this process we obtain a sequence of nested intervals $[a_k, b_k]$

such that $b_k - a_k = (B - A)/2^{-k-1}$ and $x_{n_k} \in [a_k, b_k]$. By the Nested Intervals Lemma there exists $c \in [a_k, b_k]$ for all $k \in \mathbb{N}$ and then

$$c - (b_k - a_k) \le a_k \le x_{n_k} \le b_k \le c + (b_k - a_k).$$

The Squeeze principle implies $\lim_{k \to \infty} x_{n_k} = c$. This completes the proof for real x.

For a bounded complex x with $u_n = \Re x_n$ and $v_n = \Im x_n$ we have that the sequence $n \mapsto u_n$ is also bounded. By what we have already proved there is a convergent subsequence, $u_{n_k} \to \hat{u}$. Similarly, the bounded sequence $k \mapsto v_{n_k}$ has a convergent subsequence $v_{n_{k_j}} \to \hat{v}$. By Theorem 10.4 it follows that

$$\lim_{j \to \infty} x_{n_{k_j}} = \hat{u} + \imath \hat{v}.$$

$\square$

Remark 10.8   Although we did not mention it explicitly we used the Recursion Theorem 5.2 in defining the sequence of intervals $[a_n, b_n]$.

Exercises

(!) **Exercise 10.7.1**   Let $x_1 = 1$, $x_{n+1} = -1 + x_n^2/2$. Prove that $n \mapsto x_n$ is Cauchy and $x_n \to 1 - \sqrt{3}$. [Hint: Calculate x_2, x_3. Use induction to prove that $x_3 \le x_n \le x_2$. Then show that $|x_{n+1} - x_n| \le \frac{7}{8}|x_n - x_{n-1}|$, for $n \ge 3$. Deduce $|x_{n+1} - x_n| \le 7^{n-1}8^{1-n}$. Then use $|x_n - x_{n+1}| + \cdots + |x_m - x_{m-1}| \le 7^{n-1}8^{2-n}$ to prove that it is Cauchy. The limit l must satisfy $l = -1 + l^2/2$.]

(i) **Exercise 10.7.2**   Prove that for real sequences $n \mapsto x_n$, $n \mapsto y_n$ with $x_n \le y_n$ for all $n \in \mathbb{N}$

$$\limsup_{n \to \infty} x_n \le \limsup_{n \to \infty} y_n, \tag{10.43}$$

$$\liminf_{n \to \infty} x_n \le \liminf_{n \to \infty} y_n. \tag{10.44}$$

Give an example such that

$$\liminf_{n \to \infty} y_n < \limsup_{n \to \infty} x_n.$$

[Hint: If $\limsup_{n \to \infty} y_n = \infty$ there is nothing to prove. Otherwise set $\bar{\beta}_n = \sup\{y_n, y_{n+1}, \ldots\}$ and use $\beta_n \le \bar{\beta}_n$. Similarly for $\liminf$. For the example set $y_n = 1 + (-1)^n$, $x_n = y_n/2$.]

(!) (i) **Exercise 10.7.3**   Let $n \mapsto x_n$ be convergent and $n \mapsto y_n$ bounded. Prove that

$$\limsup_{n \to \infty}(x_n + y_n) = \lim_{n \to \infty} x_n + \limsup_{n \to \infty} y_n.$$

State and prove a similar theorem for limit inferior. [Hint: Let $x_n \to l$. For $n > N$ obtain $l - \varepsilon + y_n \le x_n + y_n \le l + \varepsilon + y_n$ then pass to the limit superior.]

(!) (i) **Exercise 10.7.4**   Prove that for bounded sequences

$$\liminf_{n \to \infty} x_n + \liminf_{n \to \infty} y_n \le \liminf_{n \to \infty} (x_n + y_n) \le \liminf_{n \to \infty} x_n + \limsup_{n \to \infty} y_n$$

$$\limsup_{n \to \infty} (x_n + y_n) \le \limsup_{n \to \infty} x_n + \limsup_{n \to \infty} y_n. \quad (10.45)$$

Give examples showing that all inequalities may be strict. [Hint: For lim sup obtain $\sup \{x_k + y_k; k \ge n\} \le \sup \{x_k; k \ge n\} + \sup \{y_k; k \ge n\}$ then pass to the limit as $n \to \infty$. For an example set $x_n = -y_n = (-1)^n$.]

(!) **Exercise 10.7.5**   Let $x_1 = 1$, $x_2 = 2$ and $x_{n+1} = \sqrt{x_n x_{n-1}}$. Prove that the sequence x is Cauchy and find its limit. [Hint: $1 \le x_n \le 2$, $|x_{n+1} - x_n| = \sqrt{x_n}(\sqrt{x_{n-1}} + \sqrt{x_n})^{-1}|x_n - x_{n-1}| \le |x_n - x_{n-1}|/\sqrt{2}$, $x_{n+1}\sqrt{x_n} = x_n\sqrt{x_{n-1}} = \cdots = 2$.]

(i) **Exercise 10.7.6**   Intervals $\left]0, \dfrac{1}{n}\right[$, $n \in \mathbb{N}$ are nested but there is no

$c \in \bigcap_{n=1}^{\infty}]0, \dfrac{1}{n}[$. Similarly, intervals $[n, \infty[$ are also nested but $\bigcap_{n=1}^{\infty} [n, \infty[= \emptyset$.
Why this does not contradict Theorem 10.17.

(!) (i) **Exercise 10.7.7**   Prove: If $n \mapsto x_n$ is a bounded sequence and m is not the limit of this sequence then there exists a number $l \ne m$ and a subsequence converging to l. [Hint: There is a positive number α and a subsequence $n \mapsto x_{n_j}$ such that $|x_{n_j} - m| \ge \alpha$. Set $y_j = x_{n_j}$, by the Bolzano-Weierstrass Theorem there is a subsequence $y_{j_k} \to l$ and $|l - m| \ge \alpha$.]

10.8   Comments and supplements

A number λ is said to be an accumulation point of a sequence if there exists a subsequence converging to λ. If there is a subsequence with an infinite limit, for instance ∞, we say also that ∞ is an accumulation point of the sequence. For a bounded sequence it is clear that the set of all accumulation point is also bounded. Moreover, using the notation (10.37), (10.38) and(10.39) we have $\alpha_{n_k} \leq x_{n_k} \leq \beta_{n_k}$. It follows from Theorem 10.8 that for any accumulation point λ

$$\liminf_{n \to \infty} x_n \leq \lambda \leq \limsup_{n \to \infty} x_n. \tag{10.46}$$

We now show that $\limsup\limits_{n \to \infty} x_n$ is the largest accumulation point by defining a subsequence converging to it. For every $n \in \mathbb{N}$ there is a j_n such that

$$\alpha_n - \frac{1}{n} < x_{j_n} \leq \alpha_n.$$

By the definition of limit superior and Theorem 10.8 clearly

$$x_{j_n} \to \limsup_{n \to \infty} x_n.$$

Similarly the limit inferior is the smallest accumulation point. By the Bolzano-Weierstrass Theorem every bounded sequence has an accumulation point and if it has only one then it is convergent by Theorem 10.15. If a sequence, not necessarily bounded, has only one accumulation point then it has a limit.[4] A sequence can have many accumulation points; the sequence with the following terms

$$1, \frac{1}{2}, \frac{1}{3}, \frac{2}{3}, \frac{1}{4}, \frac{2}{4}, \frac{3}{4}, \frac{1}{5}, \frac{2}{5}, \frac{3}{5}, \frac{4}{5}, \dots$$

has any number between 0 and 1 as an accumulation point. The introduction of infinite limits superior and inferior enables simple formulation of some theorems, for instance, *a sequence has a limit if and only if its limit inferior and limit superior are equal.*

Exercise 10.8.4 asks for the proof of the following theorem: If $x_n > 0$ for all $n \in \mathbb{N}$ then

$$\liminf_{n \to \infty} \frac{x_{n+1}}{x_n} \leq \liminf_{n \to \infty} \sqrt[n]{x_n} \leq \limsup_{n \to \infty} \sqrt[n]{x_n} \leq \limsup_{n \to \infty} \frac{x_{n+1}}{x_n}. \tag{10.47}$$

In inequalities like these we interpret the symbol ∞ as larger than any real number and $-\infty$ as smaller than any real number. If, for instance $x_{2n-1} = 1$ and $x_{2n} = n$ then, with this convention, we may write

$$0 = \liminf_{n \to \infty} \frac{x_{n+1}}{x_n} < \liminf_{n \to \infty} \sqrt[n]{x_n} = 1 = \limsup_{n \to \infty} \sqrt[n]{x_n} < \limsup_{n \to \infty} \frac{x_{n+1}}{x_n} = \infty.$$

---

[4]Possibly ∞ or $-\infty$.

Exercises

ⓘ **Exercise 10.8.1**   *Define a sequence for which all real numbers are accumulation points.* [Hint: $1/2, -1/2, 1/3, -1/3, 2/3, -2/3, 4/3, -4/3, 5/3, -5/3, 1/4, -1/4, 3/4, \ldots, 11/4, -11/4, \ldots$]

⚠ **Exercise 10.8.2**   *Prove: If $y_n \to Y \geq 0$ and $n \to x_n$ is bounded then*

$$\limsup_{n\to\infty} y_n x_n \leq Y \limsup_{n\to\infty} x_n.$$

State a similar theorem for $\liminf$. [Hint: $x_n y_n \leq y_n \beta_n$. Consider separately $Y = 0$.]

⚠ **Exercise 10.8.3**   *Prove:*

$$\liminf_{n\to\infty} x_n \leq \liminf_{n\to\infty} \frac{x_1 + x_2 + \cdots + x_n}{n}$$
$$\leq \limsup_{n\to\infty} \frac{x_1 + x_2 + \cdots + x_n}{n} \leq \limsup_{n\to\infty} x_n.$$

[Hint:$(1/(n + p)) \sum_1^{n+p} x_k \leq (1/(n + p)) \left(\sum_1^n x_k + p\beta_n \right)$, let $p \to \infty$ using Exercises 10.7.3 and 10.7.4 as well as Theorem 10.15. Then let $n \to \infty$.]

⚠ **Exercise 10.8.4**   *Prove Inequalities (10.47).*

Chapter 11

Series

In this chapter we introduce infinite series and prove some basic convergence theorems. We also introduce power series—a very powerful tool in analysis.

11.1 Definition of convergence

Study of the behaviour of the terms of a sequence when they are successively added leads to infinite series. For an arbitrary sequence $n \mapsto a_n \in \mathbb{C}$ we can form another sequence by successive additions as follows:

$$s_1 = a_1$$
$$s_2 = a_1 + a_2$$
$$s_3 = a_1 + a_2 + a_3$$
$$\vdots$$
$$s_n = a_1 + a_2 + \cdots + a_n = \sum_{i=1}^{n} a_i \tag{11.1}$$

To indicate that we consider $n \mapsto s_n$ rather than $n \mapsto a_n$ we write

$$\sum a_i. \tag{11.2}$$

The symbol (11.2) is just an abbreviation for the sequence $n \mapsto s_n$, with s_n as in (11.1). We shall call (11.2) a "series" or an "infinite series"; a_n is the nth term of (11.2); s_n is the nth partial sum of (11.2).

It is usually clear from the context what the partial sums for a series like (11.2) are. On the other hand, if $a_i = \dfrac{1}{k^i}$ for some positive integer k

and every natural number i then

$$s_n = \frac{1}{k} + \frac{1}{k^2} + \cdots + \frac{1}{k^n}.$$

However the symbol $\sum k^{-i}$ is ambiguous. On occasions like this we would write $\sum k^{-i}$ rather than $\sum_i k^{-i}$. The letter i in (11.2) can be replaced by another letter without altering the meaning of the symbol; for example we can write $\sum a_j$ or $\sum a_k$ instead of $\sum a_i$. For instance, the geometric series, that is the sequence $n \mapsto s_n$ where $s_n = 1 + q + q^2 + \cdots + q^{n-1}$, is denoted by $\sum q^{i-1}$.

Definition 11.1 (Convergence)   The series (11.2) is said to be convergent to a sum S if

$$\lim_{n \to \infty} \sum_{i=1}^{n} a_i = S. \tag{11.3}$$

The notation

$$\sum_{i=1}^{\infty} a_i = S \tag{11.4}$$

is used as an abbreviation for series (11.2) to be convergent to a sum S.

Obviously, by Theorem 10.1 a series can have at most one sum, so we may speak of 'the sum' rather than 'a sum'. The phrases 'the series is convergent to the sum S', 'the series has the sum S' and 'the series converges to S', or variants of these, are used interchangeably. Equation (11.3) has, of course, the same meaning as[1]

$$\lim_{n \to \infty} s_n = S.$$

The notation and terminology for infinite series comes from the times when the theory was rather vague, but the notation has persisted ever since. The usage of the term *sum*[2] is a bit unfortunate because the sum of a series is *not* the usual sum, but the limit of the sequence of partial sums. Instead

---

[1] With s_n given by (11.1).

[2] It comes from times when the sum of a series was not well defined but regarded vaguely, intuitively and imprecisely as a sum of infinitely many terms of the series.

of (11.2) or (11.4) the following suggestive notation is often used:

$$a_1 + a_2 + a_3 + \cdots \tag{11.5}$$

or

$$a_1 + a_2 + a_3 + \cdots = S. \tag{11.6}$$

The notations of (11.5) and (11.6) have the advantage of being rather indicative of the terms of the series: for instance one gets a better impression of the series

$$\sum \left[\frac{1}{2}(1 + (-1)^k)(\frac{2}{k})^2 + \frac{1}{2}(1 + (-1)^{k+1})\, 3^{-\frac{k+1}{2}} \right] \tag{11.7}$$

by writing

$$\frac{1}{3} + 1 + \frac{1}{3^2} + \frac{1}{2^2} + \frac{1}{3^3} + \frac{1}{3^2} + \frac{1}{3^4} + \frac{1}{4^2} + \frac{1}{3^5} + \cdots$$

rather than (11.7). When using (11.6) one has to realize that the same symbol, namely

$$a_1 + a_2 + a_3 + \cdots$$

has two different meanings—it denotes the sequence of partial sums $n \mapsto \sum_{k=1}^{n} a_k$ and also $\lim_{n \to \infty} \sum_{k=1}^{n} a_k$. Perhaps we should add that most authors use the symbol $\sum_{k=1}^{\infty} a_k$ instead of our symbol $\sum a_k$, which, of course, has the same disadvantage as (11.5), namely denoting the series and its sum by the same symbol.

Since a series is nothing but the sequence of its partial sums, many definitions for sequences carry over to series. For example, a series is called *divergent, divergent to* ∞ (or $-\infty$) according to whether the sequence of its partial sums diverges, diverges to ∞ (or $-\infty$) respectively. Similarly—as with sequences—the change of finitely many terms of a series does not affect its convergence or divergence, but changes the sum in the obvious way.

In particular, the series

$$a_1 + a_2 + a_3 + \cdots$$

converges if and only if, for $N \in \mathbb{N}$

$$a_{N+1} + a_{N+2} + \cdots$$

does, and then

$$a_1 + a_2 + \cdots = (a_1 + a_2 + \cdots + a_N) + (a_{N+1} + a_{N+2} + \cdots)$$

or better still

$$\sum_{i=1}^{N} a_i + \sum_{i=1}^{\infty} a_{N+i} = \sum_{i=1}^{\infty} a_i. \tag{11.8}$$

By analogy with the finite sum we shall often write $\displaystyle\sum_{i=N+1}^{\infty} a_i$ instead of

$\displaystyle\sum_{i=1}^{\infty} a_{N+i}.$

Example 11.1   It is an immediate consequence of Definition 11.1 and the result of Example 10.14 that

$$1 + 1 + \frac{1}{2!} + \frac{1}{3!} + \frac{1}{4!} + \cdots = \mathrm{e}.$$

Example 11.2   The geometric series $\sum q^{i-1}$ converges to $\frac{1}{1-q}$ if $|q| < 1$. Indeed, $s_n = 1 + q + \cdots + q^{n-1} = \frac{1-q^n}{1-q}$ for $q \neq 1$ and $\lim_{n\to\infty} s_n = \frac{1}{1-q}$ since $\lim_{n\to\infty} q^n = 0$ for $|q| < 1$ by Example 10.13. Consequently $\sum_{i=1}^{\infty} q^{i-1} = \frac{1}{1-q}$

Example 11.3   Let us consider the series

$$\frac{1}{1\cdot 3} + \frac{1}{3\cdot 5} + \frac{1}{5\cdot 7} + \cdots .$$

We need some useful expression for the partial sum

$$s_n = \frac{1}{1\cdot 3} + \frac{1}{3\cdot 5} + \cdots + \frac{1}{(2n-1)(2n+1)} \quad .$$

By using the identity

$$\frac{1}{(2k-1)(2k+1)} = \frac{1}{2}\left(\frac{1}{2k-1} - \frac{1}{2k+1}\right),$$

we obtain

$$s_n = \frac{1}{2}\left(1 - \frac{1}{3} + \frac{1}{3} - \frac{1}{5} + \cdots + \frac{1}{2n-1} - \frac{1}{2n+1}\right) = \frac{1}{2} - \frac{1}{2(2n+1)}.$$

Since $s_n \to \frac{1}{2}$ we have

$$\sum_{k=1}^{\infty} \frac{1}{(2k-1)(2k+1)} = \frac{1}{2}$$

Example 11.4   Let $a_0 . a_1 a_2 a_3 \cdots$ be the decimal fraction for $x \in \mathbb{R}$, $x > 0$. We know from Example 5.8 that

$$a_0 + \frac{a_1}{10} + \cdots + \frac{a_n}{10^n} \leq x < a_0 + \frac{a_1}{10} + \cdots + \frac{a_n}{10^n} + \frac{1}{10^n}$$

It follows from the Squeeze Principle and Example 10.13 that

$$\lim_{n \to \infty} \sum_{i=0}^{n} \frac{a_i}{10^i} = x,$$

that is

$$\sum_{i=0}^{\infty} \frac{a_i}{10^i} = x. \tag{11.9}$$

Consequently, if $a_0 . a_1 a_2 a_3 \cdots$ is the decimal fraction for x, then x is the sum of the series (11.9).

For $x > 0$ it is customary, in contrast to what we have done in Example 5.8 to call $a_0 . a_1 a_2 a_3 \cdots$ the decimal fraction for x if (11.9) holds, where $a_0 \in \mathbb{N}_0$, and $a_i \in \{0, 1, 2 \ldots, 9\}$ for $i \in \mathbb{N}$. Every real x can then be represented by a decimal fraction, however this representation is no longer unique. For example, if $x > 0$ and $x = a_0 . a_1 a_2 a_3 \cdots a_N 999 \cdots$ with $a_N < 9$, then also $x = a_0 . a_1 a_2 a_3 \cdots \bar{a}_N 000 \cdots$ where $\bar{a}_N = a_N + 1$. This is obvious because

$$\frac{a_N}{10^N} + \sum_{k=N+1}^{\infty} \frac{9}{10^k} = \frac{a_N + 1}{10^N}.$$

Similar comments apply with appropriate changes made to binary and ternary fractions and obvious modifications should be made for negative reals, for instance the decimal expansion for $-\frac{4}{3}$ is $-1.33333 \ldots$.

Example 11.5   The series $1 + \frac{1}{\sqrt{2}} + \frac{1}{\sqrt{3}} + \cdots$ is divergent to ∞. We have

$$s_n = 1 + \frac{1}{\sqrt{2}} + \cdots + \frac{1}{\sqrt{n}} \geq \frac{n}{\sqrt{n}} = \sqrt{n}$$

and $s_n \to \infty$.

We shall see in the next section that sums of series share many properties, but not all, with ordinary sums (that is sums over a finite number of terms). For example, it is not permissible to remove (infinitely many) parentheses in a series. To see this, consider the series

$$(1 - 1) + (1 - 1) + \cdots .$$

All the terms of this series are zero, hence the series converges to zero. By removing parentheses we obtain the series

$$1 - 1 + 1 - 1 + \cdots$$

with partial sums $s_1 = 1$, $s_2 = 0$, $s_3 = 1$, ..., $s_{2k} = 0$, $s_{2k+1} = 1$, which is clearly divergent. On the other hand, inserting parentheses into a *convergent* series is permissible, because this amounts to a selection of a subsequence of partial sums, and we know that every subsequence converges to the same limit as the original sequence.

Exercises.

Exercise 11.1.1   *Decide which of the following series are convergent and which are divergent, and find the sums of the convergent series.*

(1) $(\frac{1}{2} + \frac{1}{3}) + (\frac{1}{2^2} + \frac{1}{3^2}) + (\frac{1}{2^3} + \frac{1}{3^3}) \cdots$

(2) $\dfrac{1}{1 \cdot 2} + \dfrac{1}{2 \cdot 3} + \dfrac{1}{3 \cdot 4} + \cdots ;$

(3) $\dfrac{1}{1 \cdot 4} + \dfrac{1}{4 \cdot 7} + \dfrac{1}{7 \cdot 10} + \cdots ;$

(4) $\dfrac{1}{1 \cdot 2 \cdot 3} + \dfrac{1}{2 \cdot 3 \cdot 4} + \dfrac{1}{3 \cdot 4 \cdot 5} + \cdots ;$

(5) $\dfrac{1}{1000} + \dfrac{2}{1001} + \dfrac{4}{1003} + \cdots ;$

(6) $\sum \dfrac{1}{\sqrt{k+1} + \sqrt{k}};$

(7) $x + \sum \left({}^{k+1}\!\sqrt{x} - \sqrt[k]{x} \right)$ with $x > 0.$

(i) **Exercise 11.1.2**   Prove that if $a_n = x_n - x_{n+1}$ and $\lim\limits_{n \to \infty} x_n = l$ then the series $\sum a_n$ is convergent and $\sum\limits_{n=0}^{\infty} a_n = x_0 - l.$

(i) **Exercise 11.1.3**   *The series*

$$1 - (1 - 1) - (1 - 1) - \cdots$$

converges to 1. By inserting parentheses differently we obtain the series

$$(1 - 1) + (1 - 1) + \cdots$$

which converges to zero. Explain this seeming paradox.

(i) **Exercise 11.1.4**   *If $\sum a_k$ and $\sum b_k$ converge, and $a_k \le b_k$ for $k \in \mathbb{N}$ then $\sum_{k=1}^{\infty} a_k \le \sum_{k=1}^{\infty} b_k$. Prove this result, and show that if in addition that $a_m < b_m$ for some $m \in \mathbb{N}$ then $\sum_{k=1}^{\infty} a_k < \sum_{k=1}^{\infty} b_k$.*

Exercise 11.1.5   *Find the approximate sum of the series $\sum \dfrac{k+1}{2^k k}$ to three decimal places. [Hint: use the previous exercise and (11.8); since $\dfrac{k+1}{2^k k} < \dfrac{2}{2^k}$, the sum $\sum_{k=1}^{N} \dfrac{k+1}{2^k k}$ is approximately equal to $\sum_{k=1}^{\infty} \dfrac{k+1}{2^k k}$ on three decimal places if $\sum_{k=N+1}^{\infty} \dfrac{1}{2^{k-1}} < \dfrac{1}{2} 10^{-3}$.]*

(!) **Exercise 11.1.6**   *Prove that $\sum_{i=0}^{\infty} (i+1)x^i = \dfrac{1}{(1-x)^2}$ for $|x| < 1$. [Hint: use Exercises 10.4.7, 5.3.6 and Example 10.11. (This exercise can be done easily with the help of Theorem 13.25 of Chapter 13.)]*

(i) **Exercise 11.1.7**   *Prove that*

(1) *Partial sums of a convergent series form a bounded sequence;*
(2) *If $\sum a_k$ converges then for every positive ε there exists an $M \in \mathbb{N}$ such that $\left| \sum_{k=n}^{\infty} a_k \right| < \varepsilon$ for all $n > M$.*

11.2   Basic theorems

The following theorem is an immediate consequence of Definition 11.1 and theorems on limits of sequences, especially Theorem 10.6.

Theorem 11.1

(i) If $\sum a_k$ converges, and $c \in \mathbb{C}$, then so does $\sum ca_k$, and

$$\sum_{k=1}^{\infty} ca_k = c \sum_{k=1}^{\infty} a_k$$

(ii) If both series $\sum a_k$ and $\sum b_k$ converge, then $\sum(a_k + b_k)$ converges, and

$$\sum_{k=1}^{\infty} a_k + \sum_{k=1}^{\infty} b_k = \sum_{k=1}^{\infty}(a_k + b_k).$$

Remark 11.1　　Using (i) with $c = -1$ and then applying (ii) gives

$$\sum_{k=1}^{\infty} a_n - \sum_{k=1}^{\infty} b_n = \sum_{k=1}^{\infty}(a_k - b_k)$$

Example 11.6　　Consider the series

$$1 - x + x^3 + x^4 + x^6 - x^7 + x^9 + x^{10} + x^{12} - \cdots \quad .$$

For $|x| < 1$ both of the series $1 + x^3 + x^6 + \cdots$ and $-x + x^4 - x^7 + \cdots$ converge and by Remark 11.1 we have

$$1 - x + x^3 + x^4 + x^6 - x^7 \cdots = 1 + x^3 + x^6 + \cdots - x + x^4 - x^7 + \cdots$$
$$= \frac{1}{1 - x^3} - \frac{x}{1 + x^3}.$$

Theorem 11.2　　If $\sum a_k$ converges then $\lim_{k \to \infty} a_k = 0.$

Proof.　　Let $s_n = \sum_{k=1}^{n} a_k$ and $\sum_{k=1}^{\infty} a_k = S$. Since $\lim_{n \to \infty} s_n = \lim_{n \to \infty} s_{n-1} = S$ we have $\lim_{n \to \infty} a_n = \lim_{n \to \infty}(s_n - s_{n-1}) = 0.$ □

Example 11.7　　The series $\sum_{k=1}^{\infty}(kx)^k$ converges if and only if $x = 0$. Indeed, if $x \neq 0$, then there exists a K such that $K \geq \frac{1}{|x|}$. Then for $k \geq K$ we have

$|kx| \geq 1$, therefore $\left|(kx)^k\right| \geq 1$. Consequently, $\lim\limits_{k \to \infty} (kx)^k \neq 0$, and the series must diverge by Theorem 11.2.

Remark 11.2 The converse of Theorem 11.2 is false. For example, the series $\sum \dfrac{1}{\sqrt{k}}$ from Example 11.5 diverges although $\lim\limits_{k \to \infty} \dfrac{1}{\sqrt{k}} = 0$.

Theorem 11.3 *A series with non-negative terms converges if and only if the sequence of partial sums is bounded.*

Proof. The assumption that the terms are non-negative automatically implies that they are real. The sequence of partial sums is increasing because the terms are non-negative. Hence it converges if and only if it is bounded, by Theorems 10.9 and 10.5. □

Example 11.8 The series $\sum \dfrac{1}{k^\alpha}$ converges if $\alpha > 1$, $\alpha \in \mathbb{Q}$. We have

$$
\begin{aligned}
s_{2n} &= \sum_{k=1}^{2n} \frac{1}{k^\alpha} \\
&= 1 + \frac{1}{2^\alpha} + \frac{1}{3^\alpha} + \cdots + \frac{1}{(2n)^\alpha} \\
&= 1 + \frac{1}{3^\alpha} + \frac{1}{5^\alpha} + \cdots + \frac{1}{(2n-1)^\alpha} + \frac{1}{2^\alpha}\left(1 + \frac{1}{2^\alpha} + \frac{1}{3^\alpha} + \cdots + \frac{1}{n^\alpha}\right) \\
&< 1 + \frac{1}{2^\alpha} + \frac{1}{4^\alpha} + \cdots + \frac{1}{(2n)^\alpha} + \frac{1}{2^\alpha} s_n \\
&\leq 1 + \frac{2}{2^\alpha} s_{2n}.
\end{aligned}
$$

Consequently, for $\alpha > 1$ we have

$$
s_{2n} \leq \frac{2^{\alpha-1}}{2^{\alpha-1} - 1}. \tag{11.10}
$$

Since $s_{2n-1} \leq s_{2n}$ it follows from (11.10) that the sequence of partial sums is bounded and the series converges by Theorem 11.3.

Example 11.9 The series $\sum \dfrac{1}{k}$, which is called the *harmonic series*, diverges. We will prove this indirectly, by assuming the contrary. Let $\displaystyle\sum_{k=1}^{\infty} \dfrac{1}{k} =$

S. The partial sums of $\sum \dfrac{1}{2k-1}$ are bounded above by S, hence this series also converges; let $\displaystyle\sum_{k=1}^{\infty} \dfrac{1}{2k-1} = T$. Then, and by Theorem 11.1,

$$S = 1 + \frac{1}{2} + \frac{1}{3} + \cdots = 1 + \frac{1}{3} + \frac{1}{5} + \cdots + \frac{1}{2}\left(1 + \frac{1}{2} + \frac{1}{3} + \cdots\right) = T + \frac{1}{2}S.$$

Consequently, $S = 2T$. On the other hand,

$$S = 1 + \frac{1}{2} + \frac{1}{3} + \cdots = 1 + \frac{1}{3} + \frac{1}{5} + \cdots + \frac{1}{2} + \frac{1}{4} + \frac{1}{6} \cdots$$
$$< 1 + \frac{1}{3} + \frac{1}{5} + \cdots + 1 + \frac{1}{3} + \frac{1}{5} + \cdots = 2T \quad (11.11)$$

which is a contradiction.

Theorem 11.4 (Comparison test) *If $0 \le a_k \le b_k$ for $k \in \mathbb{N}$, then the convergence of $\sum b_k$ implies the convergence of $\sum a_k$.*

Proof. If $\sum b_k$ converges, then its partial sums are bounded. This implies that the partial sums of $\sum a_k$ are also bounded, which in turn implies that $\sum a_k$ converges. □

Remark 11.3 The conclusion of Theorem 11.4 can be rephrased as *the divergence of $\sum a_k$ implies the divergence of $\sum b_k$.*

Example 11.10 The series $\sum \dfrac{1}{n^\alpha}$ diverges for $\alpha \le 1$, $\alpha \in \mathbb{Q}$. For $\alpha = 1$ this was established in Example 11.9. For $\alpha < 1$ we have $\dfrac{1}{n^\alpha} > \dfrac{1}{n}$ and divergence follows from the comparison test.

Exercises

Exercise 11.2.1 *Use the comparison test to prove the convergence or divergence of the following series.*

(1) $\displaystyle\sum \frac{1}{k^2 + 1}$;

(2) $\dfrac{1}{k^2 - 3k + 3}$;

(3) $\dfrac{1}{\sqrt{k} + 1}$.

(i) **Exercise 11.2.2**   Give examples of series $\sum a_k$ such that

    (1) $a_k \to 0$ and the series diverges;

    (2) the partial sums are bounded and the series diverges;

    (3) the partial sums are bounded, $a_k \to 0$ and the series diverges.

(!) (i) **Exercise 11.2.3**   If $a_n > 0$, $b_n > 0$ for $n \in \mathbb{N}$ and $\lim\limits_{n\to\infty} \dfrac{a_n}{b_n} = L > 0$ prove that the series $\sum a_k$ and $\sum b_k$ either both converge or both diverge. What conclusion can be drawn if $L = 0$? [Hint: There exists $N \in \mathbb{N}$ such that $|L|b_n/2 < a_n < 3|L|b_n/2$. Use the comparison test. If $L = 0$ then convergence of $\sum b_n$ implies that of $\sum a_n$ but $\sum b_n$ might diverge and $\sum a_n$ converge.]

11.3   *Maple* and infinite series

For a few series it is easy to find the sum. Generally speaking, however, it is difficult to find a sum of an infinite series exactly and on many occasions only a numerical approximation of the sum can be found.

    Maple can find sums of many series almost instantly. For instance the sum of

$$\sum_{1}^{\infty} \frac{1}{(n+1)(n+2)\cdots(n+10)}$$

can be found by the same method as in Example 11.3 but would require far more work. Finding the sum of an infinite series in *Maple* is very similar to finding a sum of finitely many terms. For example

```
>   sum(1/product((n+i),i=1..10),n=1..20);
```
$$\frac{715357}{23363003664000}$$
```
>   sum(1/product((n+i),i=1..10),n=1..infinity);
```
$$\frac{1}{32659200}$$

    Maple can find sums of infinite series for which we have not yet developed enough theory. For instance

```
>   sum(1/n^2,n=1..infinity);
```

$$\frac{1}{6}\pi^2$$

In other situations we can even obtain a result hard to understand. For example

```
>  sum(1/(n!)^2,n=1..infinity);
```

$$\text{hypergeom}([1], [2, 2], 1)$$

Maple returned the result in terms of the hypergeometric function but we need not to go into the study of this function: we can obtain a numerical approximation easily with the `evalf` command.

```
>  evalf(%);
```

$$1.279585302$$

Exercises

Exercise 11.3.1   *Find the sum of the series* $\displaystyle\sum_{i=1}^{\infty}\frac{1}{n^2-1}$ *from the first principles and by* Maple.

Exercise 11.3.2   *Find* $\displaystyle\sum_{i=1}^{\infty}\frac{1}{n^3}$ *to fifteen decimal places.*   [Hint: `Digits=15;evalf(sum(1/n^3,n=1..infinity)).`]

11.4   Absolute and conditional convergence

Theorem 10.16, the so-called Bolzano–Cauchy convergence principle, leads to

Theorem 11.5   *The series $\sum a_k$ converges if and only if for every positive ε, there exists an N such that*

$$\left| \sum_{k=n+1}^{n+p} a_k \right| < \varepsilon \qquad (11.12)$$

for every natural number p and $n > N$.

Remark 11.4   Theorem 11.5 is *also* referred to as the *Bolzano–Cauchy Theorem*, or as the *general principle of convergence*.

Proof.   Condition (11.12) is equivalent to the partial sums forming a Cauchy sequence. Hence, the sequence of partial sums converges if and only if (11.12) is satisfied.   □

Corollary 11.5.1   *If $\sum a_k$ converges, then for every positive number ε there exists an M such that for $n > M$*

$$\left| \sum_{k=n+1}^{\infty} a_k \right| < \varepsilon.$$

This result was proved directly from the definition of convergence in Exercise 11.1.7 part (ii). Here we provide another proof.

Proof.   There exists an M such that

$$\left| \sum_{k=n+1}^{n+p} a_k \right| < \frac{\varepsilon}{2}$$

for $n \in \mathbb{N}$, $p \in \mathbb{N}$, $n > M$. By taking limits as $p \to \infty$ we obtain

$$\left| \sum_{k=n+1}^{\infty} \right| \leq \frac{\varepsilon}{2} < \varepsilon.$$

□

Example 11.11   We prove again the divergence of the harmonic series $\sum \frac{1}{k}$. Clearly

$$\left| \sum_{k=n+1}^{2n} \frac{1}{k} \right| = \frac{1}{n+1} + \frac{1}{n+2} + \cdots + \frac{1}{2n} \geq \frac{n}{2n} = \frac{1}{2}$$

Consequently, the Bolzano–Cauchy condition is not satisfied for $\varepsilon = \frac{1}{2}$ and $p = n$ and the series must diverge.

Theorem 11.6 (Leibniz test)　*If $a_1 \geq a_2 \geq a_3 \geq \cdots \geq 0$ then the series $\sum(-1)^{k-1}a_k$ converges if and only if $a_k \to 0$.*

A series $\sum(-1)^{k-1}a_k$ with $a_k \geq 0$ for every $k \in \mathbb{N}$ is called an *alternating series*.

Proof.　I. The 'only if' part is an immediate consequence of Theorem 11.3.

II. Let $s_n = \sum_{k=1}^{n}(-1)^{k-1}a_k$. The sequence $n \mapsto s_{2n}$ is increasing and the sequence $n \mapsto s_{2n-1}$ is decreasing. Indeed we have $a_{2n+1} - a_{2n+2} \geq 0$ and consequently $s_{2n+2} = s_{2n} + (a_{2n+1} - a_{2n+2}) \geq s_{2n}$. Similarly, $-a_{2n} + a_{2n+1} \leq 0$ and $s_{2n+1} = s_{2n-1} - a_{2n} + a_{2n+1} \leq s_{2n-1}$. Moreover, $s_{2n+1} = s_{2n} + a_{2n+1} \geq s_{2n}$. Thus

$$s_2 \leq s_4 \leq s_6 \leq \cdots \leq s_5 \leq s_3 \leq s_1.$$

The bounded closed intervals $[s_{2n}, s_{2n+1}]$ are nested and $s_{2n+1} - s_{2n} = a_{2n+1} \to 0$. By Theorem 6.7.3 there exists an $S \in [s_{2n}, s_{2n+1}]$ for every $n \in \mathbb{N}$. Since $a_n \to 0$ for every positive ε, there exists an N such that $0 \leq a_{2N+1} < \varepsilon$. If $m > 2N + 1$ then

$$|S - s_m| \leq s_{2N+1} - s_{2N} = a_{2N+1} < \varepsilon,$$

since both $S \in [s_{2N}, s_{2N+1}]$ and $s_m \in [s_{2N}, s_{2N+1}]$.　　　　□

Remark 11.5　The sum of an alternating series with decreasing terms is trapped between s_{2n} and s_{2n-1} for every $n \in \mathbb{N}$. Consequently, the error made in replacing S by a partial sum, that is $S - s_n$, does not exceed in absolute value the first omitted term, and has the same sign.

Example 11.12　Since $n^{-\alpha} \downarrow 0$ for $\alpha > 0$, the series

$$1 - \frac{1}{2^\alpha} + \frac{1}{3^\alpha} - \frac{1}{4^\alpha} + \cdots$$

converges for $\alpha > 0$, $\alpha \in \mathbb{Q}$. Using Remark 11.5 with $\alpha = 1$ yields

$$\frac{1}{2} \leq \sum_{k=1}^{\infty}(-1)^{n+1}\frac{1}{n} \leq 1$$

or, with $\alpha = 2$ and $n = 100$

$$\sum_{n=1}^{100}(-1)^{n+1}\frac{1}{n^2} < \sum_{n=1}^{\infty}(-1)^{n+1}\frac{1}{n^2} \leq \sum_{n=1}^{100}(-1)^{n+1}\frac{1}{n^2} + \frac{1}{101^2}$$

Theorem 11.7   If $\sum|a_k|$ converges, then so does $\sum a_k$.

Proof.   By the Bolzano–Cauchy Theorem, for every positive ε there exists an N such that

$$\sum_{k=n+1}^{n+p}|a_k| < \varepsilon$$

for $n \in \mathbb{N}$, $p \in \mathbb{N}$ and $n > N$. Since

$$\left|\sum_{k=n+1}^{n+p}a_k\right| < \sum_{k=n+1}^{n+p}|a_k| < \varepsilon,$$

the series $\sum a_k$ converges, by the Bolzano–Cauchy Theorem.   □

For series with *real* terms there is another proof of Theorem 11.7 which employs an idea often used on other occasions.

Proof.   Let

$$a_k^+ = \frac{|a_k| + a_k}{2} \qquad a_k^- = \frac{|a_k| - a_k}{2}$$

so that $a_k^+ = a_k$ if $a_k \geq 0$ and $a_k^- = -a_k$ if $a_k \leq 0$. Then $0 \leq a_k^+ \leq |a_k|$, $0 \leq a_k^- \leq |a_k|$, and then the convergence of $\sum|a_k|$ implies the convergence of $\sum a_k^+$ and $\sum a_k^-$. This in turn implies convergence of $\sum a_k$, by Remark 11.1, since $a_k = a_k^+ - a_k^-$.   □

Definition 11.2 (Absolute and Conditional Convergence)   If the series $\sum|a_k|$ converges, then $\sum a_k$ is said to be *absolutely convergent*. A convergent series which does not converge absolutely is said to be *conditionally convergent*.

Remark 11.6   The alternative proof of Theorem 11.7 shows that for a conditionally convergent series $\sum a_k$ with real terms, both of the series $\sum a_k^+$ and $\sum a_k^-$ are divergent.

Example 11.13   The series $\sum (-1)^{\lfloor \frac{k}{2} \rfloor} \dfrac{1}{k^2}$ that is the series

$$1 - \frac{1}{2^2} - \frac{1}{3^2} + \frac{1}{4^2} + \frac{1}{5^2} - \cdots$$

converges absolutely because $\sum \dfrac{1}{k^2}$ converges by Example 11.8.

Example 11.14   Let $\alpha \in \mathbb{Q}$. The series $\sum (-1)^{k+1} \dfrac{1}{k^\alpha}$ converges by Example 11.12 for $\alpha > 0$ and the series $\sum \dfrac{1}{k^\alpha}$ diverges for $\alpha \leq 1$, by Example 11.13. Consequently $\sum \dfrac{(-1)^{k+1}}{k^\alpha}$ is only conditionally convergent for $0 < \alpha \leq 1$.

Absolutely convergent series have some properties in common with finite sums which are not shared by conditionally convergent series. One such property is considered in the next section.

Exercises.

Exercise 11.4.1   *Test for the absolute or the conditional convergence of the following series*

(1) $\dfrac{1}{600} - \dfrac{2}{1100} + \dfrac{3}{1600} - \dfrac{4}{2100} + \cdots$;

(2) $\dfrac{2}{2\sqrt{2}-1} - \dfrac{3}{3\sqrt{3}-1} + \dfrac{4}{4\sqrt{4}-1} - \dfrac{5}{5\sqrt{5}-1} + \cdots$;

(3) $\sum (-1)^{k+1} \dfrac{k}{k^3-2}$.

ⓘ **Exercise 11.4.2**   *If a_n are terms of a bounded sequence and $\sum b_k$ converges absolutely, prove that $\sum a_k b_k$ converges absolutely.*

Exercise 11.4.3   *If $\sum a_k$ converges absolutely prove that $\sum \dfrac{k+1}{k} a_k$ does too.*

ⓘ **Exercise 11.4.4**   *Prove that if $\sum a_k^2$ and $\sum b_k^2$ are convergent then $\sum a_k b_k$ converges absolutely. [Hint: $|a_k b_k| \leq 1/2(a_k^2 + b_k^2)$]*

ⓘ **Exercise 11.4.5**   *Prove that the series*

$$1 - \frac{1}{2} + \frac{1}{2} - \frac{1}{4} + \frac{1}{3} - \frac{1}{8} + \frac{1}{4} - \frac{1}{16} + \frac{1}{5} - \frac{1}{32} + \frac{1}{6} - \cdots$$

has unbounded partial sums and therefore diverges. *Does this contradict the Leibnitz test? If not, why not?*

Exercise 11.4.6   *The series*

$$\frac{1}{\sqrt[4]{2}-1} - \frac{1}{\sqrt[4]{2}+1} + \frac{1}{\sqrt[4]{3}-1} - \frac{1}{\sqrt[4]{3}+1} + \cdots$$

that is, the series whose terms are given by

$$a_{2k-1} = \frac{1}{\sqrt[4]{k+1}-1} \qquad a_{2k} = \frac{-1}{\sqrt[4]{k+1}+1}$$

diverges and $a_k \to 0$. Prove this, and explain why this example does not contradict the Leibniz test. [Hint: $\displaystyle\sum_{k=2}^{n}(\frac{1}{\sqrt[4]{k}-1} - \frac{1}{\sqrt[4]{k}+1}) = \sum_{k=2}^{n}\frac{2}{\sqrt{k}-1}.$]

11.5   Rearrangements

If $k \mapsto n_k$ is a bijection of $\mathbb{N}$ onto itself, then the series $\sum a_{n_k}$ is called a *rearrangement* of $\sum a_k$. The rearranged series contains all the terms of the original series and no other terms, but contains them in different order.

> **Theorem 11.8**   *Every rearrangement of an absolutely convergent series converges absolutely and has the same sum.*

Proof.   The idea of the proof is simple. If we go far enough, partial sums of both series will have a lot of terms in common, namely the first N, and the sum of all other terms in which they differ would be small by the Bolzano–Cauchy Theorem. The formal proof is as follows: We write $\sum_{k=1}^{N} a_k = s_N$ and $\sum_{k=1}^{\infty} a_k = S$. Given $\varepsilon > 0$, choose N so that $\displaystyle\sum_{k=N+1}^{\infty} |a_k| < \frac{1}{2}\varepsilon$; then $|S - s_N| < \frac{1}{2}\varepsilon$. Now there exists an m_0 such that

$$\{n_1, \ldots, n_m\} \supseteq \{1, \ldots, N\} \qquad \text{for all } m \geq m_0$$

and then $\{n_1, \ldots, n_m\} = \{1, \ldots, N, \nu_1, \ldots, \nu_K\}$ for some $K \geq 0$ where ν_i are distinct integers all greater than N. Writing $\sigma_m = \displaystyle\sum_{k=1}^{m} a_{n_k}$, we have

$$\sigma_m = \sum_{k=1}^{N} a_k + \sum_{k=1}^{K} a_{\nu_k},$$

so

$$\left|\sigma_m - s_N\right| = \left|\sum_{k=1}^{K} a_{\nu_k}\right| \le \sum_{k=1}^{K} |a_{\nu_k}| \le \sum_{k=N+1}^{\infty} |a_k| < \frac{1}{2}\varepsilon$$

and hence

$$\left|\sigma_m - S\right| = \left|\sigma_m - s_N + s_N - S\right| < \frac{1}{2}\varepsilon + \frac{1}{2}\varepsilon = \varepsilon$$

for all $m \ge m_0$, that is $\sigma_m \to S$ as $m \to \infty$. $\qquad\square$

Example 11.15 The series $\sum \dfrac{(-1)^{k+1}}{\sqrt{k}}$ converges conditionally by Example 11.14. We now show that its rearrangement

$$1 + \frac{1}{\sqrt{3}} - \frac{1}{\sqrt{2}} + \frac{1}{\sqrt{5}} + \frac{1}{\sqrt{7}} - \frac{1}{\sqrt{4}} + \cdots \qquad (11.13)$$

diverges. We have

$$\sum_{k=1}^{n}\left(\frac{1}{\sqrt{4k-3}} + \frac{1}{\sqrt{4k-1}} - \frac{1}{\sqrt{2k}}\right) \ge \sum_{k=1}^{n}\left(\frac{2}{\sqrt{4k}} - \frac{1}{\sqrt{2k}}\right)$$

$$= \left(1 - \frac{1}{\sqrt{2}}\right) \sum_{k=1}^{n} \frac{1}{\sqrt{k}}$$

$$\ge \left(1 - \frac{1}{\sqrt{2}}\right) \frac{n}{\sqrt{n}}.$$

Consequently the partial sums of (11.13) are not bounded and the series must diverge.

Exercises.

(i) **Exercise 11.5.1** Let $S = \sum_{k=1}^{\infty} \dfrac{(-1)^{k+1}}{k}$. *Show that the series*

$$1 + \frac{1}{3} - \frac{1}{2} + \frac{1}{5} + \frac{1}{7} - \frac{1}{4} + \frac{1}{9} + \frac{1}{11} - \frac{1}{6} + \frac{1}{13} + \cdots \qquad (11.14)$$

and

$$1 - \frac{1}{2} - \frac{1}{4} + \frac{1}{3} - \frac{1}{6} - \frac{1}{8} + \frac{1}{5} - \frac{1}{10} - \frac{1}{12} + \frac{1}{7} - \cdots \qquad (11.15)$$

converge and have sums $\frac{3}{2}S$ *and* $\frac{1}{2}S$, *respectively.* [Hint: For the series in (11.14) show that the series S can be written in either of the forms

$$S = \sum_{k=1}^{\infty} \left(\frac{1}{2k-1} - \frac{1}{2k} \right),$$

$$S = \sum_{k=1}^{\infty} \left(\frac{1}{4k-3} - \frac{1}{4k-2} + \frac{1}{4k-1} - \frac{1}{4k} \right),$$

and from the first of these

$$\frac{1}{2}S = \sum_{k=1}^{\infty} \left(\frac{1}{4k-2} - \frac{1}{4k} \right).$$

Now add the last two series. For the series in (11.15) proceed in a similar fashion. Do not forget to prove the convergence of all the series involved in whatever procedure you use.]

11.6   Convergence tests

Since the sum of a series cannot often be found it is important to have criteria of convergence which will allow us to establish at least that the sum exists. If we know that a series converges we can than perform operations which will help in evaluation of the sum. In this section we prove two simple theorems which are based on comparison of a given series with the geometric series. In Section 11.8 we consider several finer tests of convergence.

Theorem 11.9 (Ratio or d'Alembert's Test)   *If there exists an N and a q such that*

$$\left| \frac{a_{n+1}}{a_n} \right| \leq q < 1 \qquad (11.16)$$

for $n \geq N$, then the series $\sum a_n$ is absolutely convergent; if however,

$$\left| \frac{a_{n+1}}{a_n} \right| \geq 1 \qquad (11.17)$$

for $n \geq N$ then the series diverges.

Condition (11.16) or (11.17) is satisfied if $\lim\limits_{n \to \infty} \left| \dfrac{a_{n+1}}{a_n} \right| = l < 1$ or if

$$\lim_{n \to \infty} \left| \frac{a_{n+1}}{a_n} \right| = l > 1, \text{ respectively. Hence we have}$$

Corollary 11.9.1 (d'Alembert's Limit Test) *If* $\lim\limits_{n \to \infty} \left| \dfrac{a_{n+1}}{a_n} \right| = l$, *then* $\sum a_n$ *converges absolutely if* $l < 1$ *and diverges if* $l > 1$.

This corollary follows in a straightforward way from Theorem 11.9, so here we prove the theorem.

Proof. We may assume that $N \in \mathbb{N}$ and obtain successively from (11.16)

$$|a_{N+1}| \le |a_N| q, \qquad |a_{N+2}| \le |a_{N+1}| q \le |a_N| q^2, \dots \quad .$$

By an easy induction, $|a_{N+k}| \le |a_N| q^k$. We now have that the series $\sum |a_{N+k}|$ converges by Theorem 11.4—the comparison test. Hence $\sum a_k$ converges absolutely. If (11.17) holds, then $0 < |a_N| \le |a_{N+1}| \le |a_{N+2}|$ and it is not true that $a_n \to 0$. The series diverges by Theorem 11.2. $\square$

Example 11.16 For the series $\sum n x^n$ we have

$$\lim_{n \to \infty} \left| \frac{a_{n+1}}{a_n} \right| = \lim_{n \to \infty} \frac{n+1}{n} |x| = |x|.$$

By Corollary 11.9.1 this series converges for $|x| < 1$ and diverges for $|x| > 1$. For $|x| = 1$ the series obviously diverges. Perhaps it is worth mentioning that for $|x| = 1$ Theorem 11.9 is still applicable, whereas Corollary 11.9.1 is not.

Example 11.17 The series $\sum \dfrac{1}{n^\alpha}$, with $\alpha \in \mathbb{Q}$ converges for $\alpha > 1$ and diverges for $\alpha \le 1$ (see Examples 11.8 and 11.10).

$$\lim_{n \to \infty} \frac{a_{n+1}}{a_n} = 1 \quad \text{regardless of } \alpha.$$

This shows that no conclusion can be drawn regarding the convergence or divergence of the series $\sum a_n$ from $\lim\limits_{n \to \infty} \left| \dfrac{a_{n+1}}{a_n} \right| = 1$.

Theorem 11.10 (Root Test or Cauchy's Test) *If there exists an N such that*

$$\sqrt[n]{|a_n|} \leq q < 1 \qquad (11.18)$$

for $n \geq N$, then $\sum a_n$ converges absolutely; if

$$\sqrt[n]{|a_n|} \geq 1 \qquad (11.19)$$

for $n \geq N$, then $\sum a_n$ diverges.

Corollary 11.10.1 *If $\lim_{n \to \infty} \sqrt[n]{|a_n|} = l$, then the series $\sum a_n$ converges absolutely for $l < 1$ and diverges for $l > 1$.*

Remark 11.7 As with d'Alembert's limit test, no conclusion can be drawn regarding convergence or divergence if $l = 1$. The series from Example 11.17 illustrates this again.

We now give the proof of Theorem 11.10.

Proof. Again we may assume that $N \in \mathbb{N}$. If (11.18) holds then $|a_n| \leq q^n$ for $n \geq N$. Since $q < 1$, the series $\displaystyle\sum_{k=N}^{\infty} |a_k|$ converges by the comparison test (Theorem 11.3). If (11.19) holds, then $|a_n| \geq 1$ for $n \geq N$, and consequently $a_n \to 0$ is false and the series diverges by Theorem 11.2. $\qquad\square$

Example 11.18 The series

$$1 + \frac{1}{3} + \frac{1}{2} + \frac{1}{3^2} + \frac{1}{2^2} + \frac{1}{3^3} + \cdots$$

is obviously convergent as a sum of two geometric series. Since $\sqrt[n]{a_n}$ is $2^{-\frac{1}{2}} + \frac{1}{2n}$ for n odd and $3^{-\frac{1}{2}}$ for n even, we can apply Theorem 11.10 to assert convergence. On the other hand, as $a_{n+1}/a_n > 1$ for an even n and $a_{n+1}/a_n < 1$ for an odd n, Theorem 11.9 is not applicable.

Inequalities (10.47) actually show that whenever Corollary 11.9.1 is applicable, so is Corollary 11.10.1. Moreover, in Corollary 11.9.1, the lim can be replaced by limsup for the convergence and by liminf for divergence test. Similar comments apply to Corollary 2.

Exercises

Exercise 11.6.1   *Test the following series for convergence:*

(1) $\sum \dfrac{(-1)^n}{(2n-1)!}$;

(2) $\sum \dfrac{n}{2^n}$;

(3) $\sum \dfrac{2n+1}{(\sqrt{2})^n}$;

(4) $\sum \dfrac{n!}{n^n}$;

(5) $\sum \dfrac{e^n n!}{n^n}$;

(6) $\sum \dfrac{1 \cdot 3 \cdot 5 \cdots (2n-1)}{2 \cdot 5 \cdot 8 \cdots (3n-1)}$;

(7) $\sum \left(\dfrac{n}{2n-1}\right)^n$;

(8) $\sum \dfrac{1}{3^n}\left(\dfrac{n+1}{n}\right)^{n^2}$.

11.7   Power series

We encountered the series $\sum nx^n$ in Example 11.16. Series of the form

$$\sum c_n x^n \tag{11.20}$$

are called *power series.* A power series represents a function

$$x \mapsto \sum_{n=0}^{\infty} c_n x^n.$$

The domain of this function is the set of all x for which the series converges—the so-called *domain of convergence.* For the power series from Example 11.16 the domain of convergence was the unit disc in the complex plane. The next theorem says, roughly speaking, that the domain of convergence of any power series is always a disc in the complex plane.

Power series are, after polynomials, the simplest functions of a complex variable. In complex analysis many theorems on polynomials are extended to power series. However this goes far beyond the framework of this book.

Theorem 11.11 (Circle of convergence)   *For a power series*
$\sum a_n x^n$ *there are only three possibilities:*

(i)   *it converges only for $x = 0$;*

(ii)   *it converges absolutely for all $x \in \mathbb{C}$;*

(iii) *there exists a positive ρ such that the series converges absolutely for $|x| < \rho$ and diverges for $|x| > \rho$.*

Definition 11.3 (Radius of Convergence)   The number ρ defined in Theorem 11.11 is called the *radius of convergence* of the series $\sum c_n x^n$. It is customary to say that $\rho = 0$ in case (i) and that $\rho = \infty$ in case (iii).

Proof.   Let

$$S = \{x; \text{ there exists } K \text{ such that } |c_n x^n| \le K \text{ for every } n \in \mathbb{N}\}.$$

Obviously $0 \in S$ and if there is no other element in S then the series diverges by Theorem 11.2 for all $x \ne 0$, this is case (i). Let $\rho = \sup S$, if S is bounded. Let now x be arbitrary if S is unbounded and $|x| < \rho$ if S is bounded. In either case it is possible to choose $r \in S$ with $|x| < r$ and we select such r. Then we obtain

$$|c_n x^n| = |c_n r^n| \left(\frac{|x|}{r}\right)^n \le K \left(\frac{|r|}{r}\right)^n$$

The series $\sum c_n x^n$ now converges absolutely, by the comparison test. Hence we established convergence in cases (ii) and (iii). If $|x| > \rho$ the series cannot converge because the sequence $n \mapsto c_n x^n$ is unbounded.   □

Remark 11.8   Nothing can be asserted concerning the behaviour of the power series for $|x| = \rho$. The series $\sum n x^n$ diverges for all complex x with $|x| = \rho = 1$, the series $\sum \dfrac{1}{n^2} x^n$ converges absolutely for all complex x with $|x| = \rho = 1$, the series $\sum (-1)^{n+1} \dfrac{x^{2n}}{n}$ converges conditionally for both $x = 1$ and $x = -1$ and the series $\sum \dfrac{1}{n} x^n$ converges conditionally for $x = -1$ and diverges for $x = 1$. The convergence or divergence of a given power series for complex x with $|x| = \rho$ is a rather delicate matter which we shall not attempt to solve.

The series of the form

$$\sum c_n (x - x_0)^n \qquad (11.21)$$

are also called power series, x_0 is the *centre* of this *power series*. All statements for series (1) lead immediately to similar statements for series (11.21) via the substitution $y = x - x_0$. For instance, the analogue of Theorem 11.11 says that the series (11.21) converges for $x = x_0$ only, or for all x, or there is a positive ρ such that (11.21) converges for $|x - x_0| < \rho$ and diverges for $|x - x_0| > \rho$. The number ρ is then the radius of convergence of (11.21).

Exercises.

Exercise 11.7.1   *Find the radius of convergence ρ of the following power series, and investigate the convergence for $x = \pm\rho$.*

(1) $\displaystyle\sum \frac{k!\, x^k}{k^k}$;

(2) $\displaystyle\sum \frac{x^{2k+1}}{2k+1}$;

(3) $\displaystyle\sum \frac{1}{2k+1} x^k$;

(4) $\displaystyle\sum \frac{x^{3k}}{k^2}$;

(5) $\displaystyle\sum k^5 x^k$.

Exercise 11.7.2   *If $\lim_{n\to\infty} |c_n/c_{n+1}|$ exists, prove that the radius of convergence of (11.20) is this limit. (Show that this is also true for $\lim_{n\to\infty} |c_n/c_{n+1}| = 0$ or ∞). Give an example of a power series with positive (and finite) radius of convergence for which $\lim_{n\to\infty} |c_n/c_{n+1}|$ does not exist.*

Exercise 11.7.3   *Let $\limsup_{n\to\infty} \sqrt[n]{|c_n|} = \alpha$. Prove that the radius of convergence ρ is given by:*

(1) $\rho = \dfrac{1}{\alpha}$ *if $0 < \alpha$, $\alpha \in \mathbb{R}$;*

(2) $\rho = 0$ *if $\alpha = \infty$;*

(3) $\rho = \infty$ *if $\alpha = 0$.*

This theorem was originally proved by A. L. Cauchy and then forgotten until it was re-discovered by another French mathematician, J. Hadamard. The theorem became rather important afterwards.

(i) **Exercise 11.7.4**   *Using the previous result show that the series $\sum c_k x^k$ and $\sum k c_k x^k$ have the same radius of convergence.*

(i) **Exercise 11.7.5**   *If there exists $K \in \mathbb{R}$ and $m \in \mathbb{N}$ such that $|c_n| \leq K n^m$ for all $n \in \mathbb{N}$, prove that the radius of convergence of (11.20) is at least 1. Prove also that the radius of convergence of the series $\sum n^m x^n$ is 1. [Hint: The radius of convergence of the last series is 1 by Exercise 11.7.2; for the first part use the the comparison test.]*

11.8   Comments and supplements

To find the sum of an infinite series in a simple closed form is often difficult or impossible. In Sections 11.1 and 11.2 we found sums of only a few series. It can be shown that

$$1 + \frac{1}{2^2} + \frac{1}{3^2} + \cdots = \frac{\pi^2}{6} \qquad (11.22)$$

and that

$$1 - \frac{1}{2} + \frac{1}{3} - \frac{1}{4} + \cdots = \log 2 \qquad (11.23)$$

However, this cannot be done by merely finding the limit of the partial sums; more theory is needed before we can attempt to establish (11.22) and (11.23). (Perhaps we should mention that so far we have not even defined π or the function $x \mapsto \log x$.) We have already pointed out that in cases where it is difficult or impossible to find the sum of a series it is important to establish that the series converges. The tests we have so far, namely the root and ratio tests, are rather crude, and many other tests for convergence exist. These tests are based on comparison with another series known to be convergent. The slower the convergence of this latter series, the finer the test derived. We mention here a few more tests.

11.8.1   *More convergence tests.*

For series with non-negative and decreasing terms the following theorem is rather strong but has a simple elementary proof.

Theorem 11.12 (Cauchy's condensation test) *If* $a_1 \geq a_2 \geq a_3 \geq \cdots \geq 0$, *then the series* $\sum a_i$ *converges if and only if* $\sum 2^i a_{2^i}$ *does.*

Proof.

I. Assume $\sum 2^i a_{2^i}$ converges. We obviously have

$$a_1 + (a_2 + a_3) + (a_4 + \cdots + a_7) + \cdots + (a_{2^k} + \cdots + a_{2^{k+1}} + \cdots + a_{2^{k+1}-1})$$
$$\leq a_1 + 2a_2 + 4a_4 + \cdots + 2^k a_{2^k}$$
$$\leq \sum_{i=0}^{\infty} 2^i a_{2^i}$$

Consequently the partial sums of $\sum a_i$ are bounded and this series converges.

II. Assume $\sum a_i$ converges and put $\displaystyle\sum_{i=1}^{\infty} a_i = S$. Then

$$S \geq a_1 + a_2 + (a_3 + a_4) + (a_5 + \cdots + a_8) + \cdots + (a_{2^k+1} + \cdots + a_{2^{k+1}})$$
$$\geq a_1 + a_2 + 2a_4 + 4a_8 + \cdots + 2^{k-1} a_{2^k} + 2^k a_{2^{k+1}}$$
$$\geq a_1 + \frac{1}{2} \sum_{i=2}^{k+1} 2^i a_{2^i}$$
$$\geq \frac{1}{2} \sum_{i=0}^{k+1} 2^i a_{2^i}.$$

Consequently the partial sums of $\sum 2^i a_{2^i}$ are bounded and this series converges. $\qquad\square$

Example 11.19   We consider again $\sum n^{-\alpha}$ for $\alpha \in \mathbb{Q}$ (Examples 11.8, 11.9 and 11.10). Applying Cauchy's condensation test yields the result that $\sum n^{-\alpha}$ converges if and only if $\sum 2^{n(1-\alpha)}$ does. This latter series converges if and only if $2^{1-\alpha} < 1$, that is, if and only if $\alpha > 1$.

Theorem 11.13 (Raabe's test) *Let $a_n > 0$. If there exists an $r > 1$ and $N \in \mathbb{N}$ such that*

$$n \left(\frac{a_n}{a_{n+1}} - 1 \right) \geq r \qquad (11.24)$$

for all $n > N$, then the series $\sum a_i$ is convergent. If, however, there exists an $N_1 \in \mathbb{N}$ such that

$$n \left(\frac{a_n}{a_{n+1}} - 1 \right) \leq 1 \qquad (11.25)$$

for all $n > N_1$ then $\sum a_i$ is divergent.

Proof. It follows from (11.24) that

$$na_n \geq (r + n)a_{n+1} > (n + 1)a_{n+1},$$

and

$$na_n - (n + 1)a_{n+1} \geq (r - 1)a_{n+1}. \qquad (11.26)$$

Hence the sequence $n \mapsto na_n$ is strictly decreasing after its Nth term and it is also bounded below. Consequently it converges, which by the result of Exercise 11.1.2 implies the convergence of the series

$$\sum (ka_k - (k + 1)a_{k+1}).$$

Since $r - 1 > 0$ it follows from (11.26) by the Comparison Test that the series $\sum a_{k+1}$ also converges.

Turning our attention to divergence, we see that Inequality (11.25) implies that $na_n \leq (n + 1)a_{n+1}$ and consequently $(N + 1)a_{N+1} \leq na_n$ for $n \geq N$. It follows that there is a c such that $0 < c < na_n$ for all n. By comparing $\sum a_k$ with a multiple of the harmonic series we see that the former series diverges. $\qquad \square$

Corollary 11.13.1 *If $a_n > 0$ and $\lim\limits_{n \to \infty} n \left(\dfrac{a_n}{a_{n+1}} - 1 \right) > 1$, then $\sum a_i$ converges. If this limit is less than 1 then this series diverges.*

We know from Example 11.8 that the series $\sum 1/n^2$ converges. The ratio test is not conclusive for this series. On the other hand,

$$n\left(\frac{a_n}{a_{n+1}} - 1\right) = n\left[\left(\frac{n+1}{n}\right)^2 - 1\right] = 2 + \frac{1}{n} \geq 2$$

and Raabe's test yields convergence. Theorem 11.13 can, of course, be used for testing absolute convergence of a series; if however $n(|a_n|/|a_{n+1}| - 1) \leq 1$, then it follows only that $\sum a_i$ is not absolutely convergent. For instance $\sum (-1)^{n+1}/\sqrt{n}$ converges but $n(|a_n/a_{n+1}| - 1) = \sqrt{n}/(\sqrt{n+1} + \sqrt{n}) \leq 1$.

We had only one test for non-absolute (conditional) convergence, namely the Leibniz test. We note that this test requires the terms to alternate in sign and to decrease in absolute value. There are many more general tests, and we refer the interested reader to Knopp (1956) for Abel's, Dirichlet's and Dedekind's tests.

11.8.2 *Rearrangements revisited.*

We know from Section 11.5 that absolutely convergent series can be rearranged and the sum is not altered. We also had examples showing that a rearrangement of a conditionally convergent series may diverge or have a different sum. There is a famous theorem due to the German mathematician B. Riemann which states that a conditionally convergent series *with real terms* can be rearranged to have any sum.

> **Theorem 11.14 (Riemann's theorem on conditional convergence)** *If $A, a_i \in \mathbb{R}$ and the series $\sum a_i$ converges conditionally, then there exists a rearrangement $\sum_i a_{n_i}$ such that $\sum_{i=1}^{\infty} a_{n_i} = A$. There also exists another rearrangement which diverges.*

We shall not prove this theorem, but indicate the main idea of the proof instead. We add up the positive terms of the series until we first exceed A; hence we have $a_{n_1} + a_{n_2} + \cdots + a_{n_k} > A$. This can be done because the series of positive terms diverges to ∞. Then we add successively negative terms until the sum is smaller than A for the first time. Thus

$$a_{n_1} + a_{n_2} + \cdots + a_{n_k} > A > a_{n_1} + a_{n_2} + \cdots + a_{n_k} + a_{n_{k+1}} + \cdots + a_{n_s}.$$

Then we add positive terms to exceed A again, and by continuing with this process of jumping up and down over A we obtain the required rearrangement.

For series with complex terms the Riemann Theorem is *not* valid. This is clear from the following: If a_n, $b_n \in \mathbb{R}$, $\sum a_n$ converges absolutely and $\sum b_n$ converges conditionally then the series $\sum(a_n + \imath b_n)$ converges conditionally but by Theorem 11.8 the real part of the rearranged series will have the same sum as the real part of the original series. On the other hand we have the following corollary to the Riemann Theorem.

Corollary 11.14.1 (Rearrangement of series with complex terms) *If a_n, $b_n \in \mathbb{R}$ and the series $\sum(a_n + \imath b_n)$ converges conditionally to S then there is a rearrangement of the series which either has a sum distinct from S or diverges.*

Proof. One of the series $\sum a_n$, $\sum b_n$ must be only conditionally convergent, hence by Riemann's Theorem can be rearranged to a different sum. If we apply the same rearrangement to $\sum(a_n + \imath b_n)$ then the rearranged series cannot have S as its sum. □ ✂

11.8.3 *Multiplication of series.*

Infinite series which converge absolutely can be multiplied together. More precisely, we have the following theorem.

Theorem 11.15 (Multiplication of series) *If $\sum a_\imath$ and $\sum b_\imath$ converge absolutely, $(m, n) \mapsto k$ is a bijection from $\mathbb{N} \times \mathbb{N}$ onto $\mathbb{N}$, and $d_k = a_n b_m$ then the series $\sum d_k$ converges absolutely and*

$$\left(\sum_{n=1}^{\infty} a_n\right)\left(\sum_{m=1}^{\infty} b_m\right) = \sum_{k=1}^{\infty} d_k. \tag{11.27}$$

Remark 11.9 The content of this theorem is easy to remember: one multiplies the series $\sum a_n$ and $\sum b_m$ by multiplying each term of the first series by every term of the second series and then arranges the series with terms $a_n b_m$ arbitrarily.

Example 11.20 If we multiply the series $1 + x + x^2 + \cdots$ with itself and arrange the resulting series by increasing powers of x we obtain

$$(1 + x + x^2 + \cdots)(1 + x + x^2 + \cdots) = 1 + x + x + x^2 + x^2 + x^2 + \cdots$$
$$= 1 + 2x + 3x^2 + \cdots$$

for $|x| < 1$. Consequently

$$1 + 2x + 3x^2 + \cdots = \frac{1}{(1-x)^2}.$$

It is often convenient to arrange the product series in the following form

$$\sum_{k=1}^{\infty} c_k = a_1 b_1 + (a_2 b_1 + a_1 b_2) + (a_3 b_1 + a_2 b_2 + a_1 b_3) + \cdots .$$

The series $\sum c_j$ with $c_j = a_j b_1 + a_{j-1} b_2 + \cdots + a_1 b_j$ is sometimes called the Cauchy product of the series $\sum a_n$ and $\sum b_m$. In Example 11.20 the Cauchy product of $\sum x^{k-1}$ with itself was $\sum k x^{k-1}$.

Example 11.21   The Cauchy product of the series $= \sum \dfrac{(-1)^{n+1}}{\sqrt{n}}$ with itself is $\sum c_n$ where

$$c_n = (-1)^{n+1} \left(\frac{1}{\sqrt{n}} + \frac{1}{\sqrt{n-1}\sqrt{2}} + \cdots + \frac{1}{\sqrt{2}\sqrt{n-1}} + \frac{1}{\sqrt{n}} \right).$$

This series diverges, indeed[3] $\dfrac{1}{\sqrt{n-k}\sqrt{k+1}} \geq \dfrac{2}{n+1}$ for $0 \leq k < n-1$, and consequently $|c_n| \geq 1$, therefore $\lim\limits_{n \to \infty} c_n$ does not exist and the series $\sum c_n$ must diverge. This example shows that the Cauchy product of two conditionally convergent series may diverge. The so called Merten's Theorem asserts that the Cauchy product of an absolutely convergent series with a conditionally convergent series is convergent and has the obvious sum. For the proof see Stromberg (1981) or Knopp (1956).

After these remarks and examples we turn to the proof of Theorem 11.15.

Proof.   Let $A = \displaystyle\sum_{n=1}^{\infty} |a_n|$ and $B = \displaystyle\sum_{n=1}^{\infty} |b_n|$. Consider a sum

$$|d_1| + |d_2| + \cdots + |d_K|,$$

where each d_k above is of the form $a_n b_m$. Let N be the maximum of these n and M be the maximum of such m. Then clearly

$$|d_1| + |d_2| + \cdots + |d_K| \leq \sum_{n=1}^{N} |a_n| \sum_{m=1}^{M} |b_m| \leq AB.$$

---

[3] By $2ab \leq a^2 + b^2$.

The partial sums of $\sum |d_k|$ are bounded and hence $\sum d_k$ converges absolutely. To find the sum of $\sum d_k$ it is sufficient to find a sum of a particular rearrangement. Consider

$$a_1 b_1 + (a_2 b_1 + a_1 b_2) + \cdots$$

$$+ (a_n b_1 + a_{n-1} b_2 + \cdots a_1 b_n) \leq \sum_{i=1}^{n} a_i \sum_{j=1}^{n} b_j \quad (11.28)$$

Letting $n \to \infty$ completes the proof of Equation (11.27). $\qquad \square$

We were led to infinite series by successive addition of terms of a sequence. Successive multiplication leads to infinite products. An example of an infinite product is in Exercise 11.8.3. Some other examples of successive operations on the terms of a sequence and the taking of limits are in Exercises 11.8.4 and 11.8.5.

11.8.4 *Concluding comments*

In conclusion we add a few historical comments. It was already Archimedes who realized that the sum of a series must be defined and not be left as (intuitively) obvious. His definition anticipated the modern one given by Cauchy one and a half millennia later. Archimedes considered only (simple) series with positive terms but for these his definition was equivalent to Definition 11.1. Johann Bernoulli (1667–1748) is often credited with the divergence of the harmonic series; his proof is close to ours given in Example 11.9. The first who realized that the harmonic series diverges was most likely Nicole Oresme (1323–1382). He employed the inequality we used in Example 11.11. Oresme found the sum of

$$\frac{1}{2} + \frac{2}{4} + \frac{3}{8} + \frac{4}{16} + \cdots \quad (11.29)$$

by the following clever trick. He calculated the area T of the infinite tower in Figure 11.1

in two different ways. The areas of rectangles resting one on another and leaning on the y-axis with top right hand corners at[4] $(1,1), (\frac{1}{2}, 2), (\frac{1}{4}, 3)$ are $1, \frac{1}{2}, \frac{1}{4} \cdots$. Consequently

$$T = 1 + \frac{1}{2} + \frac{1}{4} + \cdots = 2.$$

---

[4] The first rectangle is shaded in Figure 11.1

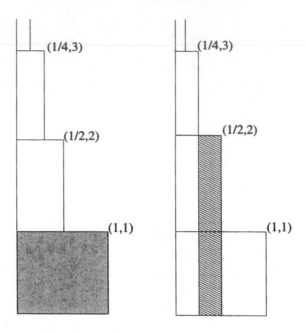

Fig. 11.1   Oresme's tower

On the other hand the areas of the 'steps'[5] above the intervals $[2^{-n}, 2^{-n+1}]$ are $n2^{-n}$; hence the area of T is also

$$T = \frac{1}{2} + \frac{2}{4} + \frac{3}{8} + \frac{4}{16} + \cdots .$$

Consequently

$$\frac{1}{2} + \frac{2}{4} + \frac{3}{8} + \frac{4}{16} + \cdots = 2.$$

For us to find the sum of the series in (11.29) is easy, it suffices to put $x = 1/2$ in Example 11.20 (and multiply by 1/2). This example illustrates two points: the intellectual power of the mathematicians of the distant past and that the wisdom accumulated in mathematical theories makes problems easier to solve.

The history of infinite series is interesting, unfortunately we have to leave it now.

---

[5]The second step is shaded differently in Figure 11.1

Exercises.

Exercise 11.8.1 *For which values of α is the series*

$$\sum (-1)^k \binom{\alpha}{k}$$

convergent? The symbol $\binom{\alpha}{k}$ *is defined as* $\dfrac{\alpha(\alpha-1)(\alpha-2)\ldots(\alpha-k+1)}{k!}$
for $\alpha \in \mathbb{R}$, $k \in \mathbb{N}$.

Exercise 11.8.2 *Using Raabe's test prove the convergence of*

$$\sum \frac{1 \cdot 3 \cdots \cdots (2n-1)}{2 \cdot 4 \cdots \cdots 2n} \cdot \frac{1}{2n+1}.$$

(i) **Exercise 11.8.3** *Prove that*

$$\lim_{n\to\infty} \prod_{i=1}^{n} \left(1 - \frac{1}{(i+1)^2}\right) = \frac{1}{2}.$$

(i) **Exercise 11.8.4** *Let* $a_1 = 2$, $a_2 = 2 + \dfrac{1}{2}$, $a_3 = 2 + \dfrac{1}{2 + \dfrac{1}{2}}$, $\cdots$. *Give*
a recursive definition of a_n. *Prove that* $\lim_{n\to\infty} a_n$ *exists and find it. (The*
sequence $n \mapsto a_n$ *is an example of a continued fraction. Continued fractions*
play an important part in number theory and numerical analysis.)

Exercise 11.8.5 *Let*

$$x_1 = \sqrt{6}, \quad x_2 = \sqrt{6 + \sqrt{6}}, \quad x_3 = \sqrt{6 + \sqrt{6 + \sqrt{6}}}, \ldots$$

Give a recursive definition of x_n. *Prove that* $\lim_{n\to\infty} x_n$ *exists and find it.*

Chapter 12

Limits and Continuity of Functions

Limits of functions are defined in terms of limits of sequences. With a function f continuous on an interval we associate the intuitive idea of the graph f being drawn without lifting the pencil from the drawing paper. Mathematical treatment of continuity starts with the definition of a function continuous at a point; this definition is given here in terms of a limit of a function at a point. In this chapter we shall develop the theory of limits of functions, study continuous functions, and particularly functions continuous on closed bounded intervals. At the end of the chapter we touch upon the concept of limit superior and inferior of a function.

12.1 Limits

Looking at the graph of $f : x \mapsto \dfrac{x}{|x|}(1-x)$ (Figure 12.1), it is natural to say that the function value approaches 1 as x approaches 0 from the right. Formally we define:

Definition 12.1   A function f, with $\operatorname{dom} f \subset \mathbb{R}$, is said to have a limit l at $\hat{x}$ from the right if for every sequence $n \mapsto x_n$ for which $x_n \to \hat{x}$ and $x_n > \hat{x}$ it follows that $f(x_n) \to l$. If f has a limit l at $\hat{x}$ from the right, we write $\lim_{x \downarrow \hat{x}} f(x) = l$.

Definition 12.2   If the condition $x_n > \hat{x}$ is replaced by $x_n < \hat{x}$ one obtains the definition of the limit of f at $\hat{x}$ from the left. The limit of f at $\hat{x}$ from the left is denoted by $\lim_{x \uparrow \hat{x}} f(x)$.

Remark 12.1   The symbols $x \downarrow \hat{x}$ and $x \uparrow \hat{x}$ can be read as "x decreases

313

to $\hat{x}$" and "x increases to $\hat{x}$", respectively.

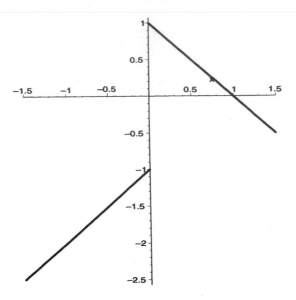

$$\text{Fig. 12.1} \quad \text{Graph of } f(x) = \frac{x}{|x|}(1-x)$$

Example 12.1   We now prove, according to Definition 12.1, the limit from the beginning of this section; that is,

$$\lim_{x \downarrow 0} \frac{x}{|x|}(1-x) = 1.$$

If $x_n \to 0$ and $x_n > 0$ then $\dfrac{x_n}{|x_n|}(1-x_n) = (1-x_n) \to 1$. Similarly, $\lim\limits_{x \uparrow 0} \dfrac{x}{|x|}(1-x) = -1$ because $\dfrac{x_n}{|x_n|}(1-x_n) = x_n - 1$ for $x_n < 0$ and since $x_n - 1 \to -1$ if $x_n \to 0$.

Remark 12.2   It was natural to say in Example 12.1 that f has a limit from the right at 0 even though f was not defined at 0. And it still would be natural to say so if f was defined at 0 no matter what $f(0)$ was. It should be remembered that Definition 12.1 (or 12.2) ignores the value of f at $\hat{x}$.

Remark 12.3   It is clear from Definition 12.1 (Definition 12.2) that two functions f and g which differ at most for $x \le \hat{x}$ ($x \ge \hat{x}$) have the same limit from the right (from the left) at $\hat{x}$ or they have no limits at all.

Example 12.2   The function $f : x \mapsto \text{saw}(1/x)$ has no limit from the right at 0. (The saw function is defined in Exercise 4.6.8 in Chapter 4.) A partial graph of f is sketched in Figure 12.2. It consists of triangles of height $1/2$ whose lateral sides are slightly curved[1] and base is formed by the interval $[1/(n+1), 1/n]$ of the x-axis. Clearly, if we choose to take as points of the sequence $x_{2k-1} = 1/k$ and $x_{2k} = 2/(2k+1)$, for $k \in \mathbb{N}$, then the values of $\text{saw}(1/x_n)$ alternate between 0 and $1/2$. Obviously the sequence $n \mapsto f(x_n)$ does not converge and the limit $\lim_{x\uparrow 0} \text{saw}(1/x)$ does not exist.

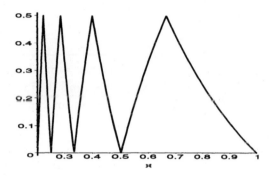

Fig. 12.2   Graph of $f(x) = \text{saw}(1/x)$

Theorem 12.1   *A function f has at most one limit from the right at $\hat{x}$ (at most one limit from the left at $\hat{x}$).*

Proof.   If l and L are limits of f at $\hat{x}$ from the right, $x_n \to \hat{x}$ and $x_n > \hat{x}$, then by Definition 12.1 $f(x_n) \to l$ and also $f(x_n) \to L$. Consequently $l = L$ (by Theorem 10.1).   □

The great advantage of Definitions 12.1 and 12.2 is that they make it possible to use the already established theory of limits of sequences (see Chapter 10) for limits of functions. This was clearly exploited in the proof of Theorem 12.1 and we shall exploit it further soon. However, we state two more definitions first.

---

[1]They are congruent to pieces of the graph of $x \mapsto 1/x$

Definition 12.3   A function f is said to have a limit l at $\hat{x}$ if for every sequence $n \mapsto x_n$ with $x_n \neq \hat{x}$ and $x_n \to \hat{x}$ it follows that $f(x_n) \to l$. The limit of f at $\hat{x}$ is denoted by $\lim_{x \to \hat{x}} f(x)$.

Remark 12.4   We left the wording of Definition 12.3 intentionally a little ambiguous. The meaning is changed according to whether $x_n \in \mathbb{R}$ or $\mathbb{C}$. If nothing is said to the contrary we shall assume that $x_n \in \mathbb{R}$ and if $x_n \in \mathbb{C}$ we shall explicitly say so. We also may make the distinction verbally by saying that the limit is in the real or the complex domain.

Example 12.3 $\lim_{x \to 2} x^2 = 4$. The proof is simple: if $x_n \to 2$ then $f(x_n) = x_n^2$ converges to 4.

Definition 12.4   If $\operatorname{dom} f \subset \mathbb{R}$ and the symbol $\hat{x}$ in Definition 12.3 is replaced by the symbol $+\infty$ or $-\infty$ (and the inequality $x_n \neq \hat{x}$ omitted) then one obtains the definition for the limit of f at $+\infty$ or $-\infty$ respectively. The limit of f at $+\infty$ or $-\infty$ is denoted by $\lim_{x \to +\infty} f(x)$ or $\lim_{x \to -\infty} f(x)$ respectively.

The letter x in any one of the symbols $\lim_{x \to \hat{x}} f(x)$, $\lim_{x \downarrow \hat{x}} f(x)$, $\lim_{x \uparrow \hat{x}} f(x)$, $\lim_{x \to +\infty} f(x)$ or $\lim_{x \to -\infty} f(x)$ can be replaced by any other letter without changing the meaning of that symbol. For example $\lim_{t \to 2} t^2 = \lim_{h \to 2} h^2 = 4$.

The next theorem can be proved in similar way to Theorem 12.1.

Theorem 12.2   *A function f has at most one limit at each of $\hat{x}$, $+\infty$ and $-\infty$.*

Example 12.4 $\lim_{x \to +\infty} \dfrac{1}{x} = 0$. Again the proof is simple: if $x_n \to +\infty$ then $\dfrac{1}{x_n} \to 0$ by Exercise 10.5.5.

Example 12.5   We find $\lim_{x \to -\infty} (x + \sqrt{x^2 - x})$. If $x_n \to -\infty$ then $\sqrt{x_n^2 - x_n} \to +\infty$ and Theorem 10.6 concerning the limit of a sum is not applicable. We first write $x + \sqrt{x^2 - x} = \dfrac{x^2 - x^2 + x}{x - \sqrt{x^2 - x}} = \dfrac{x}{x - |x|\sqrt{1 - 1/x}}$. This last expression is equal to $\dfrac{1}{1 + \sqrt{1 - 1/x}}$ for $x < 0$. Now if $x_n \to -\infty$ then

$1/x_n \to 0$ and $\dfrac{1}{1 + \sqrt{1 - 1/x_n}} \to \dfrac{1}{2}$. Consequently $\lim\limits_{x \to -\infty} (x + \sqrt{x^2 - x}) =$ $1/2$.

Theorem 12.3   If $\lim\limits_{x \to \hat{x}} f(x) = l$ and $\lim\limits_{x \to \hat{x}} g(x) = k$ then

(i) $\lim\limits_{x \to \hat{x}} |f(x)| = |l|$,

(ii) $\lim\limits_{x \to \hat{x}} (f(x) \pm g(x)) = l \pm k$,

(iii) $\lim\limits_{x \to \hat{x}} f(x)g(x) = lk$,

(iv) if $k \neq 0$ then $\lim\limits_{x \to \hat{x}} \dfrac{f(x)}{g(x)} = \dfrac{l}{k}$,

(v) if $l > 0$, $r \in \mathbb{Q}$ then $\lim\limits_{x \to \hat{x}} [f(x)]^r = l^r$,

(vi) $\lim\limits_{x \to \hat{x}} \mathrm{Max}(f(x), g(x)) = \mathrm{Max}(l, k)$,

$\lim\limits_{x \to \hat{x}} \mathrm{Min}(f(x), g(x)) = \mathrm{Min}(l, k)$.

Proof.   All assertions (i)–(vi) follow directly from Definition 12.3 and corresponding theorems for limits of sequences. As a sample we prove (iii). Let $x_n \to \hat{x}$ and $x_n \neq \hat{x}$, then $f(x_n) \to l$ and $g(x_n) \to k$ by Definition 12.3. Using Theorem 10.6 for the limit of a product of sequences we have $f(x_n)g(x_n) \to lk$ which in turn implies by Definition 12.4 that $\lim\limits_{x \to \hat{x}} f(x)g(x) - lk$.   □

Remark 12.5   An important case of assertion (iii) is $g(x) = k$ for all x. Then $\lim\limits_{x \to \hat{x}} kf(x) = k \lim\limits_{x \to \hat{x}} f(x)$. For $k = -1$ we have $\lim\limits_{x \to \hat{x}} [-f(x)] = -\lim\limits_{x \to \hat{x}} f(x)$.

Remark 12.6   All assertions in Theorem 12.3 remain valid if

(a) $\hat{x}$ is replaced everywhere either by $+\infty$ or by $-\infty$;

(b) $x \to \hat{x}$ is replaced everywhere by either $x \downarrow \hat{x}$ or by $x \uparrow \hat{x}$.

Example 12.6   We wish to find $\lim\limits_{x \to -2} \dfrac{x^2 + 3x + 2}{x^2 - 4}$. We cannot use (iv) of Theorem 12.3 directly since $\lim\limits_{x \to -2} (x^2 - 4) = 0$. Since $\dfrac{x^2 + 3x + 2}{x^2 - 4} = \dfrac{x + 1}{x - 2}$ for $x \neq -2$ we have $\lim\limits_{x \to -2} \dfrac{x^2 + 3x + 2}{x^2 - 1} = \lim\limits_{x \to -2} \dfrac{x + 1}{x - 2} = \dfrac{1}{4}$.

Remark 12.7   In the previous example we used an analogue of Remark 12.3: if two functions differ at most when $x = \hat{x}$ and one has a limit l at $\hat{x}$ then so does the other. More generally, if there exists $\Delta > 0$ such that $f(x) = g(x)$ for $0 < |x - \hat{x}| < \Delta$ then $\lim_{x \to \hat{x}} f(x) = l$ implies that $\lim_{x \to \hat{x}} g(x) = l$. This is clear from the definition of a limit.

The Squeeze Principle (see Theorem 10.3) is also valid for limits of functions.

Theorem 12.4　　*If there exists $\delta > 0$ such that*

$$g(x) \leq f(x) \leq h(x) \tag{12.1}$$

for $0 < |x - \hat{x}| < \delta$ and $\lim_{x \to \hat{x}} g(x) = \lim_{x \to \hat{x}} h(x) = l$ then f has a limit at $\hat{x}$ and $\lim_{x \to \hat{x}} f(x) = l$.

Since the formulation of the theorem involves inequalities, the functions g, f, h are automatically assumed to be real valued.

Proof.　　The proof follows the same pattern as the proof of Theorem 12.3. If $x_n \to \hat{x}$ and $x_n \neq \hat{x}$ then there exists N such that $0 < |x_n - \hat{x}| < \delta$ for $n > N$. Consequently

$$g(x_n) \leq f(x_n) \leq h(x_n)$$

for $n > N$ and it follows from the Squeeze Principle for sequences that $f(x_n) \to l$. That in turn implies that $\lim_{x \to \hat{x}} f(x) = l$. $\qquad \square$

Example 12.7　　We wish to calculate $\lim_{x \to 0} \dfrac{\sqrt[10]{1 + 3x^2} - 1}{x^2}$. In order to apply the theorem we note that, by Exercises 5.8.5 and 5.8.6 with x replaced by $3x^2$ and $n = 10$,

$$1 + \frac{3x^2}{10 \sqrt[10]{(1 + 3x^2)^9}} \leq \sqrt[10]{1 + 3x^2} \leq 1 + \frac{3x^2}{10}.$$

Consequently

$$\frac{3}{10 \sqrt[10]{(1 + 3x^2)^9}} \leq \frac{\sqrt[10]{1 + 3x^2} - 1}{x^2} \leq \frac{3}{10}.$$

Now simple application of Theorem 12.4 shows that the limit is $3/10$.

Remark 12.8　　Theorem 12.4 remains valid if:

(a) the inequality $0 < |x - \hat{x}| < \delta$ is replaced by $\hat{x} < x < x + \delta$ ($x - \delta < x < \hat{x}$), and limit is replaced by limit from the right (limit from the left);

(b) one assumes the existence of K (k) such that inequality (12.1) is valid for $x > K$ ($x < k$) and limits are replaced everywhere by limits at $+\infty$ ($-\infty$).

There is a simple relation between the limit at $+\infty$ and the limit from the right at 0.

Theorem 12.5   *We have*

$$\lim_{x \to +\infty} f(x) = l \quad \text{if and only if} \quad \lim_{y \downarrow 0} f(1/y) = l.$$

Proof.

I. Assume that $\lim\limits_{x \to +\infty} f(x) = l$ and let $y_n > 0$ and $y_n \to 0$. Then $1/y_n \to +\infty$ by Exercise 10.5.5 and therefore $f(1/y_n) \to l$; that is, $\lim\limits_{y \downarrow 0} f(1/y) = l$.

II. Assume $\lim\limits_{y \downarrow 0} f(1/y) = l$ and let $x_n \to +\infty$. Then there exists n such that $x_n > 0$ for $n > N$. For $n > N$ set $y_n = 1/x_n$. Then $y_n \to 0$ again by Exercise 10.5.5 and consequently $f(1/y_n) \to l$. However $f(1/y_n) = f(x_n)$ and hence $f(x_n) \to l$ which proves $\lim\limits_{x \to +\infty} f(x) = l$. $\square$

12.1.1   *Limits of functions in* Maple

Using *Maple* for evaluating limits is easy. The form of the command is

```
limit(fvalue,x=relevantpoint,direction);
```

fvalue stands for the value of the function for which the limit is sought, relevantpoint can be infinity or -infinity as well as some numerical value, direction can be left, right or complex. If the limit is in the real domain in the sense of Definition 12.3, two-sided so to speak, then direction should be omitted. The following examples illustrate some possibilities.

```
>  limit(2*x+floor(-x),x=3,right);
```
$$2$$

```
>  limit(x/abs(x),x=0,left);
```
$$-1$$

```
>  limit((x^2-5*x+6)/(x^2-x-2),x=2);
```

$$\frac{-1}{3}$$

```
>   limit((x^6+1)/(x^(10)+1),x=I,complex);
```

$$\frac{3}{5}$$

```
>   limit((x+1)/(x+5),x=infinity);
```

$$1$$

```
>   limit(x+sqrt(x^2-3*x),x=-infinity);
```

$$\frac{3}{2}$$

Sometimes *Maple* needs a little help

```
>   limit(1/floor(x),x=infinity);
```

$$\lim_{x \to \infty} \frac{1}{\text{floor}(x)}$$

```
>   evalf(%);
```

$$0.$$

Now there is an obvious question: is this limit exactly 0? Since

$$0 \le \frac{1}{\lfloor x \rfloor} \le \frac{1}{x-1}$$

for $x > 1$, it follows from the Squeeze Principle that, indeed, the limit is exactly 0.

Exercises

Exercise 12.1.1   *Find the following limits by paper and pencil method and by* Maple.

(1) $\displaystyle\lim_{x \to 1} \frac{2x^2 - 3x + 1}{x^2 - 4}$;

(2) $\displaystyle\lim_{x \to 2} \frac{2x^2 - 3x + 1}{x^2 - 4}$;

(3) $\displaystyle\lim_{x \to 1} \frac{2x^3 - 2x^2 + x - 1}{x^3 - x^2 + 3x - 3}$;

(4) $\displaystyle\lim_{x \to 1} \frac{x^2 - \sqrt{x}}{\sqrt{x} - 1}$;

(5) $\lim\limits_{x \to -1} \dfrac{(x+1)\sqrt{2-x}}{x^2-1}$;

(6) $\lim\limits_{x \to 0} \dfrac{\sqrt{1+x^2}-1}{x}$;

(7) $\lim\limits_{x \to 0} \dfrac{\sqrt[3]{1+x^2}-1}{x^2}$;

(8) $\lim\limits_{x \to 0} \dfrac{\sqrt[3]{1+x^2}-\sqrt[4]{1-2x}}{x+x^2}$;

(9) $\lim\limits_{x \to +\infty} \dfrac{(x+1)^5+(x+2)^5+\cdots+(x+1000)^5}{x^5+100^{20}}$;

(10) $\lim\limits_{x \to +\infty} x\left(\sqrt{x^2+1}-x\right)$;

(11) $\lim\limits_{x \to +\infty} x\left(\sqrt{x^2+\sqrt{x^4+1}}-x\sqrt{2}\right)$;

(12) $\lim\limits_{x \to -\infty} x\left(\sqrt{x^2+1}+x\right)$.

Exercise 12.1.2   *Use the Squeeze principle to prove:*

(1) $\lim\limits_{x \to 0} \lfloor 1/x \rfloor x = 1$;

(2) $\lim\limits_{x \to -\infty} \dfrac{x^2+1}{|x|\lfloor x \rfloor} = -1$;

(3) $\lim\limits_{x \to 0} \dfrac{\sqrt[100]{1+x^2}-1}{x}$;

(4) $\lim\limits_{x \to 0} x\, 1_{\mathbb{Q}}(x)$.

[Hint: For (3) use Exercise 5.8.5; for (4) use $|x\, 1_{\mathbb{Q}}(x)| \le |x|$.]

The following exercises are stated as theorems—the task is to prove them.

ⓘ **Exercise 12.1.3**   If $a \ne 0$ and $\hat{y} = a\hat{x}+b$ then $\lim\limits_{x \to \hat{x}} f(ax+b) = l$ if and only if $\lim\limits_{y \to \hat{y}} f(y) = l$. *Is this statement true if limits are replaced everywhere by limits from the right? From the left?* [Hint: No, but remains true if $a > 0$.]

ⓘ **Exercise 12.1.4**   If $a > 0$ then $\lim\limits_{x \to +\infty} f(ax+b) = l \Leftrightarrow \lim\limits_{y \to +\infty} f(y) = l$. *What happens if $a < 0$?*

ⓘ **Exercise 12.1.5** $\lim\limits_{x \to +\infty} f(x) = l \Leftrightarrow \lim\limits_{y \to -\infty} f(-y) = l$.

ⓘ **Exercise 12.1.6** $\lim\limits_{x \to \hat{x}} f(x) = l \Leftrightarrow \lim\limits_{h \to 0} f(\hat{x}+h) = l$. *State similar theorems for the limit from the right and for the limit from the left.*

(i) **Exercise 12.1.7** $\lim\limits_{x \downarrow \hat{x}} f(x) = \lim\limits_{t \to +\infty} f(\hat{x} + \frac{1}{t})$.

(i) **Exercise 12.1.8**   If $\lim\limits_{x \to c} f(x) = 0$ and there are positive δ and M such that $|g(x)| \leq M$ for $0 < |x - c| < \delta$ then $\lim\limits_{x \to c} f(x)g(x) = 0$.

12.2   The Cauchy definition

The great advantage of Definition 12.3 is that the theory of limits of sequences carries over easily to limits of functions. Sometimes it is a little inconvenient to prove that $f(x) \to l$ for *every* sequence $n \mapsto x_n$ with $x_n \to \hat{x}$ and $x_n \neq \hat{x}$. This inconvenience is often overcome by using Theorem 12.6 below, which is similar to the ε–N definition of a limit of a sequence.

If $\hat{x}, l \in \mathbb{R}$ and ε, δ are positive numbers then we call the set

$$\{(x, y); \ x \in \mathbb{R}, \ y \in \mathbb{R}, \ |x - \hat{x}| < \delta, \ |y - l| < \varepsilon\}$$

an *epsilon-delta box* centred at $(\hat{x}, l)$ (or briefly, an ε–δ box: see Figure 12.3). This ε–δ box is the interior of a rectangle with vertices at $(\hat{x} \pm \delta, l \pm \varepsilon)$. Theorem 12.6 below states that f has a limit l at $\hat{x}$ if for **every** $\varepsilon > 0$ there exists $\delta > 0$ such that the graph of f near the point $(\hat{x}, l)$ with the possible exception of $(\hat{x}, f(\hat{x}))$ lies in the ε–δ box centred on $(\hat{x}, l)$ (see Figure 12.3). The graph of f enters the ε–δ box at the left vertical side and leaves it at the right vertical sight of the box. This happens for *every* positive ε no matter how small ε is.

Theorem 12.6   *The function f has a limit l at $\hat{x}$ if and only if for every $\varepsilon > 0$ there exists $\delta > 0$ such that*

$$0 < |x - \hat{x}| < \delta \Rightarrow |f(x) - l| < \varepsilon. \tag{12.2}$$

Proof.
I. Let $\varepsilon > 0$ and assume there exists $\delta > 0$ such that the implication (12.2) holds. Let $x_n \to \hat{x}$ and $x_n \neq \hat{x}$, then there exists N such that $0 < |x_n - \hat{x}| < \delta$, and consequently $|f(\hat{x}) - l| < \varepsilon$ for $n > N$. This proves $f(x_n) \to l$ and hence $\lim\limits_{x \to \hat{x}} f(x) = l$.
II. Assume f has limit l but that it is not true that for every $\varepsilon > 0$ there exists δ such that the implication (12.2) holds. This means there exists $\varepsilon_0 > 0$ for which there is no suitable δ, and in particular $\delta = 1/n$ for $n \in \mathbb{N}$ is not suitable. In other words, for every $n \in \mathbb{N}$ there exists a number x_n

such that $0 < |x_n - \hat{x}| < 1/n$ but $|f(x_n) - l| \geq \varepsilon_0$. Now we have a sequence $n \mapsto x_n$ such that $x_n \neq \hat{x}$ and $x_n \to \hat{x}$ by the Squeeze principle,[2] but the sequence $n \mapsto f(x_n)$ does not converge to l. Hence l is not the limit of f at $\hat{x}$, a contradiction, and so there must exist a δ for every ε such that the implication (12.2) holds. □

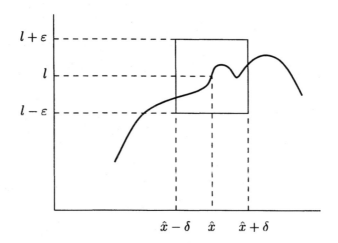

Fig. 12.3   Interpretation of Theorem 12.6

Remark 12.9   The interpretation of Theorem 12.6 with the ε δ box makes sense only if dom $f \subset \mathbb{R}$ and $l \in \mathbb{R}$. However the theorem itself is valid in the complex domain as well.

Remark 12.10   Theorem 12.6 remains valid for the limit from the right and the limit from the left if the inequality $0 < |x - \hat{x}| < \delta$ is replaced by the inequality $\hat{x} < x < \hat{x} + \delta$ and $\hat{x} - \delta < x < \hat{x}$, respectively.

Remark 12.11   The modification of Theorem 12.6 for limits at $+\infty$ or $-\infty$ reads: f has a limit l at $+\infty$ or $-\infty$ if and only if for every $\varepsilon > 0$ there exists K or k such that

$$x > K \Rightarrow |f(x) - l| < \varepsilon \quad \text{or} \quad x < k \Rightarrow |f(x) - l| < \varepsilon,$$

respectively.

---

[2]More precisely by Corollary 10.3.1.

Example 12.8   As an illustration of Theorem 12.6 we prove that $\lim\limits_{x \to -1} x^3 = -1$. In this case $f(x) - l = x^3 + 1$ and we rewrite this expression in a more convenient[3] form: $(x+1-1)^3 + 1 = (x+1)^3 - 3(x+1)^2 + 3(x+1)$. Consequently $|x^3 + 1| \leq |x + 1|(|x + 1|^2 + 3|x + 1| + 3)$. Therefore, if $|x + 1| < 1$ then $|x^3 + 1| \leq 7|x + 1|$. For $\varepsilon > 0$ we can choose $\delta = \text{Min}(\varepsilon/7, 1)$ and we have $|x + 1| < \delta \Rightarrow |x^3 + 1| \leq 7\delta < \varepsilon$.

Some authors use the implication (12.2) as the definition of a limit of f at $\hat{x}$. More precisely they say: 'f has a limit l by definition if for every positive ε there exists a positive δ such that implication (12.2) holds.' Such a definition is often referred to as Cauchy's definition of the limit. Often it is also called the ε–δ definition of the limit. Definition 12.3 is named after the German mathematician Heine.

Theorem 12.6, in conjunction with Remark 12.10, makes the following theorem fairly obvious.

Theorem 12.7   *Let I be an open interval and $f : I \mapsto \mathbb{R}$ has a limit at $\hat{x}$ if and only if $\lim\limits_{x \uparrow \hat{x}} f(x) = \lim\limits_{x \downarrow \hat{x}} f(x)$ and then $\lim\limits_{x \uparrow \hat{x}} f(x) = \lim\limits_{x \downarrow \hat{x}} f(x) = \lim\limits_{x \to \hat{x}} f(x)$.*

Proof.
I. If $\lim\limits_{x \to \hat{x}} f(x) = l$ then for every $\varepsilon > 0$ there exists $\delta > 0$ such that

$$0 < |x - \hat{x}| < \delta \Rightarrow |f(x) - l| < \varepsilon;$$

that is,

$$\hat{x} < x < \hat{x} + \delta \Rightarrow |f(x) - l| < \varepsilon,$$

and consequently $\lim\limits_{x \downarrow \hat{x}} f(x) = l$. The proof that $\lim\limits_{x \uparrow \hat{x}} f(x) = l$ is similar.
II. If $\lim\limits_{x \uparrow \hat{x}} f(x) = \lim\limits_{x \downarrow \hat{x}} f(x) = l$ then for every $\varepsilon > 0$ there exists $\delta_1 > 0$ and $\delta_2 > 0$ such that

$$\hat{x} - \delta_1 < x < \hat{x} \Rightarrow |f(x) - l| < \varepsilon, \tag{12.3}$$

and

$$\hat{x} < x < \hat{x} + \delta_2 \Rightarrow |f(x) - l| < \varepsilon, \tag{12.4}$$

---

[3]This really amounts to the use of the Taylor polynomial (see Section 6.5) but the situation is simple enough for some elementary algebra.

respectively. Let $\delta = \text{Min}(\delta_1, \delta_2)$. Then we have: if $0 < |x - \hat{x}| < \delta$ then either $\hat{x} - \delta_1 < x < \hat{x}$ or $\hat{x} < x < \hat{x} + \delta_2$, and in either case we have

$$|f(x) - l| < \varepsilon$$

by (12.3) or (12.4). $\square$

As another application of Theorem 12.6 we prove:

Theorem 12.8   Let $\text{rg}\, f \subset \mathbb{R}$ and $\text{rg}\, g \subset \mathbb{R}$. If $\lim\limits_{x \to \hat{x}} f(x) = l$, $\lim\limits_{x \to \hat{x}} g(x) = k$ and $l < k$ then there exists $\Delta > 0$ such that

$$f(x) < g(x)$$

when $0 < |x - \hat{x}| < \Delta$.

Proof.   The proof is analogous to the proof of Theorem 10.7. We choose $\varepsilon = \dfrac{k - l}{2}$. There exist $\delta_1 > 0$ and $\delta_2 > 0$ such that

$$0 < |x - \hat{x}| < \delta_1 \Rightarrow f(x) < l + \varepsilon = \frac{l + k}{2} \tag{12.5}$$

and

$$0 < |x - \hat{x}| < \delta_2 \Rightarrow \frac{l + k}{2} = k - \varepsilon < g(x). \tag{12.6}$$

Let $\Delta = \text{Min}(\delta_1, \delta_2)$, then if $0 < |x - \hat{x}| < \Delta$ then also $0 < |x - \hat{x}| < \delta_1$ and $0 < |x - \hat{x}| < \delta_2$, and then by (12.5) and (12.6)

$$f(x) < \frac{l + k}{2} < g(x)$$

follows. $\square$

Corollary 12.8.1   If $\lim\limits_{x \to \hat{x}} f(x) = l < k$ then there exists $\Delta > 0$ such that $f(x) < k$ when $0 < |x - \hat{x}| < \Delta$. Also, if $\lim\limits_{x \to \hat{x}} g(x) = k > l$ then there exists $\Delta > 0$ such that $g(x) > l$ when $0 < |x - \hat{x}| < \Delta$.

Particularly important cases of this corollary are $l = 0$ or $k = 0$; if the limit of f is positive at $\hat{x}$ then f is positive near $\hat{x}$ too (but not necessarily at $\hat{x}$ itself).

Theorem 12.9   *If* $\lim_{x \to \hat{x}} f(x) = l$ *and* $\lim_{x \to \hat{x}} g(x) = k$ *and if there exists a positive* σ *such that*

$$f(x) \leq g(x) \tag{12.7}$$

for $0 < |x - \hat{x}| < \sigma$, *then* $l \leq k$.

Proof.   Assume contrary to what we want to prove that $l > k$. Then by Theorem 12.8 (with the roles of f and g interchanged) $f(x) > g(x)$ for $0 < |x - \hat{x}| < \Delta$ for some $\Delta > 0$. Hence (12.7) does not hold if $0 < |x - \hat{x}| < \text{Min}(\sigma, \Delta)$, and this is a clear contradiction. $\square$

Analogues of Theorems 12.8 and 12.9 as well as Corollary 12.8.1 are valid for limits from the right, from the left, at $+\infty$ and at $-\infty$. Proofs of these are left for the exercises.

Bounded monotonic sequences always have limits. A similar theorem is valid for functions. The Cauchy definition is a convenient tool for the proof.

Theorem 12.10 (Limits of monotonic functions)
If $f :]a, b[\mapsto \mathbb{R}$ *is monotonic and bounded then*

$$\lim_{x \downarrow a} f(x) \quad and \quad \lim_{x \uparrow b} f(x)$$

exist.

Proof.   We shall assume that f is increasing: if it is not, the theorem for an increasing function can be applied to $-f$. We prove that the first limit is equal to

$$l = \inf \left\{ f(x); x \in]a, b[\right\}.$$

The proof that

$$\lim_{x \uparrow b} f(x) = \sup \left\{ f(x); x \in]a, b[\right\}$$

is very similar. Let $\varepsilon > 0$. By the definition of the greatest lower bound there exists $\underline{x} \in]a, b[$ such that

$$l \leq f(\underline{x}) < l + \varepsilon.$$

For $\underline{x} = a + \delta$ we now have, by monotonicity of f,

$$l \le f(x) \le f(\underline{x}) < l + \varepsilon,$$

for all x with $a < x < a + \delta$. ☐

A monotonic function, although having limits from right and left, need not have a limit. For example $\lim_{x \uparrow 0} \operatorname{sgn}(x) = -1$ and $\lim_{x \downarrow 0} \operatorname{sgn}(x) = 1$. Consequently there is no limit at 0.

There is also an analogue to Theorem 12.6 for limits at $+\infty$ or $-\infty$:

Theorem 12.11 *The function f has a limit l at ∞,*

$$\lim_{x \to +\infty} f(x) = l,$$

if and only if for every $\varepsilon > 0$ there exists K such that $x > K \Rightarrow$ $|f(x) - l| < \varepsilon$.

Proof. By Theorem 12.5 $\lim_{x \to +\infty} f(x) = l$ if and only if $\lim_{y \downarrow 0} f(1/y) = l$. This latter equation holds if and only if for every $\varepsilon > 0$ there exists $\delta > 0$ such that

$$0 < y < \delta \Rightarrow |f(1/y) - l| < \varepsilon.$$

We now set $K = 1/\delta$ and $x = 1/y$ and see that the above implication is equivalent to $x > K \Rightarrow |f(x) - l| < \varepsilon$. ☐

Remark 12.12 For the limit at $-\infty$ we have $\lim_{x \to -\infty} f(x) = l$ if and only if for every $\varepsilon > 0$ there exists k such that $x < k \Rightarrow |f(x) - l| < \varepsilon$.

Exercises

Exercise 12.2.1 *Using Theorem 12.6 prove that*

(1) $\lim_{x \to \hat{x}} (2x + 3) = 2\hat{x} + 3$,

(2) $\lim_{x \to a} x^2 = a^2$,

(3) $\lim_{x \to c} 1/x = 1/c$ *provided $c \ne 0$.*

[Hint: For (2) use the same trick as in Example 12.7; for (3) use the inequality $|1/x - 1/c| \le |x - c|/|xc| \le 2|x - c|/c^2$ for $|x| > |c/2|$.]

Exercise 12.2.2 *Using Theorem 12.6 prove that*

(1) $\lim\limits_{x\uparrow 0} \operatorname{sgn} x = -1$; (see Exercise 3.2.3 for the definition of $\operatorname{sgn} x$)

(2) $\lim\limits_{x\downarrow 0} \operatorname{sgn} x = 1$;

(3) $\lim\limits_{x\downarrow n} \lfloor x \rfloor = n \quad$ and $\lim\limits_{x\uparrow n} \lfloor x \rfloor = n - 1 \quad$ for $n \in \mathbb{Z}$;

(4) $\lim\limits_{x\downarrow n} \lfloor -x \rfloor = -n - 1 \quad$ and $\lim\limits_{x\uparrow n} \lfloor -x \rfloor = -n \quad$ for $n \in \mathbb{Z}$;

(5) $\lim\limits_{x\to 7} (\lfloor x \rfloor + \lfloor -x \rfloor) = -1$;

[Hint: For (5) you can use the results of (3) and (4) together with Theorem 12.7, but a direct proof is also easy.]

The following exercises are stated as theorems. The task is to prove them.

ⓘ **Exercise 12.2.3** If $\lim\limits_{x\downarrow \hat{x}} f(x) = l$, $\lim\limits_{x\downarrow \hat{x}} g(x) = k$ and $l < k$, then there exists $\delta > 0$ such that

$$\hat{x} < x < \hat{x} + \delta \Rightarrow f(x) < g(x).$$

State and prove an analogous theorem for the limit from the left.

ⓘ **Exercise 12.2.4** If $\lim\limits_{x\to +\infty} f(x) = l$, $\lim\limits_{x\to +\infty} g(x) = k$ and $l < k$ then there exists K such that

$$x > K \Rightarrow f(x) < g(x).$$

State and prove an analogous theorem for the limit at $-\infty$.

ⓘ **Exercise 12.2.5** If $\lim\limits_{x\to \hat{x}} f(x) = l$ then there exists $\delta > 0$ such that f is bounded on $(\hat{x} - \delta, \hat{x} + \delta)$. State and prove similar theorems for the limits from the right, from the left and for limits at $+\infty$ or $-\infty$.

ⓘ **Exercise 12.2.6** Theorem 12.9 remains valid if $x \to \hat{x}$ is replaced by $x \uparrow \hat{x}$ or by $x \downarrow \hat{x}$ and the inequalities $0 < |x - \hat{x}| < \sigma$ by $\hat{x} - \sigma < x < \hat{x}$ or by $\hat{x} < x < \hat{x} + \sigma$ respectively.

ⓘ **Exercise 12.2.7** If $\lim\limits_{x\to +\infty} f(x) = l$, $\lim\limits_{x\to +\infty} g(x) = k$ and if there exists K such that

$$x > K \Rightarrow f(x) \le g(x)$$

then $l \le k$.

⚠ **Exercise 12.2.8** If $x \in \mathbb{Q}$ let $x = p/q$ with $p \in \mathbb{Z}$ and $q \in \mathbb{N}$ and p and q relatively prime (the only positive integer dividing both is 1). Define *f as follows:* $f(x) = 1/q$ *if* $x \in \mathbb{Q}$ *and* $f(x) = 0$ *if* $x \notin \mathbb{Q}$. *Prove that* $\lim_{x \to \hat{x}} f(x) = 0$ *for* $\hat{x} \in \mathbb{Q}$.

12.3   Infinite limits

Functions, like sequences, may have infinite limits. The following definition covers the possibilities of a function having $+\infty$ or $-\infty$ as a limit.

> **Definition 12.5**   If in the definitions of a limit from the right at $\hat{x}$, limit from the left at $\hat{x}$, limit at $\hat{x}$, limit at $+\infty$ or limit at $-\infty$, the letter l denoting a real number is replaced by the symbol $+\infty$ or $-\infty$ one obtains the definition of an infinite limit. If f has a limit $+\infty$ from the right at $\hat{x}$, from the left at $\hat{x}$, at $\hat{x}$, at $+\infty$ or at $-\infty$, we write
> $$\lim_{x \downarrow \hat{x}} f(x) = +\infty, \ \lim_{x \uparrow \hat{x}} f(x) = +\infty, \ \lim_{x \to \hat{x}} f(x) = +\infty, \ \lim_{x \to +\infty} f(x) = +\infty$$
> or $\lim_{x \to -\infty} f(x) = +\infty$, respectively. A similar notation is used for the infinite limit $-\infty$.

Typical behaviour of functions which have infinite limits at $\hat{x}$ is sketched in Figure 12.4. For the function f shown

$$\lim_{x \to -\infty} f(x) = +\infty, \ \lim_{x \to \infty} f(x) = -\infty, \ \lim_{x \to 0} f(x) = \infty,$$

while

$$\lim_{x \uparrow 4} f(x) = \infty, \ \lim_{x \downarrow 4} f(x) = -\infty.$$

Limits from Definitions 12.1–12.4 are sometimes called finite limits, particularly if one wants to emphasize that the limit in question is not an infinite limit.

Example 12.9   Clearly $\lim_{x \to +\infty} x = +\infty$, $\lim_{x \to -\infty} x = -\infty$, $\lim_{x \to -\infty} x^2 = +\infty$,

$\lim_{x \downarrow 0} \frac{1}{x} = +\infty$, $\lim_{x \uparrow 0} \frac{1}{x} = -\infty$, $\lim_{x \to 0} \frac{1}{x^2} = +\infty$ but $\lim_{x \to 0} \frac{1}{x}$ does not exist, finite or infinite.

Many theorems concerning finite limits can be extended to infinite limits, but not all can and care is needed. For instance, if $\lim_{x \to \hat{x}} f(x) = +\infty$ then $\lim_{x \to \hat{x}} af(x) = +\infty$ if $a > 0$ and $\lim_{x \to \hat{x}} af(x) = -\infty$ if $a < 0$ and obviously $\lim_{x \to \hat{x}} af(x) = 0$ if $a = 0$. Some extensions of Theorem 12.3 are discussed in Exercises 12.3.3, 12.3.5 and 12.3.7. Generally speaking, no

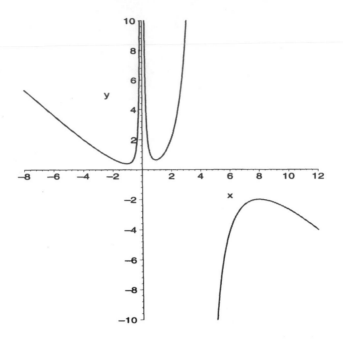

Fig. 12.4   Behaviour of functions with infinite limits

conclusion can be made concerning $\lim\limits_{x \to \hat{x}} (f(x) + g(x))$ if $\lim\limits_{x \to \hat{x}} f(x) = +\infty$ and $\lim\limits_{x \to \hat{x}} g(x) = -\infty$. For questions of this kind see Exercises 12.3.4, 12.3.6 and 12.3.7.

An analogue of the Cauchy definition for infinite limits is the following:

Theorem 12.12   Let $\operatorname{rg} f \subset \mathbb{R}$. Then $\lim\limits_{x \to \hat{x}} f(x) = +\infty$ if and only if for any K there exists $\delta > 0$ such that

$$0 < |x - \hat{x}| < \delta \Rightarrow f(x) > K. \tag{12.8}$$

Proof.

I. Take any K and assume that there exists $\delta > 0$ such that (12.8) holds. Let $x_n \neq \hat{x}$ and $x_n \to \hat{x}$, then there exists N such that $0 < |x_n - \hat{x}| < \delta$ for $n > N$. By (12.8), $f(x_n) > K$ for $n > N$. This proves that $\lim\limits_{x \to \hat{x}} f(x) = +\infty$.

II. Assume $\lim\limits_{x \to \hat{x}} f(x) = +\infty$ but that for some K_0 there does not exist $\delta > 0$ such that (12.8) holds. Then for every $n \in \mathbb{N}$ there exists a number x_n such that $0 < |x_n - \hat{x}| < 1/n$ but $f(x_n) \leq K_0$. Then $x_n \neq \hat{x}$ and $x_n \to \hat{x}$ but

$f(x_n) \to +\infty$ is false. Consequently $\lim_{x \to \hat{x}} f(x) = +\infty$ is false, which is a contradiction. $\qquad\qquad\qquad\qquad\qquad\qquad\qquad\qquad\qquad\qquad\qquad\square$

Remark 12.13 Theorem 12.12 remains valid for the limit from the right and for the limit from the left if the inequality $0 < |x - \hat{x}| < \delta$ is replaced by $\hat{x} - \delta < x < \hat{x}$ and by $\hat{x} < x < \hat{x} + \delta$ respectively.

Remark 12.14 It is an immediate consequence of Theorem 12.12 that if $f \leq g$ and $\lim_{x \to \hat{x}} f(x) = +\infty$ then $\lim_{x \to \hat{x}} g(x) = +\infty$. This remains valid for limits from the right, from the left, and limits at $+\infty$ and $-\infty$.

Example 12.10 We have $\lim_{x \to +\infty} (2x - \lfloor x \rfloor) = +\infty$. It follows from the definition of the floor function that $x \geq \lfloor x \rfloor$ and therefore $2x - \lfloor x \rfloor \geq x$. The result then follows from Example 12.9 and Remark 12.14.

Exercises

Exercise 12.3.1 *Find the following limits.*

(1) $\displaystyle\lim_{x \downarrow \sqrt{2}} \frac{1}{x^2 - 2}$;

(2) $\displaystyle\lim_{x \uparrow \sqrt{2}} \frac{1}{x^2 - 2}$;

(3) $\displaystyle\lim_{x \to -\infty} (\sqrt{x^2 + 1} - x)$;

(4) $\displaystyle\lim_{x | -1} \frac{2x + 1}{x^3 + 1}$;

(5) $\displaystyle\lim_{x \to -\infty} x^n, \quad n \in \mathbb{N}$;

(6) $\displaystyle\lim_{x \uparrow 0} \frac{1}{x^n}, \quad \lim_{x \downarrow 0} \frac{1}{x^n}, \quad n \in \mathbb{N}$;

(7) $\displaystyle\lim_{x \to +\infty} \frac{x^3 + 1}{x^2 + 5x - 7}$;

(8) $\displaystyle\lim_{x \to -\infty} \frac{x^3 + 1}{x^2 + 5x - 7}$;

(9) $\displaystyle\lim_{x \uparrow 0} 2^{\lfloor 1/x \rfloor}$;

(10) $\displaystyle\lim_{x \downarrow 0} 2^{\lfloor 1/x \rfloor}$;

(11) $\displaystyle\lim_{x \uparrow 0} \frac{1 + 2^{\lfloor 1/x \rfloor}}{2^{\lfloor 1/x \rfloor}}$;

(12) $\displaystyle\lim_{x \downarrow 0} \frac{2^{\lfloor 1/x \rfloor}}{1 + 2^{\lfloor 1/x \rfloor}}$.

ⓘ **Exercise 12.3.2**   *Prove:* $\lim\limits_{x \to \hat{x}} \dfrac{1}{f(x)} = 0 \Leftrightarrow \lim\limits_{x \to \hat{x}} |f(x)| = +\infty$. *Also prove this for* $\lim\limits_{x \uparrow \hat{x}}$ *and* $\lim\limits_{x \downarrow \hat{x}}$. [*Hint: Use Exercise 10.5.5.*]

ⓘ **Exercise 12.3.3**   *Prove: If* $\lim\limits_{x \uparrow \hat{x}} f(x) = +\infty$ *and* g *is bounded below on some interval* $(\hat{x} - \delta, \hat{x})$ *then* $\lim\limits_{x \uparrow \hat{x}} (f(x) + g(x)) = +\infty$. *Deduce that* $\lim\limits_{x \uparrow \hat{x}} (f(x) + g(x)) = +\infty$ *if* $\lim\limits_{x \uparrow \hat{x}} f(x) = +\infty$ *and* $\lim\limits_{x \uparrow \hat{x}} g(x) = l$ *or* $+\infty$.

ⓘ **Exercise 12.3.4**   *Let* F *be an arbitrary function with* $\operatorname{dom} F = (-\infty, \hat{x})$. *Find functions* f *and* g *such that* $\lim\limits_{x \uparrow \hat{x}} f(x) = +\infty$, $\lim\limits_{x \uparrow \hat{x}} g(x) = -\infty$ *and* $f + g = F$. *Deduce that* $\lim\limits_{x \uparrow \hat{x}} (f(x) + g(x))$ *can be any number,* $+\infty$, *or* $-\infty$, *or need not exist at all.* [*Hint:* $2f(x) = |F(x)| + F(x) + \dfrac{1}{\hat{x} - x}$, $g(x) = F(x) - f(x)$.]

ⓘ **Exercise 12.3.5**   *Prove: If* $\lim\limits_{x \uparrow \hat{x}} f(x) = +\infty$ *and there are positive* m *and* δ *such that* $\hat{x} - \delta < x < \hat{x} \Rightarrow g(x) \geq m$ *then* $\lim\limits_{x \uparrow \hat{x}} f(x)g(x) = +\infty$.

ⓘ **Exercise 12.3.6**   *Let* F *be an arbitrary function with* $\operatorname{dom} F = (-\infty, \hat{x})$. *Find* f *and* g *such that* $\lim\limits_{x \uparrow \hat{x}} f(x) = +\infty$, $\lim\limits_{x \uparrow \hat{x}} g(x) = 0$ *and* $fg = F$. *What can you say about* $\lim\limits_{x \uparrow \hat{x}} f(x)g(x)$? [*Hint:* $f(x) = \dfrac{|F(x)| + 1}{\hat{x} - x}$, $g(x) = \dfrac{F(x)}{f(x)}$. *The limit need not exist but if it does, it can equal anything.*]

ⓘ **Exercise 12.3.7**   *Prove: If* f *is bounded on some interval* $\hat{x} - \delta < x < \hat{x}$ *and* $\lim\limits_{x \uparrow \hat{x}} g(x) = +\infty$ *then* $\lim\limits_{x \uparrow \hat{x}} \dfrac{f(x)}{g(x)} = 0$, *and if moreover* $f(x) > 0$ *for* $\hat{x} - \delta < x < \hat{x}$ *then* $\lim\limits_{x \uparrow \hat{x}} \dfrac{g(x)}{f(x)} = +\infty$.

ⓘ **Exercise 12.3.8**   *State analogues of Exercises 12.3.3–12.3.7 for limits, limits from the right, and limits at* $+\infty$ *and* $-\infty$.

12.4 Continuity at a point

A function f is continuous at a point $\hat{x}$ if it is possible to find an approximate value of $f(\hat{x})$ from an approximate value of $\hat{x}$. Formally, we make the following definition:

Definition 12.6 (Continuity) A function f is said to be continuous at $\hat{x}$, continuous from the right at $\hat{x}$ or continuous from the left at $\hat{x}$ if

$$\lim_{x \to \hat{x}} f(x) = f(\hat{x}), \ \lim_{x \downarrow \hat{x}} f(x) = f(\hat{x}), \ \text{or} \ \lim_{x \uparrow \hat{x}} f(x) = f(\hat{x}),$$

respectively.

Example 12.11 By Example 12.8 the function $x \mapsto x^3$ is continuous at 3, the *floor* function is obviously continuous from the right at $n \in \mathbb{N}$ since $\lim_{x \downarrow n} \lfloor (x) \rfloor = n = \lfloor n \rfloor$ and similarly the function *ceil* is continuous from the left at $n \in \mathbb{N}$. More generally, every polynomial is continuous at every $\hat{x}$, the greatest integer function is continuous from the right at every point and the *ceil* function is continuous from the left at every point.

Usually, we say discontinuous instead of not continuous. For example the floor function is discontinuous from the left at any integer and the function *ceil* is discontinuous from the right at any $n \in \mathbb{Z}$. In mathematics it is sometimes convenient to consider functions which have values ∞ or $-\infty$. We do not do this in this book and, in particular, in Definition 12.6 the limit is understood as a finite limit. However, even in situations where infinite values are allowed, the function is considered to be discontinuous where $\lim_{x \to \hat{x}} f(x) = f(\hat{x}) = \pm\infty$.

In the past, mathematicians believed that for a function to be continuous everywhere was normal and to be discontinuous was an exception. To a certain extent this is somewhat tenable even today in the sense that most functions encountered in applications are continuous apart from a few points. However, today some functions discontinuous at many or even all points of the domain of definition play an important part in mathematics and in applications. Some examples of functions with discontinuities are discussed in Exercises 12.4.4–12.4.6. Obviously there are many functions continuous everywhere, for instance polynomials. The exercises contain examples of functions continuous at one point only, functions discontinuous everywhere and functions continuous at all irrational points and discontinuous at all rational points. It is an interesting fact that there is no function continuous at all rational points and discontinuous at all irrational points. The proof of this is beyond our reach at this stage.

If f is discontinuous at $\hat{x}$ and $\lim_{x \to \hat{x}} f(x)$ exists, the discontinuity at $\hat{x}$ is called removable. This is because the definition of f can be amended by declaring $f(\hat{x}) = \lim_{x \to \hat{x}} f(x)$ and thus making the function continuous at $\hat{x}$.

The changed function is, strictly speaking, a new function but it is customary to keep denoting it as f and identify it with the old f. Understanding of this point sometimes helps in dealing with some computer system algebra. For instance in *Maple* substituting $x = 2$ is, quite naturally, not allowed in $h(x) = (x - 2)^2/(x^2 - 5x + 6)$ but the *Maple* solution of $h(x) = 0$ is $x = 2$. *Maple* automatically removes the singularity at $x = 2$ by setting $h(2) = \lim_{x \to 2} h(x)$ and solves the equation after that.

Almost all statements about limits of functions translate into statements about continuity. Here we illustrate this by several corollaries to theorems or definitions which we need for future reference. One corollary to Theorem 12.7 is:

Corollary 12.7.1 (to Theorem 12.7)   *A function f is continuous at $\hat{x}$ if and only if it is continuous both from the left and from the right at $\hat{x}$.*

The Cauchy definition of a limit translates into the Cauchy definition of continuity as follows.

Corollary 12.6.1 (to Theorem 12.6)   *f is continuous at $\hat{x}$ if and only if for every $\varepsilon > 0$ there exists $\delta > 0$ such that*

$$|x - \hat{x}| < \delta \Rightarrow |f(x) - f(\hat{x})| < \varepsilon. \qquad (12.9)$$

It is a detail but an important one: the inequality $0 < |x - \hat{x}|$ from Theorem 12.6 is now omitted.

The Corollary remains valid if the word continuity is replaced by continuity from the left or continuity from the right and the inequality $|x - \hat{x}| < \delta$ by $\hat{x} - \delta < x \leq \hat{x}$ or by $\hat{x} \leq x \leq \hat{x} + \delta$, respectively.

Corollary 12.2.1 (to Definitions 12.1 and 12.2)   *f is continuous or continuous from the right or continuous from the left at $\hat{x}$ if and only if for every $x_n \to \hat{x}$ or $x_n \to \hat{x}$, $x_n \geq \hat{x}$ or $x_n \to \hat{x}$, $x_n \leq \hat{x}$, respectively, it is true that $f(\hat{x}_n) \to f(\hat{x})$.*

The only if part is not entirely obvious since now $x_n = \hat{x}$ for some n is possible. However the above Corollary can be proved using the Cauchy definition of continuity and the method of part II of the proof of Theorem 12.6. The next Corollary which is a counterpart of Theorem 12.3 is very useful for an instant recognition of whether or not a given function is continuous at some point. This helps in graphing functions in *Maple*. To obtain a correct graph of a function with a discontinuity use the option discont=true. The following *Maple* session illustrates this.

```
>   plot(signum(x-1),x=-1..3);
>   plot(signum(x-1),x=-1..3,discont=true);
```

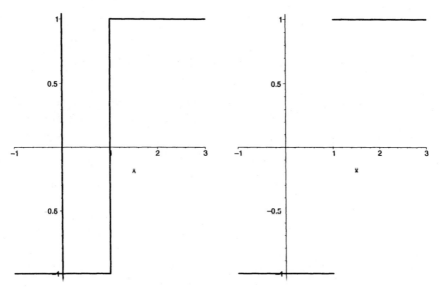

Fig. 12.5 *Maple* plotting discontinuities

Corollary 12.3.1 (to Theorem 12.3)

If f and g are continuous (continuous from the right, continuous from the left) at $\hat{x}$ then so are the functions

    (i) $x \mapsto |f(x)|$,

    (ii) $x \mapsto (f(x) \pm g(x))$,

    (iii) $x \mapsto f(x)g(x)$,

    (iv) $x \mapsto \dfrac{f(x)}{g(x)}$ *provided that $g(\hat{x}) \neq 0$*,

    (v) $x \mapsto [f(x)]^r$ *provided $f(\hat{x}) \neq 0$ and $r \in \mathbb{Q}$*,

    (vi) $x \mapsto \mathrm{Max}(f(x), g(x))$,

          $x \mapsto \mathrm{Min}(f(x), g(x))$.

Corollary 12.8.2 (to Theorem 12.8)   *Let* $\operatorname{rg} f \subset \mathbb{R}$ *and* $\operatorname{rg} g \subset \mathbb{R}$. *If* f *and* g *are continuous at* $\hat{x}$ *and* $f(\hat{x}) < g(\hat{x})$ *then there exists a positive* Δ *such that*

$$f(x) < g(x)$$

when $|x - \hat{x}| < \Delta$.

Remark 12.15   Important consequences of this Corollary are the cases when f or g are constant. For instance, if $f(\hat{x}) < M$ or $g(\hat{x}) > 0$ then $f(x) < M$ or $g(x) > 0$, respectively, for $|x - \hat{x}| < \Delta$. Similar statements hold for continuity from the right and from the left, for instance if f is continuous from the right at $\hat{x}$ and $f(\hat{x}) < M$ then there is a $\Delta > 0$ such that $f(x) < M$ for $\hat{x} \leq x < \hat{x} + \Delta$.

A very important tool for establishing continuity is the next theorem on the continuity of composite functions. It says that a composition of two continuous functions is continuous. More precisely, we have

---

Theorem 12.13   *If* g *is continuous at* $\hat{x}$ *and* f *is continuous at* $\hat{y} = g(\hat{x})$ *then* $f \circ g$ *is continuous at* $\hat{x}$.

Proof.   Let $x_n \to \hat{x}$. By Corollary 12.2.1 and continuity of g at $\hat{x}$ we have that $y_n = g(x_n) \to g(\hat{x})$, by the same Corollary and continuity of f at $\hat{y}$ we also have that $y_n \to f(\hat{y})$. In other words if $\hat{x}_n \to \hat{x}$ then $f \circ g(x_n) \to f \circ g(\hat{x})$. This, by Corollary 12.2.1, proves the theorem. $\square$

Example 12.12   A polynomial P is continuous everywhere and the square root function is continuous at every $\hat{y} > 0$. Hence the function $x :\mapsto \sqrt{P(x)}$ is continuous at every $\hat{x}$ for which $P(\hat{x}) > 0$.

Remark 12.16   Theorem 12.13 is important and has a simple wording and a simple proof. In contrast to theorems about continuity considered so far it is *NOT* valid if continuity is replaced by continuity from the right or left throughout: see Exercise 12.4.7. Also, in contrast to previously stated theorems and corollaries it does not translate directly to limits: see Exercise 12.4.8.

Exercises

ⓘ **Exercise 12.4.1** *It follows from Theorem 12.3 that a polynomial is continuous at every point $y \in \mathbb{R}$. Give a direct $\varepsilon - \delta$-proof.* [Hint: $P(x) - P(y) = q(x,y)(x-y)$ with $q(x,y)$ bounded for $|x-y| < 1$, say by K. Then, given $\varepsilon > 0$, choose $\delta = \varepsilon/K$.]

ⓘ **Exercise 12.4.2** *Why are the functions f_1, f_2 and f_3 with $f_1(x) = x^3/|x|$, $f_2(x) = x^3/|x|$ and $f_2(0) = 1$, $f_3(x) = x|x|$ all distinct? Show that if removable singularities are cleared off then all three become equal.* [Hint: All differ at 0 and two functions are equal if and only if they have the same domain of definition and have the same values for the same x. All three are equal for $x \neq 0$ and f_3 is continuous at 0.]

Exercise 12.4.3 *Decide which of the following functions have removable singularities at 0.*

(1) $f(x) = \dfrac{(1+x)^{10} - 1}{x}$;

(2) $g(x) = \dfrac{(1+x)^{10} - 1}{x^2}$;

(3) $h(x) = \dfrac{\sqrt{1-x} - 1}{x}$;

(4) $F(x) = \lceil x \rceil - \lfloor x \rfloor$;

(5) $G(x) = \lfloor x \rfloor - x$;

(6) $H(x) = x(x - \lfloor x \rfloor)$.

Prove the statements in the next three Exercises.

ⓘ **Exercise 12.4.4** *The function $x \mapsto x1_\mathbb{Q}(x)$ is continuous only at 0.*

ⓘ **Exercise 12.4.5** *The function $1_\mathbb{Q}$ is discontinuous everywhere.* (*It is called the Dirichlet's function.*)

Exercise 12.4.6 *Let p, q denote relatively prime positive integers, with $p < q$. For $x = p/q$ let $f(x) = 1/q$, $f(0) = f(1) = 1$ and $f(x) = 0$ for all other $x \in [0,1]$. This function is continuous at all irrational points and discontinuous at all rational points of $[0,1]$.* [Hint: If $\hat{x}$ is rational then by Exercise 4.6.6 there is a sequence $n \mapsto x_n \notin \mathbb{Q}$ with $x_n \to \hat{x}$ and $f(x_n) = 0$. On the other hand, if $\hat{x}$ is irrational, for any $\varepsilon > 0$ there are only finitely many rational points p/q with $1/q \geq \varepsilon$. Consequently, there is a $\delta > 0$ such that the set $\{\hat{x}; |x - \hat{x}| < \delta\}$ does not contain any of these points. For $|x - \hat{x}| < \delta$ either $f(x) = 0$ or $0 < f(x) = f(p/q) < 1/q < \varepsilon$.]

ⓘ **Exercise 12.4.7**   *Both of the floor function and* $g : x \mapsto -x^2$ *are contin-uous from the right at* 0 *and at* $g(0) = 0$ *but* $f \circ g$ *is not continuous from the right at* 0. *Show this. Give an example of* F, G *continuous from the left at* $G(1)$ *and* 1, *respectively, such that* $F \circ G$ *is not continuous from the left at* 1. [*Hint:* $f \circ g$ *is* -1 *on* $[0, 1[$ *and* $f \circ g(0) = 0$. $F(x) = \lceil x \rceil$, $G(x) = 2 - x$.]

‼ ⓘ **Exercise 12.4.8**   *The direct translation of Theorem 12.13 into limits is false. Show that if* $\lim\limits_{x \to \hat{x}} g(x) = \hat{y}$ *and* $\lim\limits_{y \to \hat{y}} f(y) = l$ *then* $\lim\limits_{x \to \hat{x}} f \circ g(x)$ *need not be* l. [*Hint:* $f(x) = \lceil x \rceil - \lfloor x \rfloor$, $g(x) = 0$, $\lim\limits_{x \to 2} f \circ g(x) = 0$, $\lim\limits_{y \to 0} f(y) = 1$.]

12.5   Continuity of functions on closed bounded intervals

> **Definition 12.7**   A function f is said to be continuous on the closed interval $[a, b]$ if it is continuous at every point c such that $a < c < b$, continuous from the left at b and continuous from the right at a.

This definition can be rephrased by saying that f is continuous on $[a, b]$ if it is continuous from the left at every point $c \in]a, b]$ and continuous from the right at every point $c \in [a, b[$.

When drawing a graph of a function f continuous on $[a, b]$ with $f(a) < 0$ and $f(b) > 0$ one has to cross the x-axis or lift the pencil from the paper. This leads to the following theorem.

> **Theorem 12.14**   *If* $f : [a, b] \mapsto \mathbb{R}$ *is continuous with* $f(a) < 0$ *and* $f(b) > 0$ *then there exists a point* $c \in]a, b[$ *such that* $f(c) = 0$.

Proof.   We denote $[a, b]$ by $[a_0, b_0]$ and consider the intervals $\left[a, \dfrac{a + b}{2}\right]$ and $\left[\dfrac{a + b}{2}, b\right]$. If $f\left(\dfrac{a + b}{2}\right) = 0$ the proof ends, otherwise one of the intervals will have the property that f is negative at the left end point and positive at the right end point. We denote this interval $[a_1, b_1]$ and continue the process. Either we find a point c with $f(c) = 0$ in a finite number of steps, or we have constructed a sequence of intervals $[a_n, b_n]$ such that $[a_n, b_n] \supset [a_{n+1}, b_{n+1}]$, and $b_n - a_n = \dfrac{b - a}{2^n} \to 0$. By the Nested Intervals Lemma (Theorem 10.17) there exists a number c such that $a_n \uparrow c$

and $b_n \downarrow c$. By construction

$$f(a_n) < 0 \qquad (12.10)$$

and

$$f(b_n) > 0. \qquad (12.11)$$

By taking limits as $n \to \infty$ in (12.11) and (12.10) we have $f(c) \le 0$ and $f(c) \ge 0$, and consequently $f(c) = 0$. $\qquad\square$

Remark 12.17 Despite its plausibility, Theorem 12.14 does require a proof. We have to realize that intuition and the concept of continuity do not fully agree; for example there are continuous functions which behave so badly they cannot be graphed. Moreover, Theorem 12.14 has to be proved as a matter of principle: we are obliged to prove it from the definition of continuity (and its established consequences) and not rely on drawings or our feelings.

The idea of the proof can be used to find the point c approximately. If we want to find c with an error not exceeding ε, we find n so large that $b_n - a_n < \varepsilon$ and then the approximate value $\bar{c} = (a_n + b_n)/2$ differs from c by less than ε. This can be easily programmed on a computer, and the method has the advantage that we have good control over the error. See Example A.3 in the Appendix.

Example 12.13 For $f : x \mapsto x^5 + x + 1$ we have $f(-1) = -1$ and $f(0) = 1$. Since f is continuous by Example 12.8, the equation $x^5 + x + 1$ has a solution in the interval $[-1, 0]$. Since f is obviously strictly increasing this is the only solution.

Example 12.14 Every cubic equation

$$f(x) = x^3 + ax^2 + bx + c = 0$$

with $a, b, c \in \mathbb{R}$ has at least one real solution. Since

$$\lim_{x \to -\infty} \left(1 + \frac{a}{x} + \frac{b}{x^2} + \frac{c}{x^3} \right) = 1,$$

there exists $k < 0$ such that

$$1 + \frac{a}{x} + \frac{b}{x^2} + \frac{c}{x^3} > \frac{1}{2},$$

for $x \le k$, and consequently

$$f(k) < \frac{1}{2} k^3 < 0.$$

Similarly there exists $K > 0$ such that

$$f(K) > \frac{1}{2}K^3 > 0.$$

By Theorem 12.14 there exists $c \in [k, K]$ such that $f(c) = 0$.

The following generalization of Theorem 12.14 is often used.

Theorem 12.15 (The Intermediate Value Theorem)
If $f : [a, b] \mapsto \mathbb{R}$ is continuous and α lies between $f(a)$ and $f(b)$ then there exists a point $c \in [a, b]$ such that $f(c) = \alpha$.

Proof. If $f(a) = f(b)$ then $\alpha = f(a)$ and the theorem is obvious. If $f(a) < f(b)$ we define $F : x \mapsto f(x) - \alpha$. Clearly F is continuous, $F(a) < 0$ and $F(b) > 0$. Therefore by Theorem 12.14 there exists a number c such that $F(c) = 0$; that is, $f(c) - \alpha = 0$. If $f(a) > f(b)$ one can apply the already established result on $-f$.   □

Example 12.15   The equation

$$f(x) = x^3 - 3\sqrt{x} + 1 = 0$$

has a solution in $[1, 2]$. Indeed $f(1) < 0$ and $f(2) > 0$ and the Intermediate value theorem applies with $\alpha = 0$.

The intermediate value theorem plays an important role in our next theorem on continuity of the inverse function. We have:

Theorem 12.16   *If f is strictly increasing and continuous on an interval I then*

  (i) *the inverse f_{-1} is strictly increasing on $f(I)$;*
  (ii) *$f(I)$ is an interval;*
  (iii) *f_{-1} is continuous on $f(I)$.*

Remark 12.18   The theorem remains valid if the phrase *strictly increasing* is replaced everywhere by *strictly decreasing*.

Proof.   Let $y_1, y_2 \in \operatorname{rg} f = \operatorname{dom} f_{-1}$ with $y_1 < y_2$. Assuming $f_{-1}(y_1) \geq f_{-1}(y_2)$ leads to

$$y_1 = f(f_{-1}(y_1)) \geq y_2 = f(f_{-1}(y_2)),$$

which is a contradiction. This proves (i).

For every $n \in \mathbb{N}$ let $b_n = b$ if b, the right-hand end-point of I, belongs to I. Otherwise let $b_n \uparrow b$ if I is bounded from above and $b_n \to \infty$ if I is not bounded above. Define a_n similarly at the left of I. Then by the intermediate value theorem each $[f(a_n, f(b_n)]$ is a part of $f(I)$ and moreover

$$f(I) = \bigcup_{n=1}^{\infty} [f(a_n), f(b_n)]$$

is an interval.

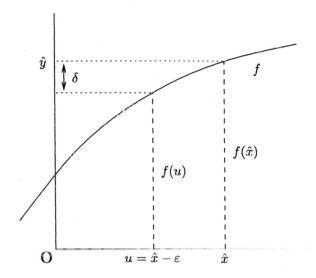

Fig. 12.6   Continuity of the inverse from the left.

To prove (iii) we denote by $\hat{y}$ a point of $f(I)$ which is not a left-hand end point of $f(I)$ and $\hat{y} = f(\hat{x})$, that is $\hat{x} = f_{-1}(\hat{y})$. For a positive ε let $\delta = f(\hat{x} - \varepsilon) - f(\hat{x})$. (See Figure 12.6, which is provided for illustration, not as a proof.) For $\hat{y} - \delta < y \le \hat{y}$ we have, by the monotonicity of f_{-1}, that

$$f_{-1}(y) > f_{-1}(\hat{y} - \delta) = \hat{x} - \delta = f_{-1}(\hat{y}) - \varepsilon, \tag{12.12}$$
$$f_{-1}(y) \le f_{-1}(\hat{y}) < f(\hat{y}) + \varepsilon. \tag{12.13}$$

The continuity from the right can be proved similarly. $\square$

Example 12.16   By Exercise 5.5.6 the function $x \mapsto x^n$, for an odd $n \in \mathbb{N}$, is strictly increasing on $\mathbb{R}$. Obviously it is continuous everywhere. Hence the inverse function $x \mapsto x^{\frac{1}{n}}$ is continuous everywhere where it is defined, and that is on $\mathbb{R}$. An alternative notation commonly used is $x^{\frac{1}{n}} = \sqrt[n]{x}$. In *Maple* it is necessary to use surd(x), but the other two symbols can have different meanings in *Maple*.

The theorem asserts that $f(I)$ is an interval and if I is bounded and closed then so is $f(I)$. However, if I is bounded but not closed, the set $f(I)$ need not be bounded. The function $x \mapsto 1/x^2, x \in]0,1]$ has as the inverse $y \mapsto 1/\sqrt{y}$, $y \in [1,\infty[$. The original function had bounded domain but the domain of the inverse was unbounded. Exercise 12.5.5 shows the importance of the assumption that the domain of f was an interval.

The next few theorems extend local properties of functions to global properties. Typical of these theorems is Theorem 12.18, proof of which establishes boundedness of the function, a global property, from continuity of the function at every point, a local property. An important tool for the proofs is the Cousin Lemma below.

A finite sequence of points t_k, $k = 0, 1, \ldots, n$ forms a division T of an interval $[a, b]$ if

$$a = t_0 < t_1 < t_2 < \cdots < t_n = b. \tag{12.14}$$

The intervals $[t_k, t_{k+1}]$ are called subintervals of the division. If there is an X_k in every subinterval $[t_{k-1}, t_k]$ for $k = 1, \ldots, n$ then we call

$$a = t_0 \leq X_1 \leq t_1 \leq X_2 \leq t_2 < \cdots \leq t_{n-1} \leq X_n \leq t_n = b \tag{12.15}$$

a tagged division of $[a, b]$. The points X_k are called tags, and X_k tags the interval $[t_{k-1}, t_k]$. We shall denote the tagged division 12.15 with dividing points $T = \{t_0, t_1, \ldots, t_n\}$ and tags $X = \{X_1, X_2, \ldots, X_n\}$ by TX. A function is called a gauge if it is positive everywhere where it is defined. If δ is a gauge on $[a, b]$ then the tagged division TX is called δ-fine if

$$X_k - \delta(X_k) < t_{k-1} < t_k < X_k + \delta(X_k) \quad \text{for all} \quad k = 1, 2, \ldots, n. \tag{12.16}$$

In proofs, a gauge is often used to control the size of subintervals of a tagged division. Of course, any other letter can be used in place of δ but use of this letter is traditional and a gauge usually enters a proof from some related ε–δ definition. We are now ready for[4]

---

[4]Lemma means an auxiliary theorem. History of mathematics knows a number of examples where a lemma becomes so important that it gains a status of a Theorem.

Theorem 12.17 (The Cousin Lemma) *For every gauge δ there is a δ-fine tagged division of a closed bounded interval $[a, b]$.*

Proof. We proceed indirectly. If there is no δ-fine tagged division of $[a, b]$ then there is no δ-fine tagged division of one of the intervals $[a, (a + b)/2]$, $[(a + b)/2, b]$. If it were then merging these two tagged divisions together would define a δ-fine tagged division of $[a, b]$. Denote this interval $[a_1, b_1]$ and continue the process indefinitely. We obtain a sequence of nested closed bounded intervals $[a_n, b_n]$ and none has a δ-fine tagged division. Since $b_n - a_n = (b - a)/2^{-n} \to 0$ there is a point $c \in [a_n, b_n]$ for every $\mathbb{N}$. For $\varepsilon = \delta(c) > 0$ there is $m \in \mathbb{N}$ such that $b_m - a_m < \delta(c)$. Consequently

$$c - \delta(c) < a_m < b_m < c + \delta(c).$$

This means that $a_m \leq c \leq b_m$ is a δ-fine tagged division of $[a_m, b_m]$, contradicting the definition of $[a_m, b_m]$. $\square$

Cousin's Lemma looks obvious, but see Exercise 12.5.6. The first application of Cousin's Lemma is:

Theorem 12.18 (Continuous functions are bounded) *If f is continuous on a closed and bounded interval $[a, b]$ then the function f is bounded on $[a, b]$.*

Proof. By Corollary 12.3.1 the function $|f|$ is continuous on $[a, b]$. By Remark 12.15 there is a positive δ such that $|f(x)| < |f(X)| + 1$ for $|x - X| < \delta$. Let TX be a δ-fine tagged division 12.15 and $M = 1 + \text{Max}(f(X_1), f(X_2), \ldots f(X_n))$. If $x \in [a, b]$ then for some natural k the point x lies in the subinterval $[t_{k-1}, t_k]$ and consequently $|x - X_k| < \delta$. It follows that

$$|f(x)| < |f(X_k)| + 1 \leq \text{Max}(f(X_1), f(X_2), \ldots f(X_n)) + 1 = M.$$

$\square$

We emphasized that $[a, b]$ was bounded and closed. The importance of this assumption is illuminated in Exercise 12.5.7. The next theorem says that a continuous f attains its maximum and minimum on $[a, b]$.

This is so here.

Theorem 12.19 (Weierstrass on extreme values) *If f is continuous on $[a, b]$ then there exist c, d in this interval such that*

$$f(c) \leq f(x) \leq f(d)$$

for all x in $[a, b]$.

Proof. We denote by $m = \inf \{f(x); x \in [a, b]\}$. For every natural n there is a point x_n such that $f(x_n) < m + 1/n$. By the Bolzano-Weierstrass Theorem there is a convergent subsequence, say $x_{n_k} \to c$ as $k \to \infty$. By Corollary 12.2.1 we obtain $f(x_{n_k}) \to f(c)$ and by Theorem 12.4 that $f(x_{n_k}) \to m$. Hence $f(c) = m$, and by Definition of the greatest lower bound $f(x) \geq m = f(c)$. The proof for d is similar or one can apply the already established part of the theorem to $-f$. □

This Weierstrass Theorem is theoretical in its character. However we shall see in the next chapter that the mere knowledge of the existence of, say, a maximum value, enables to find this extremum.

The next theorem shows that continuous functions can be well approximated by some simple functions. We shall call a function l piecewise linear on $[a, b]$ if there is a division D such that l is linear on all intervals $[x_{k-1}, x_k]$ for $k = 1, 2, \ldots x_n$.[5]

Theorem 12.20 *If $f : [a, b] \mapsto \mathbb{R}$ is continuous on a bounded and closed interval $[a, b]$ then for every positive ε there exists a piecewise linear function l such that, for all $x \in [a, b]$,*

$$|f(x) - l(x)| < \varepsilon.$$

Proof. By continuity, for every $X \in [a, b]$ we find $\delta = \delta(X)$ such that $|f(x) - f(X)| < \varepsilon/2$ for $|x - X| < \delta(X)$. Let

$$a = x_0 \leq X_1 \leq x_1 \leq \cdots \leq x_n = b$$

be a δ-fine tagged division of $[a, b]$ and l the piecewise linear function which agrees with f at all the points x_k (see Figure 12.7). If $x \in [x_{x-1}, x_k]$ then $f(x)$ lies in the interval $J_k =]f(X_k) - \varepsilon/2, f(X_k) + \varepsilon/2[$ and so does $l(x)$ because it lies between $f(x_k)$ and $f(x_{k+1})$. Consequently $|f(x) - l(x)|$ is less then the length of J_k, which is ε. □

[5]It is a consequence of the definition that a piecewise linear function l is continuous on $[a, b]$.

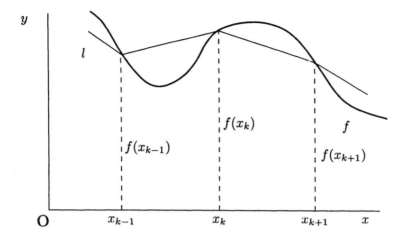

Fig. 12.7 A detail of a piecewise linear approximation.

A graph of a linear approximation to a function is depicted in Figure 12.8. It was made with *Maple*: for making such graphs see Exercise 12.5.11.

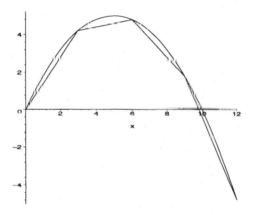

Fig. 12.8 Piecewise linear approximation

The theorem has a practical application in *Maple*. When making a plot of a continuous function, the picture on the screen is a graph of a linear approximation to the function. It has a large number of subintervals, so large that the eye sees a continuous curve rather than a polygonal line.

If f is continuous at a point $y \in [a, b]$ then for every $\varepsilon > 0$ there exists a $\delta > 0$ such that

$$|x - y| < \delta \Rightarrow |f(x) - f(y)| < \varepsilon \qquad (12.17)$$

This δ depends generally speaking on y (and also ε). For instance for the function $f : x \mapsto x^2$ a suitable δ satisfying the implication 12.17 is $\delta_0 = \sqrt{\varepsilon + y^2} - |y|$. This is easy to prove.

$$|x^2 - y^2| = \left|(x - y)^2 + 2y(x - y)\right|$$
$$\leq (x - y)^2 + 2|y||x - y| < \delta_0^2 + 2|y|\delta_0 = \varepsilon.$$

The implication 12.17 does not determine δ uniquely: any positive number smaller than δ_0 can be used as a δ in 12.17. In particular

$$\delta_1 = \frac{\varepsilon}{2\sqrt{\varepsilon + y^2}} \leq \frac{\varepsilon}{\sqrt{\varepsilon + y^2} + |y|} = \delta_0$$

Now if y is restricted to the interval $[0, 1]$ then

$$\delta_1 \geq \varepsilon/(2\sqrt{\varepsilon + 1}) = \boldsymbol{\delta}$$

and so this $\boldsymbol{\delta}$ can be used in implication (12.17). We have found something important: $\boldsymbol{\delta}$ is the same for all y. As a consequence, with $\delta = \boldsymbol{\delta}$ the implication 12.17 holds for *all* y in $[0, 1]$.[6]

Definition 12.8   We shall say that a function f is *uniformly continuous* on a set S if for every positive ε there exists a positive δ such that implication 12.17 holds for all $x, y \in S$.

Remark 12.19   The δ in Definition 12.8 may (and usually does) depend on ε, but does *not* depend on x (or y).

Theorem 12.21 (Uniform continuity)   *A function continuous on a closed bounded interval K is uniformly continuous on K.*

Proof.   By continuity of f at X for every η there exists $\delta_1(X)$ such that $|f(t) - f(X)| < \eta/2$ for $|t - X| < \delta$. Then

$$(|u - X| < \delta_1 \quad \text{and} \quad |v - X| < \delta_1) \Rightarrow |f(u) - f(v)| < \eta.$$

Let δ_2 be chosen so that the above implication holds for δ_1 and η replaced by δ_2 and $\varepsilon/2$, respectively. By Cousin's Lemma there is a tagged division

---

[6]Not merely for one particular y as before.

TX which is δ_2-fine. Define

$$\delta = \frac{1}{2}\text{Min}(\delta_2(X_k); k = 1, \ldots n)$$

If $|u-v| < \delta$ and u, v belongs to the same subinterval of T then by definition of δ

$$|f(u) - f(v)| < \frac{\varepsilon}{2} < \varepsilon.$$

If u, v do not belong to the same subinterval of T then the smaller one, say u, belongs to $[t_{k-1}, t_k]$ for some k and $v \in [t_k, t_{k+1}]$ and then

$$|f(u) - f(v)| \le |f(u) - f(t_k)| + |f(t_k) - f(v)| < \varepsilon. \qquad \square$$

Conceptually and computationally, polynomials are the simplest functions. The next theorem, very important from a theoretical point of view but also with some practical applications, says that continuous functions are close to polynomials.

Theorem 12.22 (Weierstrass theorem on polynomial approximation) *If $f : [a, b] \mapsto \mathbb{R}$ is continuous and $\varepsilon > 0$ then there exists a polynomial P such that*

$$|f(x) - P(x)| < \varepsilon \qquad (12.18)$$

for all $x \in [a, b]$.

The main idea of the proof consists in showing that if f can be approximated by a polynomial on some interval that it can also be approximated by a polynomial on a slightly larger interval. For the proof we also need a polynomial approximation U of a function which is 1 on a part of the interval and 0 on another part of $[a, b]$. We state this as a lemma.

Lemma 12.1 *If $a < c - k < c \le b$ then for every $\eta > 0$ there exists a polynomial U such that*

$$1 - \eta < U(x) \le 1 \quad \text{for} \ \ a \le x \le c - k \qquad (12.19)$$

$$0 \le U(x) \le 1 \quad \text{for} \ \ c - k < x < c \qquad (12.20)$$

$$0 \le U(x) \le \eta \quad \text{for} \ \ c \le x \le b \qquad (12.21)$$

Proof. We denote $l = b - a$ and $d = c - k/2$. First we find a polynomial p which is between 0 and 1/2 on $[a, d]$ and between 1/2 and 1 on $[d, b]$.

This is easy

$$p(x) = \frac{1}{2} + \frac{1}{2l}(x - d).$$

Then we define

$$U(x) = \left(1 - [p(x)]^n\right)^{2^n}$$

with some $n \in \mathbb{N}$ which we choose suitably later. Obviously

$$0 \le U(x) \le 1 \tag{12.22}$$

for $a \le x \le b$. Employing the Bernoulli inequality, from Equation (5.36), where n is replaced by 2^n, gives

$$U(x) \ge 1 - [2p(x)]^n \ge 1 - [2p(c - k)]^n \tag{12.23}$$

for $a \le x \le c - k$. On the other hand we have for $c \le x \le b$

$$U(x) \le \frac{1}{[2p(x)]^n} U(x) \left(1 + [2p(x)]^n\right)$$

$$\le \frac{1}{[2p(x)]^n} \left(1 - [p(x)]^{2n}\right)^{2^n} \le \frac{1}{[2p(c)]^n}. \tag{12.24}$$

Since both $2p(c - k) < 1$ and $2p(c) > 1$ it follows that both $[2p(c - k)]^n \to 0$ and $[2p(c)]^{-n} \to 0$. We can therefore find $n \in \mathbb{N}$ such that

$$[2p(c - k)]^n < \eta \text{ and } \frac{1}{[2p(c)]^n} < \eta.$$

Equations (12.19), (12.20) and (12.21) now follow from Equations (12.22), (12.23) and (12.24), respectively.   □

We now proceed with the proof proper of the Weierstrass theorem.

Proof.   For a given $\varepsilon > 0$ let S_ε be the set of all $t \le b$ such that there exists a polynomial P_ε with the property that

$$|f(x) - P_\varepsilon(x)| < \varepsilon \tag{12.25}$$

for $a \le x \le t$. By continuity of f at a there exists t_0 such that

$$|f(x) - f(a)| \le \varepsilon$$

for $a \le x \le t_0$. Consequently f can be approximated by a constant $f(a)$ on $[a, t_0]$ and $S_\varepsilon \neq \emptyset$. Let $s = \sup S_\varepsilon$. Clearly $a < s \le b$. By continuity of

f at s there is $\delta > 0$ such that

$$|f(x) - f(s)| \leq \frac{\varepsilon}{3} \qquad (12.26)$$

for $s - \delta \leq x \leq s + \delta$ and $x \leq b$. By the definition of the least upper bound there is a c with $s - \delta < c \leq s$ and $s \in S_\varepsilon$. This means there is a polynomial P_ε satisfying Equation (12.25) for $a \leq x \leq c$. Let

$$m = \text{Max}\{|f(x) - P_\varepsilon(x)|; a \leq x \leq c\}$$

and M so large that

$$M > |f(x) - P_\varepsilon(x)| + |f(x) - f(s)|$$

for all $x \in [a, b]$. We apply the lemma for $c - k = s - \delta$ to find the function U with $0 < \eta < 1$ so small that

$$m + M\eta < \varepsilon, \qquad (12.27)$$

$$M\eta < \frac{2\varepsilon}{3} \qquad (12.28)$$

This is possible since $m < \varepsilon$. Now we define

$$P(x) = f(s) + [P_\varepsilon(x) - f(s)]U(x),$$

and show that it satisfies (12.18) on $[a, b]$. First we have

$$|f(x) - P(x)| \leq |f(x) - P_\varepsilon(x)|U(x) + |f(x) - f(s)|(1 - U(x)).$$

It follows from (12.19) and (12.27) that on the interval $[a, s - \delta]$

$$|f(x) - P(x)| \leq m + M\eta < \varepsilon.$$

On the interval $[s - \delta, c]$ clearly

$$|f(x) - P(x)| \leq \varepsilon U(x) + \frac{\varepsilon}{3}(1 - U(x)) < \varepsilon.$$

Finally, using (12.21), (12.26) and (12.28) we have

$$|f(x) - P(x)| \leq MU(x) + \frac{\varepsilon}{3}(1 - U(x)) < \varepsilon,$$

on $[c, s + \delta] \cap [c, b]$. This proves that $s = b$, otherwise the inequality (12.18) would hold on $[a, s + \delta]$, contrary to the definition of s. Hence Equation (12.18) holds on $[a, s] = [a, b]$ and the proof is complete. $\qquad \square$

A function f is said to be Lipschitz on the interval $[a, b]$ if there exists a constant L such that, for all $x, y \in [a, b]$,

$$|f(x) - f(y)| \leq L|x - y|. \tag{12.29}$$

A Lipschitz function is continuous but the property of being Lipschitz is stronger: Lipschitz functions behave even better than continuous functions. The geometrical meaning is illustrated in Figure 12.29: the graph of f lies within the angle formed by lines of slopes L and $-L$ with vertex at $(u, f(u))$. It is also important to realize that this happens for every $u \in [a, b]$.

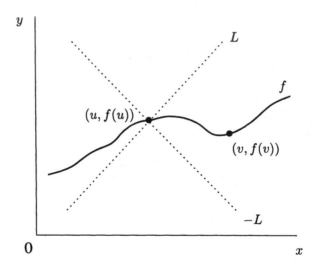

Fig. 12.9   Geometrical illustration of a Lipschitz function

For instance, if $f(x) = \sqrt{x}$ then

$$|\sqrt{x} - \sqrt{y}| \leq \frac{|x - y|}{\sqrt{x} + \sqrt{y}} \leq \frac{|x - y|}{\sqrt{y}}. \tag{12.30}$$

Consequently, for $y \geq a$, this function is Lipschitz with $L = 1/\sqrt{a}$. We leave it as Exercise 12.5.12 to show that this function is not Lipschitz on $[0, 1]$.

Every polynomial is Lipschitz on any bounded interval. Indeed

$$P(x) - P(y) = q(x, y)(x - y). \tag{12.31}$$

The expression for $q(x, y)$ is obtained by the long division algorithm from the coefficients of P and from x, y by addition and multiplication, so if x and y are restricted to a bounded interval q is bounded, say by L, and then

$$|P(x) - P(y)| \leq L|x - y|. \tag{12.32}$$

As an application of the Weierstrass Theorem we now give a second proof of the theorem on uniform continuity.

Proof. For given positive ε we find, by the Weierstrass approximation theorem a polynomial P such that for all $x \in K$

$$|f(x) - P(x)| < \frac{\varepsilon}{3}.$$

Using Inequality (12.32) we have

$$|f(x) - f(y)| \leq |f(x) - P(x)| + |P(x) - P(y)| + |P(y) - f(y)|$$
$$\leq \frac{\varepsilon}{3} + L|x - y| + \frac{\varepsilon}{3}. \tag{12.33}$$

From this inequality it is clear that implication 12.17 is satisfied for $\delta = \dfrac{\varepsilon}{3L}$ and *all* $x, y \in K$. $\square$

Exercises

Exercise 12.5.1   *Show that the equation*

$$x^{17} + \frac{199}{1 + x^4 + \mathrm{saw}(x)} = 200$$

has a solution.

(i) **Exercise 12.5.2**   *Prove that a polynomial with real coefficients and of odd degree has a real root.*

(i) **Exercise 12.5.3**   *Let $P(x) = x^n + a_1 x^{n-1} + \cdots + a_n$. If the coefficients of P are real and $a_n < 0$ then P has a positive root. Prove it.*

(!) (i) **Exercise 12.5.4**   *Prove the following. If f is a continuous bijection of $[a, b]$ onto $[c, d]$ then f is either strictly increasing or strictly decreasing on $[a, b]$. Therefore the assumption that f is strictly increasing in Theorem 12.16 can be replaced by a more general assumption that f is one-to-one. [Hint: Use an indirect proof. If $x_1 < x_2$ with $f(x_1) < f(x_2)$ and $x_3 < x_4$ with $f(x_3) > f(x_4)$ use the auxiliary function $F : t \mapsto f(x_1 + t(x_3 - x_1)) - f(x_2 + t(x_4 - x_2)), t \in [0, 1]$, and Theorem 12.15.]*

ⓘ ⓘ **Exercise 12.5.5**   *Prove the following. The assumption that* dom f *is an interval in Theorem 12.16 is innocuous but essential. Let* $f(x) = x$ *for* $x < 0$ *and* $f(x) = x - 1$ *for* $x \geq 1$ *then the inverse* f_{-1} *is discontinuous at* 0. [Hint: $f_{-1}(y) = y$ *for* $y < 0$ *and* $f_{-1}(y) = y + 1$ *for* $y \geq 0$.]

ⓘ ⓘ **Exercise 12.5.6**   *Prove the following. For* $d(x) = x$ *there is no d-fine tagged division of* $[0, 1]$. [Hint: If there was then $x_1 - 0 < d(X_1) = X_1 \leq x_1$.]

ⓘ **Exercise 12.5.7**   *The function* $x \mapsto 1/x, x \in]0, 1]$ *is not bounded. The* id *function is continuous on the closed interval* $\mathbb{R}$ *and is not bounded. Reconcile this with Theorem 12.18.*

ⓘ **Exercise 12.5.8**   *Give an example of a function continuous on all of* $\mathbb{R}$ *which does not have a maximum or minimum value.* [Hint: id.]

ⓘ **Exercise 12.5.9**   *If* P *is a polynomial show that the function* $|P|$ *attains its minimum on* $\mathbb{R}$. [Hint: Since $|P| \to \infty$ for $x \to \pm\infty$ there is a finite closed interval $I = [-a, a]$ such that $|P(x)| > |P(0)|$ for $x \notin I$. The minimum on I is the minimum on $\mathbb{R}$.]

ⓘ **Exercise 12.5.10**   *Suppose that* $f(a) > 0$ *for some* $a \in \mathbb{R}$ *and* $\lim_{x \to \infty} f(x) = \lim_{x \to -\infty} f(x) = 0$. *Prove that* f *attains its maximum on* $\mathbb{R}$. [Hint: There is a closed bounded interval I such that $f(x) < f(a)$ outside I. The maximum on I is the maximum on $\mathbb{R}$.]

ⓘ **Exercise 12.5.11**   *Given two lists* $X = [x_1, x_2, \ldots, x_n]$ *and* $Y = [y_1, y_2, \ldots, y_n]$ *with* x_k *strictly increasing, the piecewise linear function which takes the value* y_k *at* x_k *is called a linear spline. It can be made by the Maple command* `spline(X,Y,t,linear)`. *Here,* t *indicates the variable, and the option* `linear` *is needed, as there are other splines of higher degree. We shall encounter some in the next chapter. Produce a graph of* sin *together with a linear spline approximation for* $X := [0, Pi/3, 2 * Pi/3, Pi, 3 * Pi/2, 2 * Pi]$.

ⓘ **Exercise 12.5.12**   *Show that the function* $x \mapsto \sqrt{x}$ *is not Lipschitz on the interval* $[0, 1]$. [Hint: Indirect proof. If it were then $\sqrt{x} \leq Lx$ for $0 \leq x \leq 1$, a contradiction.]

Exercise 12.5.13   *Show directly that* $x \mapsto 2|x|$ *is uniformly continuous on* $\mathbb{R}$ *with* δ *from implication 12.17 equal to* $\varepsilon/2$.

Exercise 12.5.14   *Show that* $x \mapsto 1/x$ *and* $x \mapsto x^2$ *are not uniformly continuous on the intervals* $]0, 1[$ *and* $[0, \infty[$, *respectively.* [Hint: The proofs are indirect. For $\varepsilon = 1$ let $x = \delta$, $y = \delta/2$ and $x = 1/\delta$, $y = x + \delta/2$, respectively.]

12.6 Comments and supplements

Similarly to sequences, there are lim sup and lim inf for functions. There are six of these 'limits': from the left, from the right and two-sided. We state the definition for the limit superior from the left and leave it to our readers to formulate by analogy the remaining five. Assume that $f :]a, b[\mapsto \mathbb{R}$ is bounded.

$$M(\delta) = \sup\{f(x); \, b - \delta < x < b\}.$$

The function M is decreasing and bounded, since f is bounded. Consequently it has a limit from the right at 0 and this limit is, by definition, the limit superior from the left of f at b. In symbols,

$$\limsup_{x \uparrow b} f(x) = \lim_{\delta \downarrow 0} M(\delta).$$

The notation for the other five limits is (with $c \in]a, b[$):

$$\limsup_{x \downarrow a} f(x),$$

$$\limsup_{x \to c} f(x),$$

$$\liminf_{x \to c} f(x),$$

$$\liminf_{x \uparrow b} f(x),$$

$$\liminf_{x \downarrow a} f(x).$$

To illustrate these limits consider the function

$$f(x) : x \mapsto 2(\text{sgn}(x) + \frac{1}{2}) \, \text{saw}(1/x).$$

It plausible that

$$\limsup_{x \downarrow 0} f(x) = \frac{3}{2},$$

$$\limsup_{x \uparrow 0} f(x) = 0,$$

$$\liminf_{x \uparrow 0} f(x) = -\frac{1}{2},$$

$$\liminf_{x \downarrow 0} f(x) = 0.$$

Similarly as with sequences, the limit exists if and only if the limit superior and inferior are equal and that is so also when the limit is taken from the right or left. It is not difficult to prove that

$$\limsup_{x \to c} f(x) = \text{Max} \left(\limsup_{x \downarrow c} f(x), \limsup_{x \uparrow c} f(x) \right),$$

$$\liminf_{x \to c} f(x) = \text{Min} \left(\liminf_{x \downarrow c} f(x), \liminf_{x \uparrow c} f(x) \right).$$

Limit superior and limit inferior make sense only for real valued functions. In contrast, the Bolzano–Cauchy theorem applies to complex valued functions and also for limits in the complex domain.

Theorem 12.23 (Bolzano–Cauchy) *The function f has a limit from the left at b if and only if for every $\varepsilon > 0$ there is a positive δ such that*

$$|f(u) - f(v)| < \varepsilon \qquad (12.34)$$

whenever $b - \delta < u < b$ and $b - \delta < v < b$.

Proof. If f has a limit from the left, say l, then for every positive ε there is a positive δ such that, for all x with $b - \delta < x < b$,

$$|f(x) - l| < \frac{\varepsilon}{2}.$$

Taking x to be u and also v in the above inequality we have

$$|f(u) - f(v)| \leq |f(u) - l| + |f(v) - l| < \frac{\varepsilon}{2} + \frac{\varepsilon}{2},$$

whenever $b - \delta < u < b$ and $b - \delta < v < b$. Turning to the second part of the proof, let $x_n \to b$ and $x_n < b$. There is a N such that $b - \delta < x_n$ for $n > N$. This, together with Inequality (12.34), implies that $|f(x_n) - f(x_m)| < \varepsilon$, for $n, m > N$. By the Bolzano–Cauchy theorem for sequences 10.16, this implies that the sequence $n \mapsto f(x_n)$ is convergent, say to l. It remains to be shown that if $y_n \to b$ and $y_n < b$ then also $f(y_n) \to l$. By what we have already proved the sequence with terms

$$f(x_1), f(y_1), f(x_2), f(y_2), f(x_3), f(y_3), \ldots$$

also converges. Its limit is l since for the subsequence $f(x_n) \to l$. Every subsequence has the same limit, hence $f(y_n) \to l$. □

Remark 12.20 Similarly as with sequences, the Bolzano–Cauchy theorem makes it possible to prove that a function has a limit without the actual knowledge of what the value of this limit is. The Bolzano–Cauchy theorem holds also for the limit from the right, limit and limit in the complex domain. The inequalities $b - \delta < u < b$ and $b - \delta < v < b$ must be replaced by $a < u < a + \delta$ and $a < v < a + \delta$, or $0 < |c - u| < \delta$ and $0 < |c - v| < \delta$, respectively. For the limit in the complex domain the last inequalities are then required for complex u, v. For limit of f at ∞, the Bolzano–Cauchy theorem reads: a finite limit of f at ∞ exists if and only if for every positive ε there exists K such that $u, v > K$ imply $|f(u) - f(v)| < \varepsilon$.

The fact that monotonic functions have limits from the left and from the right has an interesting consequence, namely, a monotonic function can be discontinuous only at countably many points. For an increasing and bounded function f we denote, for brevity, by $f(c-)$ and $f(c+)$ the limits from the left and from the right, respectively. If f is discontinuous at c then $f(c-) < f(c+)$. If c, d are points of discontinuity of f then the open intervals $]f(c-), f(c+)[$ and $]f(d-), f(d+)[$ are disjoint. Indeed, if, for instance, $c < d$, then $f(c+) \leq f(d-)$. Now we can associate with every point c of discontinuity a rational point $r(c) \in]f(c-), f(c+)[$. This mapping is one to one: for distinct values of c, the corresponding values of $r(c)$ are also distinct, because they lie in distinct intervals. So r is a bijection of the set of discontinuities onto a subset of rationals and hence the assertion.

The function f is said to have the intermediate value property, or the Darboux property, on the interval $[\alpha, \beta]$ if it attains every value between $f(\alpha)$ and $f(\beta)$ somewhere in the interval. We have proved in Theorem 12.15 that a continuous function has the intermediate value property. The Darboux property was in the past confused with continuity but it does *not* characterize continuity. The function f equal to saw$(1/x)$ for $x \neq 0$ and $f(0) = 0$ has the intermediate value property on every subinterval of $[-1, 1]$ but is discontinuous at 0. There is a more dramatic example in Boas (1972,

p. 71) of a function which has the intermediate value property on every subinterval of $[0, 1]$ but is discontinuous at every point of $[0, 1]$.

Mathematicians of the 17th and 18th centuries had an idea of continuity but were unable to formulate it precisely. This was done by Bolzano in 1817 in a paper whose main purpose was to prove the intermediate value property for continuous functions. The Bolzano definition was what we called the Cauchy definition, following custom rather than historical accuracy. Bolzano's paper is remarkable. Firstly he set out his goal very clearly: he wanted to prove the theorem, as he said, scientifically. By that he meant a proof, which used only the definition and logical means, but did not contain any reference to intuition, geometry or the concept of motion. Secondly, in the course of his proof, Bolzano established the greatest lower bound theorem and what we called the Bolzano–Cauchy theorem for sequences. Bolzano was not only a mathematician but a moral philosopher and was politically persecuted. He was prevented from publishing some important results which were discovered much later by other mathematicians, in particular by Weierstrass.

Exercises

(!) (i) **Exercise 12.6.1**   *This exercise outlines the idea of Bolzano's proof of the intermediate value theorem. Let $f : [a, b] \mapsto \mathbb{R}$ with $f(a) < 0$ and $f(b) > 0$. Denote S the set of all $x \in [a, b]$ for which $f(x) > 0$. Show that*

1. $\xi = \inf S$ *exists and is in* $]a, b[$.
2. $f(\xi)$ *cannot be positive.*
3. $f(\xi)$ *cannot be negative.*

Chapter 13

Derivatives

Derivative can be described informally as a rate of change and as such it is extremely important in Science and applications. In this chapter we introduce derivatives as limits, establish their properties and use them in studying deeper properties of functions and their graphs. We also extend the Taylor Theorem from polynomials to power series and explore it for applications (within mathematics).

13.1   Introduction

Let us think of a body moving along a straight line. Let the distance of this body from a fixed point be a known function of time, say f. During the time interval $[t, t + h]$ the body travels the distance $f(t + h) - f(t)$ and the average velocity will be

$$\frac{f(t + h) - f(t)}{h}.$$

The velocity shown on the speedometer of a car or a plane is the instantaneous velocity and for our moving body this is the limit of the above expression as $h \to 0$.

Consider another example, this time from geometry; see Figure 13.1. The secant line joining the points $(x, f(x))$ and $(x + h, f(x + h))$ has the slope

$$\frac{f(x + h) - f(x)}{h}.$$

As $h \to 0$ the points $(x, f(x))$ and $(x + h, f(x + h))$ coalesce and the secant becomes a tangent. This corresponds to moving the ruler aligned along the secant line carefully so that the point $(x + h, f(x + h))$ moves towards

$(x, f(x))$ until it reaches its limit position, and then the secant becomes a tangent.

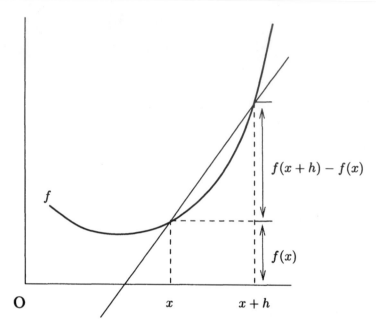

Fig. 13.1   Secant becoming a tangent

Definition 13.1   The number

$$\lim_{h \to 0} \frac{f(x+h) - f(x)}{h} \qquad (13.1)$$

is called the derivative of f at x. It is denoted by $f'(x)$ or $Df(x)$.

Geometrically, if the tangent forms an angle ϕ with the x-axis then $f'(x) = \tan \phi$.

Example 13.1   The derivative of a constant function is 0 at any point, $(c)' = 0$. Also $Dx = (x)' = 1$. For the quadratic function we have $Dx^2 = \lim_{h \to 0} (2x + h) = 2x$. Using the binomial theorem it is easy to show that $Dx^n = nx^{n-1}$ for $n \in \mathbb{N}$. Indeed,

$$\frac{(x+h)^n - x^n}{h} = nx^{n-1} + h \sum_{k=1}^{n} \binom{n}{k} h^{k-1} x^{n-k}.$$

As $h \to 0$ the second term on the right hand side tends to 0.

For $x \neq 0$ we have

$$D\frac{1}{x} = \lim_{h \to 0} \frac{1}{h}\left(\frac{1}{x+h} - \frac{1}{x}\right) = \lim_{h \to 0} \frac{-1}{(x+h)x} = -\frac{1}{x^2}. \tag{13.2}$$

Remark 13.1 The notation $f'(x)$ is more common, and *Maple* also uses $D(f)(x)$ rather than $Df(x)$. If $f(x)$ is abbreviated as y or u then $f'(x)$ is denoted by y' or u'. The derivative $Df(a)$ is sometimes calculated according to Equation (13.1) as

$$Df(a) = \lim_{x \to a} \frac{f(x) - f(a)}{x - a}.$$

The phrases 'f has a derivative at x' and 'f is differentiable at x' are used interchangeably. The limit in Equation (13.1) can be understood as a limit in the complex domain or in the real domain. These two concepts are distinct! Usually it is clear from the context which derivative is meant. In the previous example, however, all the formulae are valid for the derivative in the complex domain and then of course also for the derivative in the real domain. If the limit in Equation (13.1) is replaced by the limit from the left or right then Definition 13.1 becomes definition of the derivative of f from the left or right, respectively. The derivative from the right is denoted by $f'_+(x)$, and $f'_-(x)$. denotes the derivative from the left. Both one-sided derivatives, the derivative from the left and the derivative from the right, make sense only as derivatives in the real domain.

Example 13.2 The function $f : x \mapsto |x|$ has at zero derivative from the right 1, and from the left -1. Indeed

$$\lim_{h \downarrow 0} \frac{|h| - 0}{h} = \lim_{h \downarrow 0} 1 = 1,$$

$$\lim_{h \uparrow 0} \frac{|h| - 0}{h} = \lim_{h \uparrow 0}(-1) = -1.$$

This example shows that a continuous function need not have a derivative. Geometric intuition might suggest that a function continuous on some interval would have derivative at many points of the interval. This is not true: Bolzano was aware of this and later Weierstrass published an example of a function continuous everywhere and differentiable nowhere. We shall comment on this in the last section of this chapter.

Remark 13.2 The limit in Equation (13.1) can be infinite, and the derivative is then called an infinite derivative. In this book the word derivative means a *finite* derivative and if we allow the derivative to become infinite, we shall explicitly say so. Moreover, we shall consider infinite derivatives only for functions for which the domain of definition is a part of $\mathbb{R}$.

Example 13.3   Let $f(x) = \sqrt{x}$, $g(x) = \sqrt[3]{1-x}$. Then

$$f'_+(0) = \lim_{h \downarrow 0} \frac{\sqrt{h}}{h} = \lim_{h \downarrow 0} \frac{1}{\sqrt{h}} = \infty,$$

$$g'_-(1) = \lim_{h \to 0} \frac{\sqrt[3]{1-(1+h)} - \sqrt[3]{1-1}}{h} = \lim_{h \to 0} \frac{-1}{\sqrt[3]{h^2}} = -\infty.$$

We introduce yet another notation for the derivative: we denote $f'(x)$ by $\dfrac{d}{dx} f(x)$ or by $\dfrac{df(x)}{dx}$. The original notation is shorter and on most occasions unambiguous, but this latest notation has the advantage of indicating with respect to which variable the derivative is taken. For instance, it is clear from the notation $\dfrac{d(x^2 + tx)}{dx}$ that t should be regarded as a constant and x as a variable, that is,

$$\frac{d(x^2 + tx)}{dx} = \lim_{h \to 0} \frac{(x+h)^2 + t(x+h) - x^2 - tx}{h} = 2x + t.$$

Naturally if $f(x)$ is denoted by y then $f'(x) = y' = \dfrac{dy}{dx}$. Each notation has its advantages and disadvantages. The symbol y' hides the x, so it cannot be used for denoting, for example, $f'(1)$. Although $\dfrac{dy}{dx}$ does not hide x, it is also unsuitable for denoting the derivative at a particular point, say $x = 1$.

We now consider some differentiation with *Maple*.

```
>   restart;
>   f:=x->x^2+1/x;
```

$$f := x \rightarrow x^2 + \frac{1}{x}$$

```
>   D(f)(x);
```

$$2x - \frac{1}{x^2}$$

```
>   D(f)(1);
```

$$1$$

The command `diff` can also be used to calculate the derivative of a function with values $f(x)$.

> `diff(f(x),x);`

$$2\,x - \frac{1}{x^2}$$

> `g:=x->x*abs(x);`

$$g := x \to x\,|x|$$

> `D(g)(0);`

`Error, (in simpl/abs) abs is not differentiable at 0`

It is true that $x \mapsto |x|$ is not differentiable at 0 nevertheless g *is* differentiable at zero. *Maple* is unfortunately unable to calculate $Dg(0)$. However *Maple* can find the derivative as a limit

---

> `limit((g(h)-g(0))/h,h=0);`

$$0$$

The main advantage of using *Maple* for differentiation lies in the fact that *Maple* easily calculates derivatives which otherwise would be laborious to do, as we shall see later. Here is an example which would not be too difficult without *Maple*, but using *Maple* still has an advantage.

> `diff((1+x-x^2)/(1-x+x^2),x);`

$$\frac{1-2\,x}{1-x+x^2} - \frac{(1+x-x^2)\,(-1+2\,x)}{(1-x+x^2)^2}$$

> `simplify(%);`

$$-2\,\frac{-1+2\,x}{(1-x+x^2)^2}$$

Exercises

Exercise 13.1.1   *Prove that the function $x \mapsto \sqrt{x}$ has, for all $x > 0$, the derivative $1/(2\sqrt{x})$.*

Exercise 13.1.2   *Show that* $D\dfrac{1}{x^n} = -n\dfrac{1}{x^{n+1}}$ *for* $x \neq 0$ *and* $n \in \mathbb{N}$.
[Hint: Use equation (13.1) and the result from Example 13.1, namely that $Dx^n = nx^{n-1}$.]

Exercise 13.1.3   *Use* Maple *to find* $D(1 + \sqrt{x})/(1 - \sqrt{x})$.

13.2   Basic theorems on derivatives

If f is differentiable at x then the function

$$h \mapsto \mathcal{D}(h) = \frac{f(x+h) - f(x)}{h} \tag{13.3}$$

has a removable discontinuity at 0 and by defining $\mathcal{D}(0) = f'(x)$ becomes continuous. This leads to the next theorem, which is theoretical in character but needed at many practical occasions.

Theorem 13.1   *The function f is differentiable at x if and only if there exists a function $\mathcal{D}$ continuous at zero such that Equation (13.3) holds. If so then $\mathcal{D}(0) = f'(x)$.*

Remark 13.3   This theorem as well the next two are valid in the complex domain.

From this theorem it follows immediately:

Theorem 13.2   *If a function has a finite derivative at a point then the function is continuous at this point.*

Proof.   As $h \to 0$, Equation (13.3) yields $f(x+h) - f(x) \to \mathcal{D}(0)0 = 0$ so f is continuous at x.   □

Theorem 13.3   *If $f'(x)$ and $g'(x)$ exist and c is a constant then*

$$[cf(x)]' = cf'(x), \tag{13.4}$$
$$[f(x) + g(x)]' = f'(x) + g'(x), \tag{13.5}$$
$$[f(x)g(x)]' = f'(x)g(x) + f(x)g'(x). \tag{13.6}$$

These formulae are more easily remembered as

$$(cu)' = cu',$$
$$(u + v)' = u' + v',$$
$$(uv)' = u'v + uv'.$$

Proof. The Formulae (13.4) and (13.5) are immediate consequences of theorems on limits and Definition 13.1. It is important to note that $[-f(x)]' = -f'(x)$ follows from Equation (13.4) and then

$$[f(x) - g(x)]' = f'(x) - g'(x), \tag{13.7}$$

by Equation (13.5). Let us now turn to Equation (13.6). By Theorem 13.1

$$f(x + h) = f(x) + \mathcal{D}_1(h)h, \tag{13.8}$$
$$g(x + h) = g(x) + \mathcal{D}_2(h)h. \tag{13.9}$$

Consequently

$$f(x+h)g(x+h) - f(x)g(x) = (f(x)\mathcal{D}_2(h) + g(x)\mathcal{D}_1(h) + \mathcal{D}_1(h)\mathcal{D}_2(h)h)\,h. \tag{13.10}$$

Dividing by h and then passing to the limit as $h \to 0$ leads to 13.6, since $\mathcal{D}_1(0) = f'(x)$ and $\mathcal{D}_2(0) = g'(x)$. □

Remark 13.4 This Theorem together with the formula $Dx^n = nx^{n-1}$ makes it clear that the definition of the derivative of a polynomial as given in Equation (6.25) and the definition of a polynomial according to Definition 13.1 do agree.

Exercises

ⓘ **Exercise 13.2.1** *Prove by induction:*

$$(u_1 u_2 \ldots u_n)' = u_1' u_2 \ldots u_n + u_1 u_2' u_3 \ldots u_n + \cdots + u_1 u_2 \ldots u_{n-1} u_n'.$$

ⓘ **Exercise 13.2.2** *Show that $f + g$ and fg might have derivatives even though neither f nor g has. [Hint: $f(x) = |x|$, $g(x) = -|x|$ for $x = 0$.]*

ⓘ **Exercise 13.2.3** *The function $x \mapsto \mathrm{sgn}(x)$ has an infinite derivative $+\infty$ at 0 but is discontinuous at 0. Does this contradict Theorem 13.2 and if not why not.*

The chain rule

The next theorem concerning the derivative of a composite function is one of the most important rules for differentiation.

Theorem 13.4 (The chain rule) *If g is differentiable at x and f differentiable at $g(x)$ then $f \circ g$ is differentiable at x and*

$$(f(g(x)))' = f'(g(x))g'(x). \qquad (13.11)$$

The equation is often rewritten in the following way with $y = f(x)$ and $u = g(x)$

$$\frac{d}{dx}f(u) = f'(u)g'(x), \qquad (13.12)$$

$$\frac{dy}{dx} = \frac{dy}{du}\frac{du}{dx}. \qquad (13.13)$$

The symbols $\frac{dy}{du}$ and $\frac{du}{dx}$ are indivisible but still Equation (13.13) is easily remembered as fractions which cancel. Before we prove the theorem we illustrate it by two examples.

Example 13.4 We wish to calculate $D(1 + x^2)^{20}$. We can expand the expression by the binomial theorem and then take the derivative. Far more convenient is to use Theorem 13.4, set $u = 1 + x^2$ and $y = f(u) = u^{20}$. Then by Equation (13.13),

$$\frac{dy}{du} = 20u^{19},$$

$$\frac{du}{dx} = 2x,$$

$$\frac{dy}{dx} = 20u^{19}2x = 40x(1 + x^2)^{19}.$$

In Example 13.1 we established the formula $(x^n)' = nx^{n-1}$ for a positive integer n. We now extend the formula for negative integers.

Example 13.5 Set $m = -n$ then m is positive and

$$(x^n)' = \left[\left(\frac{1}{x}\right)^m\right]' = m\left(\frac{1}{x}\right)^{m-1}\left(\frac{-1}{x^2}\right) = nx^{n-1}.$$

Clearly this is valid for $x \neq 0$.

We now take up the proof of Theorem 13.4.

Proof. By Theorem 13.1

$$f(u+k) - f(u) = \mathcal{D}_1(k)k, \tag{13.14}$$
$$g(x+h) - g(x) = \mathcal{D}_2(h)h, \tag{13.15}$$

with $\mathcal{D}_1$ and $\mathcal{D}_2$ continuous at 0, $\mathcal{D}_1(0) = f'(u)$ and $\mathcal{D}_2(0) = g'(x)$. Setting $u = g(x)$ and $k = g(x+h) - g(x)$ leads to

$$\frac{f(g(x+h)) - f(g(x))}{h} = \mathcal{D}_1(k)\mathcal{D}_2(h). \tag{13.16}$$

Since $k \to 0$ as $h \to 0$, passing to the limit in Equation (13.16) as $h \to 0$ gives Equation 13.11. $\qquad\square$

Example 13.6 Sometimes the chain rule needs to be applied to the inner function. This happens when the function is multiply composed. We wish to find $\left(x + \sqrt{1 + x^2}\right)^5$. Let us recall from Exercise 13.1.1 that $\left(\sqrt{x}\right)' = 1/(2\sqrt{x})$.

Firstly, by Theorem 13.4

$$D\sqrt{1 + x^2} = \frac{1}{2\sqrt{1 + x^2}} 2x$$

and then

$$D\left(x + \sqrt{1 + x^2}\right)^5 = 5\left(x + \sqrt{1 + x^2}\right)^4\left(1 + \frac{x}{\sqrt{1 + x^2}}\right)$$

If $f(u) = 1/u$ then the chain rule leads to

$$\left(\frac{1}{g(x)}\right)' = -\frac{g'(x)}{g(x)^2}.$$

Applying this and Equation (13.6) from Theorem 13.3 to the product $f(x)\dfrac{1}{g(x)}$ yields

Theorem 13.5 *If f and g have derivatives at x and $g'(x) \neq 0$ then*

$$\left(\frac{f(x)}{g(x)}\right)' = \frac{f'(x)g(x) - f(x)g'(x)}{g(x)^2}. \tag{13.17}$$

This equation can be rewritten as

$$\left(\frac{u}{v}\right)' = \frac{u'v - uv'}{v^2}.$$

Exercises

Exercise 13.2.4   *Use theorems of this section to calculate the following derivatives and then check your results by* Maple.

1. $\left(\dfrac{ax + b}{a + b}\right)'$;

2. $\left(\dfrac{ax + b}{cx + d}\right)'$;

3. $\left(\dfrac{3x + 7}{(x^2 + x + 1)^2}\right)'$;

4. $\left(\dfrac{x}{x + \sqrt{1 + x^2}}\right)'$;

5. $\left(\sqrt{\dfrac{x + 1}{x - 1}}\right)'$;

ⓘ **Exercise 13.2.5**   *Prove that* $(f(ax + b))' = af'(ax + b)$. [Hint: Use Theorem 13.4.]

ⓘ **Exercise 13.2.6**   *Theorem 13.4 is not valid if the derivative is replaced by right-hand (or left-hand) derivative. Give an example of this.* [Hint: $g(x) = -x$, $f(x) = |x|$ at $x = 0$.]

ⓘ **Exercise 13.2.7**   *Prove that Theorem 13.4 remains valid if the derivative is replaced by right-hand (or left-hand) derivative, provided g is strictly increasing (decreasing).*

Derivative of the inverse function

Now we study the derivative of the inverse function. We consider a function f strictly monotonic on some interval I. Figure 13.2 provides a good motivation.

The tangent to the graph of f at $(x, f(x))$ forms an acute angle α with the x-axis. Hence the tangent to the graph of f_{-1} at $(f(x), x) = (y, f_{-1}(y))$ subtends an angle α with the y-axis and the angle $\pi/2 - \alpha$ with the x-axis. It follows that

$$f'_{-1}(y) = \tan\left(\frac{\pi}{2} - \alpha\right) = \frac{1}{\tan(\alpha)} = \frac{1}{f'(x)}.$$

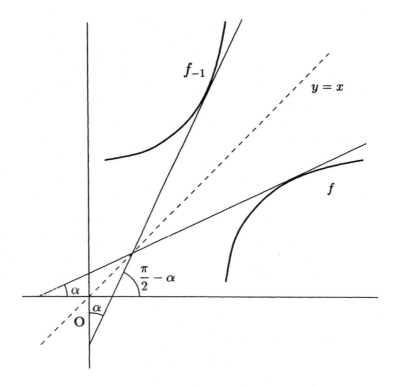

Fig. 13.2   Derivative of the inverse

We state this as

Theorem 13.6 (Derivative of the inverse function)   If f is strictly monotonic and f has a derivative $f'(x) \neq 0$ at an interior point x of the interval I then the inverse function has the derivative $f'_{-1}(y)$ at the point $y = f(x)$ and

$$f'_{-1}(y) = \frac{1}{f'(x)}. \qquad (13.18)$$

The above formula is easily remembered as

$$\frac{dy}{dx} = \frac{1}{\dfrac{dx}{dy}} \qquad (13.19)$$

with the understanding that y denotes $f(x)$.

Proof. Define $k(h) = f_{-1}(y + h) - f_{-1}(y)$. By the Theorem on the continuity of the inverse, k is continuous at 0 with $k(0) = 0$. Denote $y = f(x)$ then $x = f_{-1}(y)$ and we obtain successively:

$$f_{-1}(y + h) = x + k(h), \tag{13.20}$$

$$y + h = f(x + k(h)), \tag{13.21}$$

$$h = f(x + k(h)) - f(x). \tag{13.22}$$

By Theorem 13.1

$$f(x + k(h)) - f(x) = \mathcal{D}(k(h))k(h), \tag{13.23}$$

with $\mathcal{D}$ continuous at 0. Substituting for h from Equation (13.22) gives

$$\frac{k(h)}{h} = \frac{1}{\mathcal{D}(k(h))}. \tag{13.24}$$

Using Equation (13.20) for $k(h)$ leads to

$$\frac{f_{-1}(y + h) - f_{-1}(y)}{h} = \frac{1}{\mathcal{D}(k(h))}.$$

Taking limit as $h \to 0$ proves Equation (13.18). $\qquad\qquad\square$

A more detailed and careful examination of the proof will show that if $f'(x) = 0$ then $f'_{-1}(y) = \pm\infty$ with the $+$ sign valid for a strictly increasing f and $-$ sign for a strictly decreasing f.

Example 13.7   Let $f(x) = x^m$ with $m \in \mathbb{N}$. Then f is increasing, for m odd on all of $\mathbb{R}$ and for m even on $[0, \infty[$. The inverse[1] $f_{-1}(y) = y^{1/m}$ has the derivative

$$f'_{-1}(y) = \frac{dy^{\frac{1}{m}}}{dy} = \frac{1}{mx^{m-1}} = \frac{1}{m}y^{\frac{1}{n}-1}. \tag{13.25}$$

We know that the formula $(x^n)' = nx^{n-1}$ holds for $n \in \mathbb{Z}$. Equation (13.25) extends it to n of the form $1/m$ with $m \in \mathbb{N}$. We now use the chain rule to extend it for any rational $m = p/q$, $p \in \mathbb{Z}$ and $q \in \mathbb{N}$.

$$(x^m)' = \left(\left(x^{\frac{1}{q}}\right)^p\right)'$$

$$= p\left(x^{\frac{1}{q}}\right)^{p-1}\frac{1}{q}x^{\frac{1}{q}-1} = \frac{p}{q}x^{\frac{p-1}{q}+\frac{1}{q}-1} = \frac{p}{q}x^{\frac{p}{q}-1} = mx^{m-1}. \tag{13.26}$$

---

[1]For even m we tacitly understand that f is restricted to $[0, \infty[$.

Exercises

ⓘ **Exercise 13.2.8** The validity of Theorem 13.6 can be extended to endpoints of $I = [a, b]$. If, for instance, f is strictly increasing and $f'_+(a) \neq 0$, then $(f_{-1})'_+$ exists at $c = f(a)$ and

$$(f_{-1})'_+ (c) = \frac{1}{f'_+(b)}.$$

Prove it. [Hint: Define $F(x) = f(x)$ for $x \geq a$ and $F(x) = f(a) + f'_+(a)(x - a)$ for $x < a$. Show that Theorem 13.6 can be applied to F and prove $(f_{-1})'_+ (c) = (F_{-1})'_+ (c)$.]

ⓘ **Exercise 13.2.9** Prove: If f is strictly increasing on I and $f'(c) = +\infty$ for an interior point $c \in I$ then for $d = f(c)$ the derivative $f'_{-1}(d) = 0$

13.3 Significance of the sign of derivative.

If a function has a positive derivative then the tangent forms an acute angle with the x-axis and it is plausible that then the function increases, locally at a point or in an interval. We shall prove both theorems in this regard. We start with the case of the derivative having a constant sign on an interval.

Theorem 13.7 (On the sign of the derivative) Assume that a function f is continuous on the interval I and denote by I^o the set of interior points of I. Then

  (i) $f'(x) \geq 0$ in $I^o \Rightarrow f$ is increasing in I;
  (ii) $f'(x) \leq 0$ in $I^o \Rightarrow f$ is decreasing in I.
  (iii) $f'(x) = 0$ in $I^o \Rightarrow f$ is constant in I;
  (iv) $f'(x) > 0$ in $I^o \Rightarrow f$ is strictly increasing in I;
  (v) $f'(x) < 0$ in $I^o \Rightarrow f$ is strictly decreasing in I;

Remark 13.5 This theorem remains valid if f' is replaced by f'_+ (or f'_-) everywhere. It is this more general version which we prove.

Proof. We prove first (i) indirectly. So assume that $f'_+(x) \geq 0$ in I^o and that there are points $x_1 < x_2$ with $f(x_1) > f(x_2)$. It is easy to explicitly define a linear function l with the following properties; see Figure 13.3.

  (a) $l'(x) < 0$ for all x,

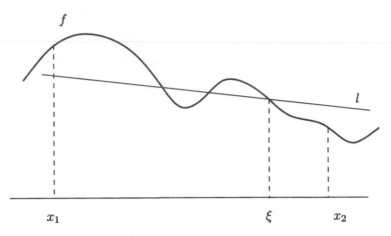

Fig. 13.3   A function which does not increase from x_1 to x_2

(b) $f(x_1) > l(x_1)$,

(c) $f(x_2) < l(x_2)$.

Let $S = \{x; f(x) - l(x) > 0, x_1 \le x \le x_2\}$ and $\xi = \sup\{S\}$. Since $f(x_2) - l(x_2) < 0$ by continuity of $f - l$ there is a positive δ such that $f(x) - l(x) < 0$ for $x_2 - \delta < x \le x_2$. Consequently $\xi < x_2$. Using (b) and continuity of $f - l$ it follows similarly that $\xi > x_1$. By the definition of the greatest lower bound, for every $n \in \mathbb{N}$, there is x_n with $\xi + 1/n > x_n \ge \xi$ and $x_n \in S$. Passing to the limit as $n \to \infty$ in the inequality $f(x_n) - l(x_n) > 0$ leads to $f(\xi) \ge l(\xi)$. Since $f(x) - l(x) < 0$ for $x > \xi$, by continuity $f(\xi) \le l(\xi)$, hence $f(\xi) = l(\xi)$ and

$$f(x) - f(\xi) < l(x) - l(\xi) \quad \text{for} \quad x > \xi. \tag{13.27}$$

Dividing by $x - \xi$ and letting $x \to \xi$ gives

$$f'_+(\xi) \le l'(\xi) < 0. \tag{13.28}$$

This contradicts (i).

To prove (ii) it is sufficient to apply (i) to the function $-f$.

After (i) and (ii) are established, (iii) becomes obvious.

The proof of (iv) is again indirect. By (i) f is increasing, so assume it is not strictly increasing. Then there are two points a, b with $a < b$ in I such that $f(a) = f(b)$. Obviously f is constant on $[a, b]$, hence $f_+(x) = 0$ for $a < x < b$, which is a contradiction.

Finally, part(v) follows from (iv) applied to $-f$. □

Before the age of computers this theorem was an indispensable tool for plotting functions. It is still important but at most occasion the intervals where f decreases or increases can be found by plotting the graph of f by *Maple*.

Example 13.8

We wish to find the intervals where the following polynomial is monotonic.

```
>   p:=x->18*x^5-125*x^3+385*x+1;
```
$$p := x \rightarrow 18\,x^5 - 125\,x^3 + 385\,x + 1$$

We look at the graph of p

```
>   plot(p,-3..3);
```

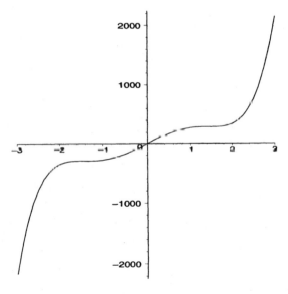

Fig. 13.4   Graph of a quintic

To get a better idea we plot p on $[0, 2]$, in Figure 13.5.

```
>   plot(p,0..2);
```

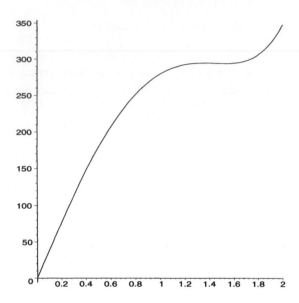

Fig. 13.5　Graph of a quintic, with restricted domain and range

The exact location of points where f changes from increasing to decreasing is still not clear. In order to apply apply Theorem 13.7 we calculate the points where the derivative is zero. Between consecutive points the derivative does not change its sign and the function is therefore monotonic.

```
>   solve(D(p)(x)=0);
```

$$\frac{1}{3}\sqrt{21}, \ -\frac{1}{3}\sqrt{21}, \ \frac{1}{6}\sqrt{66}, \ -\frac{1}{6}\sqrt{66}$$

Obviously p increases for $x < -\frac{\sqrt{66}}{6}$ then decreases till $-\frac{\sqrt{21}}{3}$ and so on. On reflection it was more convenient to use fsolve rather then solve; this is often so in examples like this.

```
>   fsolve(D(p)(x)=0);
```

$$-1.527525232, \ -1.354006401, \ 1.354006401, \ 1.527525232$$

Since we were dealing with a polynomial, fsolve found all solutions to

$p(x) = 0$ automatically. In more general situations, additional care is needed to find all solutions.

Example 13.9   The problem consists of finding a rectangle of largest area cut from a metal plate with the lower left hand corner removed; see Figure 13.6. We seek the maximum value of $A(x) = (d-x)(v-1+x)$ on the interval $[0, 1]$.

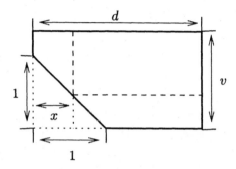

Fig. 13.6   A rectangular plate

The derivative $A'(x) = d - v + 1 - 2x$ is positive for $x < c = (d - v + 1)/2$ and negative for $c > x$. If $0 < c < 1$ then the largest value of A is attained at c and the dimensions of the largest rectangle are $d - c$ and $v - 1 + c$. If $c < 0$ then $A'(x) < 0$ on $[0, 1]$, the function is strictly decreasing and A attains its maximum value at the left end-point of $[0, 1]$ and the dimensions of the largest rectangle are d and $v - 1$. Similarly, if $c > 1$ then A has the largest value for $x = 1$. We needed the zero of A but the zero in itself was of secondary importance, the crucial importance was the sign of the derivative.

Example 13.10   Find intervals of monotonicity of the following function.

```
>   restart;
>
>   f:=x->100/(401*x^3-899*x^2+600*x);
```

$$f := x \to 100\,\frac{1}{401\,x^3 - 899\,x^2 + 600\,x}$$

```
>   D(f)(x);
```

$$-100\,\frac{1203\,x^2 - 1798\,x + 600}{(401\,x^3 - 899\,x^2 + 600\,x)^2}$$

```
>   L:=[solve(D(f)(x)=0,x)];
```

$$L := [\frac{899}{1203} - \frac{1}{1203}\sqrt{86401}, \frac{899}{1203} + \frac{1}{1203}\sqrt{86401}]$$

```
>  evalf(%);
```

$$.5029588776, .9916379636$$

```
>  a:=L[1];b=L[2];
```

$$a := \frac{899}{1203} - \frac{1}{1203}\sqrt{86401}$$

$$b = \frac{899}{1203} + \frac{1}{1203}\sqrt{86401}$$

---

f is discontinuous at 0, hence Theorem 13.7 cannot be applied to any interval containing zero. We are left with the intervals $]-\infty, 0[$ and $]0, a]$, $[a, b]$ and $[b, \infty[$. Near 0 the derivative is negative, by continuity it is negative on $]-\infty, 0[$ and $]0, a[$. Hence f is strictly decreasing on these intervals. For large x the derivative is obviously also negative, hence f strictly decreases on $[b, \infty[$. On the remaining interval f strictly increases. Our result is confirmed by the graph

---

```
>  plot(f(x),x=-1..2,y=-2..2);
```

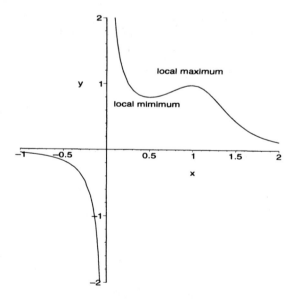

Fig. 13.7   Graph of $f(x) = 100/(401x^3 - 899x^2 + 600x)$

We observe from the graph in Figure 13.7 that at $x = \sqrt{21}/3$ the function is larger than at nearby points although obviously f is larger still for x positive and close to 0. We say that f has a local maximum at $\sqrt{21}/3$. More precisely

> **Definition 13.2**   A function $f : I \mapsto \mathbb{R}$ is said to have a local maximum at c, an interior point of I, if there exists a positive δ such that, for $0 < |x - c| < \delta$,
>
> $$f(x) \leq f(c). \tag{13.29}$$
>
> If this inequality is strict then the maximum is also strict. If inequality (13.29) is reversed then f has a local minimum and if the reversed inequality is strict then f has a strict local minimum.

The common name for a local maximum or a local minimum is a *local extremum*. The function $x \mapsto (x + |x|)(2 - x)$ has a local minimum at 0, the function $x \mapsto x^2(x - 1)$ has a strict local maximum at 0.

> **Theorem 13.8**   If f has a local extremum at c and $f'(c)$ exists then $f'(c) = 0$.

Proof.   For definiteness let us assume f has a local maximum. Then, since $f(x) - f(c) \leq 0$ to the right of c, it follows that $f'_+(c) \leq 0$. Similarly $f'_-(c) \geq 0$. Since the derivative exists $f'(c) = f'_+(c) = f'_-(c)$ and consequently $f'(c) = 0$. $\square$

It is important to be aware that the converse of Theorem 13.8 is false. The function $x \mapsto x^3$ has derivative 0 at $x = 0$ but is strictly increasing on $\mathbb{R}$ and therefore cannot have a local extremum at 0. The next Theorem guarantees the existence of a local extremum.

> **Theorem 13.9**   If f is continuous at c and
>
> $$f'(x) > 0 \tag{13.30}$$
>
> on some interval $]a, c[$ and
>
> $$f'(x) < 0 \tag{13.31}$$
>
> on some interval $]c, b[$ then f has a strict local maximum at c. If reversed inequalities hold in (13.30) and (13.31) then f has a strict local minimum at c.

This Theorem is easily remembered as: If f' changes sign at c then it has a local extremum at c.

Proof. By Theorem 13.7 and by inequality (13.30), the function f increases to the left of c. Hence it is smaller to the left of c than it is at c. Similarly, f decreases to the right of c so it is smaller there than at c. The proof for the local minimum is similar, or we can apply the already proved part of the theorem to $-f$. □

Example 13.11 Let $f(x) = (1 - x^2)^{10} x^{20}$. It is clear without any differentiation that f has strict local minima at $-1, 0, 1$. However there are other extrema. Since

$$f'(x) = 20(1 - x^2)^9 x^{19}(1 - 2x^2),$$

and f' changes sign at both $x = -1/\sqrt{2}$ and $x = 1/\sqrt{2}$ the function has strict local maxima at these points.

Many practical problems in applications require finding the largest or the smallest value of a given function on a given interval. Obviously, the largest or the smallest value is either a local extremum or the value of the function at one of the ends of the interval. The next example shows how to use *Maple* in such a situation.

Example 13.12
We wish to find the largest and smallest value of the function f, defined below, on the interval $[0, 2]$.

```
>  restart;
>  f:=x->15/(1+abs(x-1))+x^2+2*x-1;
```

$$f := x \rightarrow 15\,\frac{1}{1 + |x - 1|} + x^2 + 2x - 1$$

```
>  g:=D(f)(x);
```

$$g := -15\,\frac{\operatorname{abs}(1, x - 1)}{(1 + |x - 1|)^2} + 2x + 2$$

$\operatorname{abs}(1, x - 1)$ denotes $\operatorname{sgn}(x - 1)$. The reason why *Maple* uses this alternative notation need not concern us. However we note that f is not differentiable at $x = 1$.

```
>  L:=solve(g,x);
```

$$L := \frac{1}{6}\,(802 + 18\,\sqrt{1985})^{(1/3)} + \frac{\frac{2}{3}}{(802 + 18\,\sqrt{1985})^{(1/3)}} - \frac{1}{3},$$

$$-\frac{1}{2}\,(38 + 2\,\sqrt{345})^{(1/3)} - \frac{2}{(38 + 2\,\sqrt{345})^{(1/3)}} + 1$$

For better understanding we convert these numbers to decimal fractions.

```
>  L1:=evalf(%);
```

$$L1 := 1.674571001,\ -1.583911177$$

Now we put together the end-points, the points where the derivative does not exist (there is just one) and zeros of the derivative which lie in $[0, 2]$.

```
>  K:=0,1,2,L1[1];
```

$$K := 0,\ 1,\ 2,\ 1.674571001$$

We evaluate the function f at these points, using the *Maple* command map to map the set $\{K\}$ onto the set $\{K1\}$.

```
>  K1:=map(f,{K});
```

$$K1 := \{17,\ 14.11084811,\ \frac{13}{2},\ \frac{29}{2}\}$$

In order to apply max and min we convert K1 to an expression sequence.

```
>  K2:=op(K1);
```

$$K2 := 17,\ 14.11084811,\ \frac{13}{2},\ \frac{29}{2}$$

```
>  max(K2);min(K2);
```

$$17$$

$$\frac{13}{2}$$

17 is the value of f at $x = 1$, 13/2 is attained at 0. We had better check it:

```
>  is(17=f(1)); is(13/2=f(0));
```

$$true$$

$$true$$

We succeeded, and have found the maximum and minimum values. However, when using a computer algebra system, it is always advisable to check the results. A small typing error can cause an enormous error in the final result. Moreover, it is easy to overlook some subtle feature of *Maple*, for instance that solve or fsolve did not find all solutions to $Df(x) = 0$. Here the easiest check is to plot f.

>   plot(f,0..2);

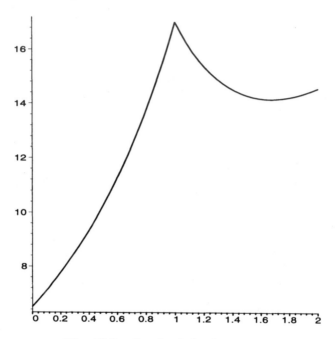

Fig. 13.8   Graph of the function f

If the derivative is 0 at some point then the function might or might not have a local extremum at that point. We get a more definite answer if the derivative is positive.

A function f is said to be *increasing at a point c* if there is a $\delta > 0$ such that

$$f(c - h) < f(c) \quad \text{and} \quad f(c + h) > f(c), \tag{13.32}$$

for $0 < h < \delta$. If the inequalities (13.32) for f are reversed the function is said to be decreasing at c.

> **Theorem 13.10**   If $Df(c) > 0$ or $Df(c) < 0$ then f is increasing at c or decreasing at c, respectively.

Proof.   By Theorem 13.1

$$f(c+h) - f(c) = \mathcal{D}(h)h, \tag{13.33}$$

with $\mathcal{D}(0) = Df(c)$ and $\mathcal{D}$ continuous at 0. If $Df(c) > 0$ then by Remark 12.15 we have $\mathcal{D}(h) > 0$ for $|h| < \delta$. Consequently Equations 13.32 hold. The proof is entirely similar if $Df(c) < 0$.   □

The derivative of a function has the intermediate property even if it is not continuous.

> **Theorem 13.11**   If f' exists on an open interval I then it has the intermediate value property.

Proof.   For sake of definiteness let $f'(x_1) < L < f'(x_2)$ with $x_1 < x_2$. Define an auxiliary function $F : x \mapsto Lx - f(x)$. On the interval $[x_1, x_2]$ the function F attains its maximum, say at c. By Theorem 13.10 F is increasing at x_1 and decreasing at x_2. Consequently $c \neq x_1$ and $c \neq x_2$ and the maximum at c is local. By Theorem 13.8 we have $F'(c) = 0$ and consequently $f'(c) = L$.   □

This Theorem is not valid for one-sided derivatives, For $f(x) = |x|$ we have $f'_+(1) = 1$, $f'_+(-1) = -1$ but $f'_+(x)$ is never zero on $]0, 1[$.

Exercises

Exercise 13.3.1   Find the local maxima and minima of $\sqrt[3]{(x^2 - 1)^2}$. [Hint: The answer is $-1, 0, 1$.]

ⓘ **Exercise 13.3.2**   Prove that if a continuous f has only one local minimum at c on an interval I and no local maximum then $f(c)$ is the smallest value of f on I. [Hint: Indirect proof, assume that $f(d) < f(c)$ and find a local maximum between c and d.]

ⓘ **Exercise 13.3.3**   The previous exercise is particularly useful if I is either infinite or open. Find the minimum value of $x + 1/x^2$ on $]0, \infty[$.

ⓘ **Exercise 13.3.4**   *Prove: If a continuous f has local minima at a and b, $b > a$, then it has at least one local maximum at some point $c \in]a, b[$ If further, f has no local minimum inside $]a, b[$ then f attains its largest value at c. Show also that the assumption that f continuous is essential.*

ⓘ **Exercise 13.3.5**   *If f is increasing on I with $\operatorname{rg} g \subset I$ then g and $f \circ g$ have the same extrema (local or otherwise). Prove it.*

Exercise 13.3.6   *A truck travels between Sydney and Brisbane, a distance of 700 km, at the uniform speed of x km per hour and does not exceed the speed limit of 100 km per hour. Assuming that the consumption of diesel for 100 km is given by $a + bx^2$ of litres, that the cost of the crew is c dollars per hour and the cost of diesel per litre is \$0.74. For what c is the limit speed of 100 km/hour the most economical?*

Exercise 13.3.7   *Among all circular cones of surface area S find the one which has the largest volume.*

Exercise 13.3.8   *Find the circular cone of largest volume inscribed in a sphere of radius R.*

ⓘ **Exercise 13.3.9**   *Use Theorem 13.7 to prove: If f' is constant on some open interval I then f is linear on I.*

13.4   Higher derivatives

If a function f has a derivative at every point of an interval I then the function $x \mapsto f'(x)$ is denoted naturally as f'. If this function has a derivative at x then it is called the second derivative of f at x and is denoted by $f''(x)$ or $D^2 f(x)$. The second derivative can again have a derivative, the third derivative, denoted by f'''. The nth derivative $f^{(n)}$ is defined inductively

$$f^{(n)}(x) = \left(f^{(n-1)} \right)'(x). \tag{13.34}$$

The notation with $D^2 = DD$ or generally D^n is also used. For instance

$$(x^7)'' = 7 \left(x^6 \right)' = 42x^5 = D^2 x^7,$$

or if $f(x) = (1 + x)^m$ with $m \in \mathbb{Q}$ then

$$D(1 + x)^m = m(1 + x)^{m-1}, \tag{13.35}$$

$$D^2(1 + x)^m = m(m - 1)(1 + x)^{m-2} \tag{13.36}$$

$$\vdots \tag{13.37}$$

$$D^n(1 + x)^m = m(m - 1) \cdots (m - n + 1)(1 + x)^{m-n}. \tag{13.38}$$

For $x = 0$ we obtain $f^{(n)}(0) = m(m - 1) \cdots (m - n + 1)$.

The Leibniz formula for the nth derivative of the product reads

$$(uv)^{(n)} = \sum_{k=0}^{n} \binom{n}{k} u^{(k)} v^{(n-k)}. \tag{13.39}$$

In this formula $u^{(0)}$, the zero derivative of u, is to be understood as u itself. Similarly $v^{(0)} = v$. This convention is not only used here but generally throughout calculus. The proof of Equation (13.39) is by induction, and is similar to the proof of the binomial theorem 5.4 and is left to the readers as an exercise.

13.4.1 *Higher derivatives in* Maple

```
>   restart;
>   f:=x->x^3;
```

$$f := x \to x^3$$

```
>   D(f)(x);
```

$$3\,x^2$$

```
>   Df:=unapply(%,x);
```

$$Df := x \to 3\,x^2$$

```
>   D(Df)(-2);
```

$$-12$$

```
>   diff(x^3,x,x);
```

$$6\,x$$

```
>   subs(x=-2,%);
```

$$-12$$

If we have to differentiate many times it is convenient to produce a sequence
of x's first.

> `dnu:=seq(x,k=1..11):diff(x^(100),dnu);`

$$5653408585997652480000\, x^{89}$$

> `dnum:=seq(x,k=1..100):diff(1/(1+x),dnum):subs(x=0,%);`

$93326215443944152681699238856266700490715968264438162146859\backslash$
$29638952175999932299156089414639761565182862536979208272\,23\backslash$
$758251185210916864000000000000000000000000000$

By Equation (13.38), for $m = -1$ this number is equal to 100!; this is easily
confirmed:

> `%/100!;`

$$1$$

13.4.2   *Significance of the second derivative*

Applying Theorems 13.9 and 13.10 to the second derivative gives

---

Theorem 13.12   *Assume $f'(c) = 0$. If $f''(c) > 0$ then f has a strict*
local minimum at c, if $f''(c) < 0$ then f has a strict local maximum at
c.

This theorem deals with the second derivative at a point. We are now
going to study the behaviour of a function if the second derivative is of a
definite sign on some interval.

A function f is said to be convex on an interval I if

$$f(x_2) \le f(x_1) + \frac{f(x_3) - f(x_1)}{x_3 - x_1}(x_2 - x_1), \qquad (13.40)$$

for $x_1 < x_2 < x_3$ in I. If the inequality in Equation (13.40) is strict the
function is strictly convex. f is said to be concave or strictly concave on
I if $-f$ is convex or strictly convex, respectively. The terms concave up
(upwards) and concave down (downwards) are used by some authors instead
of convex and concave. Equivalent forms of inequality (13.40) are

$$f(x_2)(x_3 - x_1) - f(x_1)(x_3 - x_2) - f(x_3)(x_2 - x_1) \le 0, \quad \text{and} \quad (13.41)$$

$$\frac{f(x_2) - f(x_1)}{x_2 - x_1} \leq \frac{f(x_3) - f(x_2)}{x_3 - x_2}. \tag{13.42}$$

The geometrical meaning of Equation (13.40) is: The point $(x_2, f(x_2))$ lies below the straight line joining $(x_1, f(x_1))$ with $(x_2, f(x_2))$, whereas Equation (13.42) indicates that the slope of the line joining $(x_1, f(x_1))$ to $(x_3, f(x_3))$ is smaller than the slope of the line joining $(x_3, f(x_3))$ to $(x_2, f(x_2))$. This means that the difference quotients are monotonically increasing. The next theorem extends this monotonicity to the derivative.

Theorem 13.13 *A function differentiable on an open interval I is convex if and only if f' is increasing on I. It is strictly convex if and only if f' is strictly increasing on I.*

Corollary 13.13.1 *A function differentiable on an open interval I is concave if and only if f' is decreasing on I. It is strictly concave if and only if f' is strictly decreasing on I.*

Proof. Let f be convex. Sending $x_2 \to x_1$ in inequality (13.42) gives

$$f'(x_1) \leq \frac{f(x_3) - f(x_1)}{x_3 - x_1}. \tag{13.43}$$

Taking the limit again in inequality (13.42), this time as $x_2 \to x_3$, leads to

$$\frac{f(x_3) - f(x_1)}{x_3 - x_1} \leq f'(x_3). \tag{13.44}$$

Combining Inequalities (13.43) and (13.44) leads to $f'(x_1) \leq f'(x_3)$. Since x_1 and x_3 were arbitrary, subject only to the condition $x_1 < x_3$, we have proved the monotonicity of f'.

If f is not convex then there are three points $x_1 < x_2 < x_3$ such that inequality (13.40) is false. In other words

$$f(x_2) - f(x_1) - \frac{f(x_3) - f(x_1)}{x_3 - x_1}(x_2 - x_1) > 0. \tag{13.45}$$

We denote

$$Q = \frac{f(x_3) - f(x_1)}{x_3 - x_1}$$

and chose an auxiliary function

$$F(x) = f(x) - f(x_1) - Q(x - x_1). \tag{13.46}$$

From inequality (13.45) we obtain $F(x_1) = 0 < F(x_2)$, consequently, by Theorem 13.7, there is a point ξ_{12} between x_1 and x_2 such that $f'(\xi_{12}) > Q$. Reasoning similarly leads to ξ_{23} between x_2 and x_3 for which $f'(\xi_{23}) < Q$. This proves f' is not increasing.

If f is strictly convex then f' is still increasing, if it were not strictly increasing it would be constant on some subinterval of I, and then f would be linear on this subinterval by Exercise 13.3.9, hence not strictly convex.

If f is strictly increasing but not strictly convex then F defined in Equation (13.46) would be zero at x_1, x_2, x_3 and would have local extrema in $]x_1, x_2[$ and $]x_2, x_3[$. At these points $F'(x) = 0$ and consequently at these points f' would be equal to Q. So, f' would not be strictly increasing. $\square$

This Theorem, together with Theorem 13.7, leads to the next Theorem, which is easier to apply.

Theorem 13.14   *Assume that f'' exists on an open interval I. Then*

(i) $f''(x) \geq 0$ in $I^\circ \Rightarrow f$ is convex in I;
(ii) $f''(x) \leq 0$ in $I^\circ \Rightarrow f$ is concave in I;
(iii) $f''(x) = 0$ in $I^\circ \Rightarrow f$ is linear on I;
(iv) $f''(x) > 0$ in $I^\circ \Rightarrow f$ is strictly convex in I;
(v) $f''(x) < 0$ in $I^\circ \Rightarrow f$ is strictly concave in I.

Example 13.13

```
> restart;
> f:=x->(41*x^3-119*x)^(1/3);
```

$$f := x \to (41\, x^3 - 119\, x)^{(1/3)}$$

```
> diff(f(x),x,x);
```

$$-\frac{2}{9} \frac{(123\, x^2 - 119)^2}{(41\, x^3 - 119\, x)^{(5/3)}} + \frac{82\, x}{(41\, x^3 - 119\, x)^{(2/3)}}$$

```
> f2:=normal(%);
```

$$f2 := -\frac{238}{9} \frac{123\, x^2 + 119}{(41\, x^3 - 119\, x)^{(5/3)}}$$

The numerator is positive, so the sign is determined by the denominator. To find where the derivative is positive and where it is negative we find the zeros of the denominator.

```
>  L:=[solve(41*x^3-119*x=0,x)];
```

$$L := [0, \frac{1}{41} \sqrt{4879}, -\frac{1}{41} \sqrt{4879}]$$

We denote

```
>  a:=L[3];b:=L[2];
```

$$a := -\frac{1}{41} \sqrt{4879}$$

$$b := \frac{1}{41} \sqrt{4879}$$

---

f is convex on $]0, b[$ and $] - \infty, a[$ and concave on the two remaining intervals. The graph is shown in Figure 13.9. When plotting f the command surd must be used rather than a fractional exponent because cube roots of *negative* numbers are involved.

```
>  restart;
>  plot(surd(41*x^3-119*x,3),x =-2..2);
```

A function f is said to have an *inflection point* at c if f' has a strict local extremum at c. An immediate consequence of this definition is

---

Theorem 13.15   *If f has an inflection at c and $f''(c)$ exists then $f''(c) = 0$.*

An inflection point is of significance for the graph of f since the graph of f attaches itself tightly to the tangent at the inflection point: the graph crosses the tangent from one side to the other. If, for instance, f' has a strict local maximum at c then, for some positive δ, the function

$$F :\longmapsto f(x) - f(c) - f'(c)(x - c)$$

has a positive derivative for $c - \delta < x < c$ and also for $c < x < c + \delta$. Consequently F is strictly increasing and

$$F(x) < 0 = F(c) \quad \text{for} \quad x < c, \tag{13.47}$$

$$F(x) > 0 \quad \text{for} \quad x > c. \tag{13.48}$$

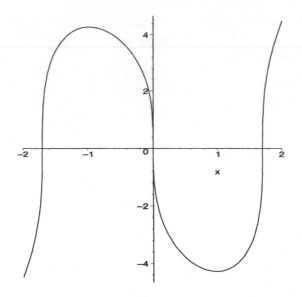

Fig. 13.9   Convexity, concavity

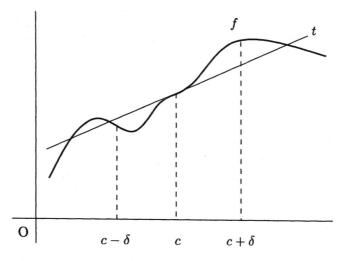

Fig. 13.10   An inflection point at $x = c$

Inequality (13.47) means that the graph of f lies first below the tangent and then by inequality (13.48) above the tangent. Similarly, if f' has a strict local minimum, then the graph crosses from above to below the tangent.

If $f''(c) = 0$ then f might have an inflection point at c but need not. $(x^4)'' = 4 \cdot 3x^2$ is zero at $x = 0$ but there is no inflection. The next Theorem guarantees the existence of an inflection point.

> **Theorem 13.16**   If $f''(c) = 0$ and $f'''(c) \neq 0$ then f has an inflection point at c.

Proof.   This is an immediate consequence of the definition of an inflection point and Theorem 13.12. $\square$

Example 13.14   The function $f : x \mapsto 7x + x^3$ has the second derivative equal to zero at $x = 0$, the third derivative $f'''(0) = 6$. Consequently f has an inflection point at $x = 0$ by Theorem 13.16.

Exercises

Exercise 13.4.1   *Find intervals where the given functions are convex and where they are concave.*

1. $f(x) = 3x^5 - 10x^3 + 6x - 7$,
2. $y(x) = (x - 3)^6 - 3x + 4$,
3. $h(x) = x^3/(x^2 + 1)$.

(!) (i) **Exercise 13.4.2**   *Prove: If f is convex, but not strictly convex, on an interval I, then there is a subinterval of I on which f is linear.*

Exercise 13.4.3   *Find a and b such that $f : x \mapsto 2ax^3 + bx^2$ has an inflection point at $x = 1$.*

(!) (i) **Exercise 13.4.4**   *Prove: A differentiable function has an inflection between two consecutive local extrema.*

(i) **Exercise 13.4.5**   *The function $f : x \mapsto x|x|$ has an inflection point at $x = 0$, however it is not true that $f''(0) = 0$. Why? [Hint: Consider $f'(x)$ for $x > 0$ and for $x < 0$.]*

(!) (i) **Exercise 13.4.6**   *Prove: f is convex on $[a, b]$ if and only if f'_+ or f'_- exists on $[a, b]$ and is increasing.*

13.5   Mean value theorems

Figure 13.11 suggests that there is a point $(c, f(c))$ on the graph of f where the tangent is parallel to the line joining the end-points, that is

$$f'(c) = \frac{f(b) - f(a)}{b - a}. \qquad (13.49)$$

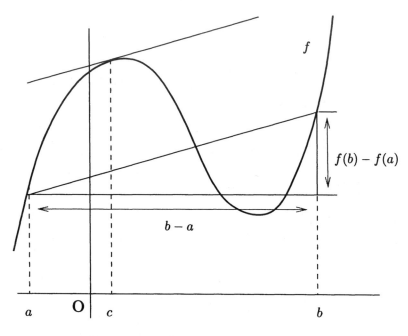

Fig. 13.11   The mean value theorem

We state this as

Theorem 13.17 (The Lagrange mean value theorem)   *If* $f :$ $[a, b] \mapsto \mathbb{R}$ *is continuous on* $[a, b]$ *and differentiable for* $a < x < b$ *then there exists a point* $c \in [a, b]$ *such that Equation (13.49) holds.*

Figure 13.12 exhibits the same phenomenon for a parametrically given curve $\{[f(t), g(t)]; a \leq t \leq b\}$. We state this as

> **Theorem 13.18 (The Cauchy mean value theorem)** If $f : [a, b] \mapsto \mathbb{R}$ and $g : [a, b] \mapsto \mathbb{R}$ are continuous on $[a, b]$ and differentiable for $a < t < b$ then there exists a point c with $a < c < b$ such that
>
> $$[f(b) - f(a)]g'(c) = [g(b) - g(a)]f'(c). \qquad (13.50)$$

The Lagrange mean value theorem is a special case of the Cauchy mean value theorem for $g(x) = x$. It suffices to prove the Cauchy theorem. The term *mean value theorems* is not really appropriate; the emphasis should lie on the increment of the function or functions, not on the mean value c, but in English-speaking mathematics the terminology is firmly established. If $g'(x) \neq 0$ for $a < x < b$ then it follows from (13.50) that[2] $g(b) - g(a) \neq 0$ and Equation (13.50) takes a more convenient form

$$\frac{f(b) - f(a)}{g(b) - g(a)} = \frac{f'(c)}{g'(c)}. \qquad (13.51)$$

Formulae (13.49), (13.50) and (13.51) are often written in a different but equivalent form. One writes x and $x+h$ instead of a and b. Then c becomes $x + \Theta h$ with $0 < \Theta < 1$. Equation (13.51) then becomes

$$\frac{f(x + h) - f(x)}{g(x + h) - g(x)} = \frac{f'(x + \Theta h)}{g'(x + \Theta h)}. \qquad (13.52)$$

The advantage in using the notation with Θ is that for negative h, equations like (13.52) hold without any need to change the inequality for Θ. The numbers c or Θ are not uniquely determined, for instance, for constant f the number c can be any number in $]a, b[$ and Θ any number in $]0, 1[$. The dependence of Θ on f, x and h in the Lagrange mean value theorem is studied in Exercises 13.5.1 and 13.5.2. The importance of the mean value theorem comes from the fact that it allows conclusions to be made about the function from knowledge of the derivative, even if nothing more is known about Θ than $0 < \Theta < 1$.

For the proof of the Cauchy theorem we denote

$$\Delta f = f(b) - f(a),$$
$$\Delta g = g(b) - g(a),$$
$$F(x) = [g(x) - g(a)]\Delta f - [f(x) - f(a)]\Delta g.$$

---

[2]By choosing $f(x) = x$.

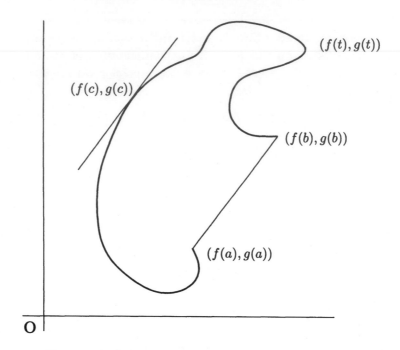

Fig. 13.12   The Cauchy mean value theorem

Readers familiar with analytic geometry will recognize that $|F(x)|$ is twice the area of a triangle with vertices $(f(a), g(a))$, $(f(x), g(x))$, $(f(b), g(b))$.

Proof. F is zero at the ends of the interval $[a, b]$. By Weierstrass' theorem 12.22 the function F attains its extrema in $[a, b]$. Since $F(a) = F(b) = 0$ either the maximum or the minimum is attained at an interior point c. By Theorem 13.8 the derivative $F'(c) = 0$ and this is equivalent to Equation (13.50). $\qquad\square$

Exercises

(i) **Exercise 13.5.1**   Show: If $f(t) = t^2$ then Θ from the Lagrange mean value theorem is $1/2$, independent of x.

(!) (i) **Exercise 13.5.2**   Find Θ from the Lagrange mean value theorem if $f(t) = t^3$. Show that $\Theta \to 1/2$ as $h \to 0$.

(i) **Exercise 13.5.3**   The Lagrange mean value theorem becomes false if f fails to be differentiable at one point. [Hint: $f(x) = |x|$, $[a, b] = [-1, 1]$.]

ⓘ **Exercise 13.5.4** *Prove: Between two real zeros of a differentiable function there is a zero of the derivative.*

ⓘ **Exercise 13.5.5** *For complex valued functions the mean value theorems are false. Consider the following counterexample to the Lagrange theorem:* $f(x) = x^3 - x + \imath(x^3 - x^2)$ *and* $[a, b] = [0, 1]$.

13.6 The Bernoulli–l'Hospital rule

Theorems for calculating limits by means of derivatives are known as l'Hospital rules. However they were discovered by J. Bernoulli who communicated them to Marquis l'Hospital, who promptly published them under his own name. Their importance for practical evaluation of limits is diminished when limits can be found easily by *Maple*. However, we need the rule in the next section. We first make a remark indicating that the rule is plausible.

Remark 13.6 If $f(a) = g(a) = 0$ then

$$\lim_{x \to a} \frac{f(x)}{g(x)} = \lim_{x \to a} \frac{\dfrac{f(x) - f(a)}{h}}{\dfrac{g(x) - g(a)}{h}} = \frac{f'(a)}{g'(a)}, \tag{13.53}$$

provided $f'(a)$, $g'(a)$ exist and $g'(a) \neq 0$. If we try to apply this result to $f(x) = \sqrt[5]{x} - 1$ and $g(x) = \sqrt[3]{x} - 1$ for $a = 1$ we encounter a difficulty: $g'(1)$ does not exist. However the limit can be calculated as

$$\lim_{x \to u} \frac{f(x)}{g(x)} = \lim_{x \to u} \frac{f'(x)}{g'(x)}, \tag{13.54}$$

with the result that

$$\lim_{x \to 1} \frac{\sqrt[5]{x} - 1}{\sqrt[3]{x} - 1} = \lim_{x \to 1} \frac{3}{5} \frac{\sqrt[3]{(x - 1)^2}}{\sqrt[5]{x^4}} = 0.$$

In the above we have two variants. In the first the required limit is equal to the ratio of the derivatives at the point in question. In the second the limit is equal to the limit of the ratio of the derivatives at that point. It is a rather subtle, but significant, point to recognise that these two procedures are distinct. The combination of these two methods is important for extending the rule to the case where the first few derivatives of f and g are zero at the point. The following theorem describes the general situation.

> **Theorem 13.19** *Let I be an interval, a an interior point of I and $f : I \mapsto \mathbb{R}$ and $g : I \mapsto \mathbb{R}$. If n is a non-negative integer, $f^{(n)}$ and $g^{(n)}$ are continuous at a,*
>
> $$f(a) = f'(a) = \cdots = f^{(n)}(a) = 0, \qquad (13.55)$$
>
> $$g(a) = g'(a) = \cdots = g^{(n)}(a) = 0, \qquad (13.56)$$
>
> *and EITHER*
>
> (i) $\lim_{x \to a} \dfrac{f^{(n+1)}(x)}{g^{(n+1)}(x)} = l$
>   OR
> (ii) *both $f^{n+1}(a)$ and $g^{n+1}(a) \neq 0$ exist*
>
> *then*
>
> $$\lim_{x \to a} \frac{f(x)}{g(x)} \qquad (13.57)$$
>
> *exists and is equal to l or to $\dfrac{f^{n+1}(a)}{g^{n+1}(a)}$, respectively.*

Proof. We commence the proof of (i) for[3] $n = 0$. Using the Cauchy mean value theorem gives

$$\frac{f(x)}{g(x)} = \frac{f(a+h) - f(a)}{g(a+h) - g(a)} = \frac{f'(a+\Theta h)}{g'(a+\Theta h)}, \qquad (13.58)$$

for some Θ with $0 < \Theta < 1$. Let $h_n \to 0$ then[4] $\Theta h_n \to 0$ and by assumption (i) and the definition of a limit, $f'(a+\Theta h_n)/g'(a+\Theta h_n) \to l$. Consequently $\lim_{x \to a} \dfrac{f(x)}{g(x)} = l$. We proceed by induction now.

$$\lim_{x \to a} \frac{f^n(x)}{g^n(x)} = \lim_{x \to a} \frac{f^{n+1}(x)}{g^{n+1}(x)} = l \qquad (13.59)$$

by assumption and by what we have already established. Since

$$\lim_{x \to a} \frac{f(x)}{g(x)} = \lim_{x \to a} \frac{f^n(x)}{g^n(x)}, \qquad (13.60)$$

by the induction hypothesis, we now have (i).

---

[3] $f^{(0)}$ and $g^{(0)}$ are understood as f and g, respectively.
[4] Θ depends on n but this dependence has no influence on the proof.

For (ii), we have by (i) that Equation (13.60) is true and by Equation (13.53) applied to f^n and g^n in place of f and g we obtain (ii). □

The next Theorem has a very similar proof. It says, roughly speaking, that the derivative cannot have "jump discontinuities".

Theorem 13.20   If $f : I \mapsto \mathbb{R}$ is continuous at an interior point of interval I and $\lim\limits_{x \to a} f'(x) = l$ then f is differentiable at a and $f'(a) = l$.

Proof.   By the Lagrange mean value theorem

$$\frac{f(a + h) - f(a)}{h} = f'(a + \Theta h)$$

with $0 < \Theta < 1$. If $h_n \to 0$ then $\Theta h_n \to 0$ and the right hand side tends (by assumption) to l. Consequently, the left hand side has also limit l and this means that the derivative exists and is l. □

The derivative can exist at a even if $\lim\limits_{x \to a} f'(x)$ does not. For instance, if $f(x) = x^2 \operatorname{saw}(1/x)$ for $x \neq 0$ and $f(0) = 0$ then f has a discontinuous derivative at 0.

The Bernoulli–l'Hospital rule can also be applied in situation when $g(x) \to \infty$. For sake of completeness we state the next theorem in which the symbol **Lim** stands consistently throughout the theorem for any of $\lim\limits_{x \to a}$ or $\lim\limits_{x \uparrow a}$ or $\lim\limits_{x \downarrow a}$ or $\lim\limits_{x \to \infty}$ or $\lim\limits_{x \to -\infty}$.

Theorem 13.21   If $\mathbf{Lim}\ \dfrac{f'(x)}{g'(x)} = l$ then

$$\mathbf{Lim}\ \frac{f(x)}{g(x)} = l$$

if EITHER

$$\mathbf{Lim}\ f(x) = \mathbf{Lim}\ g(x) = 0$$

OR

$$\mathbf{Lim}\ |g(x)| = \infty$$

This theorem remains true if l denotes ∞ or $-\infty$.

Exercises

Exercise 13.6.1   *Use the methods of this section to evaluate the limits in Exercise 12.1.1.*

(i) **Exercise 13.6.2**   *Reckless application of the Bernoulli–l'Hospital rule can lead to wrong results. For instance,* $\lim_{x\to 1}\left((x^2+2)/(3x^3-1)\right) = \lim_{x\to 1}(2x/9x^2) = 2/9$ *but clearly* $\lim_{x\to 1}\left((x^2+2)/(3x^3-1)\right) = 3/2$. *Where is the mistake.*

(!) (i) **Exercise 13.6.3**   *Show that Theorem 13.20 can be extended to one-sided limits and one-sided derivatives.*

(!) (i) **Exercise 13.6.4**   *The Bernoulli–l'Hospital rule can be extended to one-sided derivatives. Prove: If g is monotonic,* $\lim_{x\downarrow a} f(x) = \lim_{x\downarrow a} g(x) = 0$ *and* $\lim_{x\downarrow a} f'_+(x)/g'_+(x) = l$ *or* $\lim_{x\downarrow a} f'_-(x)/g'_-(x) = l$ *then* $\lim_{x\downarrow a} f(x)/g(x) = l$. [*Hint: For the proof and discussion of theorems of this type see Vyborny and Nester (1989).*]

13.7   Taylor's formula

Often it is convenient and sometimes even necessary to approximate a given function by another function, which is simpler or has better properties. It is easy to work with polynomials, which have many simple good properties. For example, they have derivatives of all orders and their function values can be computed by multiplication and addition only. Even computers can handle polynomials better than most other functions: *Maple* finds *all* zeros of a polynomial by `fsolve`, but need not do so for a more complicated function. We have already encountered polynomial approximation in Weierstrass' Theorem 12.22 which deals with a uniform approximation of a continuous function on an interval. However, the theorem does not really give any means of finding this approximation effectively and constructively. Our study in this section will deal with a different approximation. It is constructive and effective, but approximates the given function only[5] near a given point.

We encountered Taylor polynomial before in Section 6.5. If f is a poly-

---

[5] However, generally speaking, and given some luck, the approximation can be good everywhere.

nomial of degree n then

$$T_n(x) = f(a) + \frac{f'(a)}{1!}(x-a) + \frac{f''(a)}{2!}(x-a)^2 + \cdots$$
$$+ \frac{f^{(n-1)}(a)}{1!}(x-a)^{n-1} + \frac{f^{(n)}(a)}{n!}(x-a)^n$$

is a polynomial equal to f for all x. If f is n times differentiable at a then T_n can always be formed, regardless of whether or not f is a polynomial. It is natural to expect that T_n would be a good approximation of f and then the estimate of

$$R_{n+1}(x) = f(x) - T_n(x) \tag{13.61}$$

would determine how good the approximation is. In the last chapter of this book we shall derive powerful but reasonably simple estimates of R_{n+1}. In this section we shall be satisfied with the next Theorem 13.22, which says that R_n is of an order of magnitude smaller than $(x-a)^n$. We illustrate the nature of approximation by plotting some graphs in *Maple*. Let as recall that the command[6] `taylor(f(x),x=a,k):convert(%,polynom);` produces the Taylor polynomial for f of degree k and it can be then used for plotting. The following lines of *Maple* code produce the first eight approximations, t1–t8, for the function defined below.

```
> restart;
> F:=x->1/sqrt(1+x);
```

$$F := x \rightarrow \frac{1}{\sqrt{1+x}}$$

```
> k:=1; while k<9 do
> t||k:=convert(taylor(F(x),x=0,k),polynom)
> k:=k+1; end do:
```

A few approximations, together with f, are depicted in the next two Figures. In Figure 13.13 the second and third approximations are plotted; in Figure 13.14 the third and eighth approximations are graphed.

Maple uses the symbol $h(x) = o(\phi(x))$ and $g(x) = O(\phi(x))$ to indicate that

$$\frac{h(x)}{\phi(x)} \rightarrow 0 \tag{13.62}$$

---

[6]Explained earlier for polynomials but applicable generally.

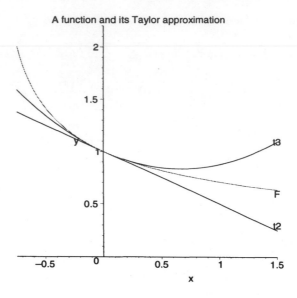

Fig. 13.13   Taylor approximations

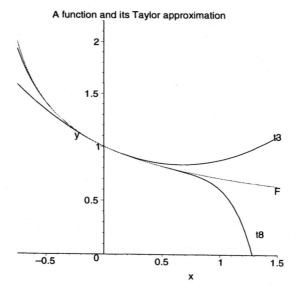

Fig. 13.14   Taylor approximations

or that $\dfrac{g(x)}{\phi(x)}$ is bounded. These symbols do not reveal the point at which the limit in Equation (13.62) is taken or the interval on which $0(\phi(x))$ is bounded. This must be understood from the context. These symbols are convenient and widely used, but unfortunately the behavior of these symbols might look bizarre to the uninitiated: for instance the relations[7] $o(x) + 0(x) = 0(x)$ or $o(x) + o(x) = o(x)$ are correct, but they *do not imply* that $o(x) = 0$. We are now ready for the main theorem of this section.

Theorem 13.22   If f has the n-th order derivative at a then

$$f(x) - T_n(x) = R_{n+1}(x) = o((x-a)^n). \qquad (13.63)$$

Proof.   Denote $g(x) = (x-a)^n$. Clearly

$$D^k R_{n+1}(a) = 0 \quad \text{for} \quad k = 1, \ldots, n,$$
$$D^k g(a) = 0 \quad \text{for} \quad k = 1, \ldots, n-1, \quad D^n g(a) = n!$$

By Theorem 13.21

$$\lim_{x \to a} \frac{R_{n+1}(x)}{(x-a)^n} = \frac{0}{n!} = 0.$$

$\square$

This theorem is called the Taylor formula with Peano remainder. We use it for a proof of a theorem on local extrema.

Theorem 13.23   Let $f : |a, b| \mapsto \mathbb{R}$ and n be even. If

$$f'(a) = f''(a) = \cdots = f^{(n-1)}(a) = 0,$$
$$f^{(n)}(a) \neq 0,$$

then f has a local extremum at a. It is a maximum if $f^n(a) < 0$ and a minimum if $f^n(a) < 0$.

Proof.   We apply Theorem 13.22 assuming $f^n(a) > 0$.

$$f(x) = f(a) + \left(\frac{f^n(a)}{n!} + \alpha(x) \right)(x-a)^n,$$

with $\alpha(x) \to 0$. By Corollary 12.8.1 there is a positive δ such that $f^n(a) + \alpha(x) > 0$ for $|x - a| < \delta$ and then $f(x) < f(a)$, which means there is a local

---

[7]The relevant point is 0.

minimum at a. The case $f^n(a) < 0$ can be handled similarly, or we can apply the already proved part to $-f$. $\qquad\square$

The next Theorem has very similar proof; we skip it.

> **Theorem 13.24** *Let $f : [a, b] \mapsto \mathbb{R}$ and n be odd. If*
> $$f''(a) = \cdots = f^{(n-1)}(a) = 0,$$
> $$f^{(n)}(a) \neq 0,$$
> *then f has an inflection at a.*

Exercises

(i) **Exercise 13.7.1** *Prove the following addition to Theorem 13.23: If n is odd then there is no local extremum at a.*

(i) **Exercise 13.7.2** *Prove the following addition to Theorem 13.24: If n is even then there is no inflection at a.*

Exercise 13.7.3 *Find the local maxima, minima and inflection points for each of the following functions:*

1. $x^2(1 - x)^3$,
2. $x^m(1 - x)^l$ *with $m \in \mathbb{N}$ and $l \in \mathbb{N}$,*
3. $1/\sqrt{1 + x^4}$.

13.8 Differentiation of power series

If a function is given by a power series

$$f(x) = c_0 + c_1(x - a) + c_2(x - a)^2 + c_3(x - a)^3 + \cdots + c_n(x - a)^n + \cdots \quad (13.64)$$

then it is easy to find the derivative, with differentiation like that for polynomials:

$$f'(x) = c_1 + 2c_2(x - a) + 3c_3(x - a)^2 + \cdots + nc_n(x - a)^{n-1} + \cdots. \quad (13.65)$$

This is plausible but it does require a proof.[8]

---

[8]It follows Vyborny (1987).

Theorem 13.25   *Let the power series (13.64) have a positive radius of convergence R (possibly ∞). Then the power series*

$$d(x) = c_1 + 2c_2(x - a) + 3c_3(x - a)^2 + \cdots + nc_n(x - a)^{n-1} + \cdots$$

has the same radius of convergence R and for all complex x with $|x - a| < R$

$$f'(x) = d(x).$$

Without loss of generality we take $a = 0$. We need

Lemma 13.1   *Let n denote a positive integer. For $x \in \mathbb{C}$, $h \in \mathbb{C}$ and $0 < |h| \le H$*

$$\left| (x + h)^n - x^n - hnx^{n-1} \right| \le \frac{|h|^2}{H^2} (|x| + H)^n \tag{13.66}$$

$$\left| nx^{n-1} \right| \le \frac{1}{H} \left[2(|x| + H)^n + |x|^n \right]. \tag{13.67}$$

Proof.   For the proof of inequality (13.66) we use the binomial Theorem 5.4.

$$\left| (x + h)^n - x^n - nx^{n-1}h \right| \le |h|^2 \sum_{k=2}^{n} \binom{n}{k} |x|^{n-k} |h|^{k-2}$$

$$\le \frac{|h|^2}{H^2} \sum_{k=2}^{n} \binom{n}{k} |x|^{n-k} H^k \le \frac{|h|^2}{H^2} (|x| + H)^n. \tag{13.68}$$

Since

$$H \left| nx^{n-1} \right| - (|x| + H)^n - |x|^n \le \left| (x + H)^n - x^n - Hnx^{n-1} \right|$$

Inequality (13.67) follows from inequality (13.66) with $h = H$. $\square$

We can now prove Theorem 13.25.

Proof.   For a given x with $|x| < R$ choose H such that $|x| + H < R$. Multiplying inequality (13.67) by $|c_n|$ gives

$$\left| nc_n x^{n-1} \right| \le \frac{1}{H} \left[2c_n (|x| + H)^n + c_n |x|^n \right].$$

By the comparison theorem the power series for d converges, this in turn implies that the radius of convergence for d is R. It follows from inequality (13.66) that

$$\left| \frac{f(x+h) - f(x)}{h} - d(x) \right| \le \frac{|h|}{H^2} \sum_{n=1}^{\infty} |c_n| \left(|x| + H \right)^n .$$

The sum on the right hand side is a well-defined finite number and as $h \to 0$ the right hand side tends to 0, consequently $f'(x) = d(x)$. $\qquad\square$

An important consequence of our theorem is that the derivative, being again a power series, is differentiable. Then the second derivative, as a power series, is again differentiable, and so on. A convergent power series has derivatives of all orders. Substituting $x = a$ into Equation (13.65) gives $c_1 = f'(a)$, differentiating and substituting again gives $c_2 = f''(a)/2$, and by differentiating n-times leads to $c_n = \dfrac{f^{(n)}(a)}{n!}$. This leads to

$$f(x) = f(a) + f'(a)(x-a) + \frac{f''(a)}{2!}(x-a)^2 + \cdots + \frac{f^n(a)}{n!}(x-a)^n + \cdots .$$
$$(13.69)$$

This series is called the *Taylor series* of f at a. The special case $a = 0$, that is

$$f(x) = f(0) + f'(0)x + \frac{f''(0)}{2!}x^2 + \cdots \frac{f^{(n)}(0)}{n!}x^n + \cdots \qquad (13.70)$$

has the name of *Maclaurin series*. It is important to realize that we obtained these formulae under the assumption that f was representable by a power series. This assumption must be justified for concrete functions. In Chapter 15 we shall prove some general theorems guaranteeing the validity of Equation (13.69). However, it might happen that a function has derivatives of all orders but *is not* equal to its Taylor series. Such a function is defined in Example 14.1. Here is an example in a positive direction.

Example 13.15   The formula

$$\frac{1}{1+x} = 1 - x + x^2 - x^3 + x^4 + \cdots + (-1)^n x^n + \cdots$$

is valid for $|x| < 1$. Theorem 13.25 gives

$$\frac{1}{(1+x)^2} = 1 - 2x + 3x^2 - 4x^3 + \cdots + (-1)^{n-1} n x^{n-1} + \cdots$$

$$\frac{-2}{(1+x)^3} = -2 + 3 \cdot 2x - 4 \cdot 3x^2 + \cdots + (-1)^{n-1} n(n-1) x^{n-1} + \cdots$$

and by differentiating $m - 1$ times

$$\frac{(-1)^{m-1}(m-1)!}{(1+x)^m} = \sum_{k=m-1}^{\infty} k(k-1)\cdots(k-m+2)x^{k-m+1}. \tag{13.71}$$

It is convenient to extend the definition of the binomial coefficient to any real n and positive integer k by

$$\binom{n}{k} = \frac{n(n-1)\cdots(n-k+1)}{k!} \tag{13.72}$$

for instance

$$\binom{-\frac{1}{2}}{3} = \frac{\left(-\frac{1}{2}\right)\cdot\left(-\frac{3}{2}\right)\cdot\left(-\frac{5}{2}\right)}{1\cdot2\cdot3} = -\frac{5}{16}.$$

These definitions allow us to rewrite the Maclaurin series for $(1+x)^{-m}$, $m \in \mathbb{N}$, in the following form

$$(1+x)^{-m} = \sum_{k=0}^{\infty} \binom{-m}{k} x^k \tag{13.73}$$

This equation is called the binomial expansion: it resembles Formula (5.30) in the Newton's binomial theorem for $a = 1$ and $b = x$. The difference is that now the formula is correct only for $|x| < 1$ and the expansion is infinite.

13.9 Comments and supplements

The concepts of right-hand and left-hand derivative can be further generalized, in a similar way as one-sided limits were to limits superior and inferior. The following four numbers

$$D^+f(c) = \limsup_{h\downarrow0} \frac{f(c+h)-f(c)}{h},$$

$$D^-f(c) = \limsup_{h\uparrow0} \frac{f(c+h)-f(c)}{h},$$

$$D_-f(c) = \liminf_{h\uparrow0} \frac{f(c+h)-f(c)}{h},$$

$$D_+f(c) = \liminf_{h\downarrow0} \frac{f(c+h)-f(c)}{h},$$

are called the *Dini derivates*. The first one is the upper right-hand Dini derivate, the third is the lower left-hand Dini derivate, with similar terminology for the other two.

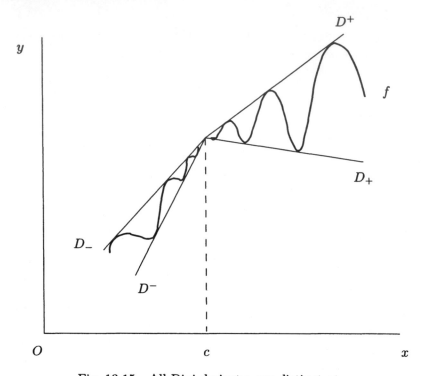

Fig. 13.15   All Dini derivates are distinct at c

The behaviour of a function with all Dini derivates distinct at c is indicated in Figure 13.9. It can be shown that, for a continuous f, the set of points c such that $D^+f(c) < D_-f(c)$ is at most countable. So is the set where $D^-f(c) < D_+f(c)$. A consequence of this is: the set where both the right-hand derivative and the left-hand derivative exist and are not equal is countable. There is another theorem in similar vein: A set of points of *strict* local maxima (or minima) of a function, continuous on some interval, is at most countable.

There is a theorem for Dini derivates similar to Theorem 13.20. If f and one of its Dini derivates is continuous on an open interval I then all Dini derivates are equal to f' on I.

Bolzano constructed a function continuous everywhere and differentiable nowhere.

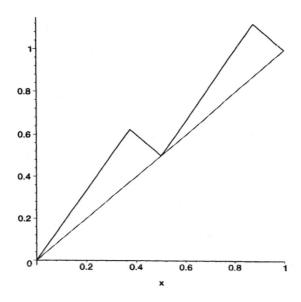

Fig. 13.16 The first approximation to the Bolzano function

Now we sketch his construction. First we define a sequence of divisions of $[0, 1]$ as follows: D_1 consists of $[0, 1]$, if u, v are consecutive points of D_n with $l = v - u > 0$ then the intervals

$$\left[u, u + \frac{3}{8}l\right] \quad \left[u + \frac{3}{8}l, u + \frac{1}{2}l\right] \quad \left[u + \frac{1}{2}l, u + \frac{7}{8}l\right] \quad \left[u + \frac{7}{8}l, v\right] \quad (13.74)$$

become subintervals of D_{n+1}. We proceed by defining a sequence of functions f_n each piecewise linear and continuous on $[0, 1]$. Let $f_1 : x \mapsto x$. If f_n has been defined and u, v are consecutive dividing points of D_n then let $f_{n+1}(u) = f_n(u)$ and

$$f'_{n+1} = \frac{5}{3} f'_n \quad \text{on the interiors of the first and third interval of (13.74),}$$

$$f'_{n+1} = -f'_n \quad \text{on the interiors of second and fourth interval of (13.74).}$$

It follows that $f_{n+1}(v) = f_n(v)$. Consequently f_{n+1} is continuous on $[0, 1]$. The graphs of f_2, f_3, f_4 are shown in Figure 13.17. The functions f_n very quickly become difficult to graph. The Bolzano's function f is the limit of

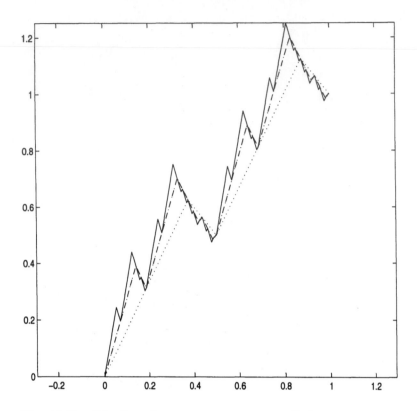

Fig. 13.17   The first three approximations to Bolzano's function

f_n, that is

$$f(x) = \lim_{n \to \infty} f_n(x)$$

for $x \in [0, 1]$. For details and proofs we refer to Výborný (2001). A famous
example is by Weierstrass[9], a relatively simple example due to Van der
Waerden, accessible to our readers, is in Spivak (1967). We consider now a
very simple example from McCarthy (1953). The function is given by

$$F(x) = \sum_{i=1}^{\infty} \frac{1}{2^n} g_n(x),$$

---

[9]The definition of the Weierstrass function is also in McCarthy (1953)

where

$$g_n(x) = \mathrm{saw}(2^{2^n} x).$$

The proof that F is continuous is easy and left for Exercise 13.9.1. Let N be a positive integer and $h_N = \pm 2^{-2} 2^{-2^N}$. The function g_N is linear on intervals of length $2^{2^N}/2$, twice as long as $|h_N|$, therefore it is possible to choose h_N positive or negative in order that x and $x + h_N$ are in the interval on which g_N is linear. Then we have

(i) $g_i(x + h_N) - g_i(x) = 0$ for $i > N$ since $2^{2^i} h_N$ is an integer[10]

(ii) $g_N(x + h_N) - g_N(x) = \dfrac{1}{4}$;

(iii)

$$\left| \sum_{i=1}^{N-1} 2^{-i} [g_i(x + h_N) - g_i(x)] \right| \leq (N-1) 2^{2^{N-1}} |h_N| \leq \frac{N-1}{4} 2^{-2^{N-1}}.$$

Using this we have

$$\left| \frac{f(x + h_N) - f(x)}{h_N} \right| \geq \frac{1}{4} \frac{1}{2^N} 4 \cdot 2^{2^N} - \frac{N-1}{4} \cdot 2^{-2^{N-1}} 4 \cdot 2^{2^N}$$

$$= 2^{-N} 2^{2^N} - (N-1) 2^{2^{N-1}}.$$

The right-hand side goes to infinity as $N \to \infty$. A finite derivative cannot exist.

Exercises

Exercise 13.9.1   *Prove that F is continuous.* [Hint: Choose k such that $\displaystyle\sum_{i=k+1}^{\infty} \frac{1}{2^n} g_n(x) < \varepsilon/3$. Then use continuity of $\displaystyle\sum_{1}^{k} \frac{1}{2^n} g_n(x)$.]

Exercise 13.9.2   *Use Maple to plot the third approximation to the Bolzano function.*

---

[10] The function saw has period 1 and hence $\mathrm{saw}(u + m) = \mathrm{saw}(u)$ for $m \in \mathbb{N}$.

Chapter 14

Elementary Functions

In this chapter we lay the proper foundations for the exponential and logarithmic functions, for trigonometric functions and their inverses. We calculate derivatives of these functions and use this for establishing important properties of these functions.

14.1 Introduction

We have already mentioned that some theorems from the previous chapter are valid for differentiation in the complex domain. Specifically, this is so for Theorem 13.1 and 13.2, for basic rules of differentiation in Theorems 13.3, 13.5 and for the chain rule, Theorem 13.4. In contrast, Theorem 13.7 makes no sense in the complex domain since the concept of an increasing function applies only to functions which have real values. However, part (iii) of Theorem 13.7 can be extended as follows.

> **Theorem 14.1**   For $a \in \mathbb{C}$ and $R > 0$ denote by S the disc $\{z; |z - a| < R\}$ or the whole complex plane $\mathbb{C}$. If $f'(z) = 0$ for all $z \in S$ then f is constant in S.

Proof.   For $t \in [0,1]$ let $Z = tz_1 + (1 - t)z_2$. If z_1 and z_2 are in S so is Z. Let $F : t \mapsto f(Z)$, $F_1(t) = \Re F(t)$ and $f_2(t) = \Im F(t)$. Then

$$F'(t) = f'(Z)(z_1 - z_2) = 0.$$

Consequently $F_1'(t) = F_2'(t) = 0$. By (iii) of Theorem 13.7 both F_1 and F_2 are constant in $[0,1]$, hence $f(z_1) = f(z_2)$ and f is constant in S.   □

An easy consequence of this Theorem is: $f : \mathbb{C} \mapsto \mathbb{C}$ is a polynomial of degree at most n if and only if $f^{(n+1)}(z) = 0$ for $z \in \mathbb{C}$.

14.2 The exponential function

A natural question to ask is: Is there a function which is not changed by differentiation? An obvious and uninteresting answer is, yes, the zero function. Hence a better question is: Does a function E exist such that $E' = E$ and $E'(a) \neq 0$ for some a? The answer is contained in the next theorem.

Theorem 14.2 *The equation*

$$E'(x) = E(x) \tag{14.1}$$

has a solution $\exp : \mathbb{C} \mapsto \mathbb{C}$ *satisfying the condition* $\exp(0) = 1$.
The function exp *has the following properties*

   (i) $\exp(1) = e$.
   (ii) *For* $z, y \in \mathbb{C}$

$$\exp(z + y) = \exp(z)\exp(y). \tag{14.2}$$

   (iii) $\exp(x) \neq 0$ *for all* $x \in \mathbb{C}$.
   (iv) *The restriction of* exp *to* $\mathbb{R}$ *is strictly increasing and for a non-negative integer* n

$$\lim_{x \to \infty} \frac{\exp(x)}{x^n} = \infty \quad \text{and} \quad \lim_{x \to \infty} \frac{x^n}{\exp(x)} = 0 \tag{14.3}$$

If E is any solution to Equation (14.1) in $\mathbb{C}$ then there is a constant c such that $E(x) = c\exp(x)$.

The graph of exp is shown in Figure 14.1

Proof. If exp existed then all derivatives at 0 would be $\exp(0) = 1$ and the Maclaurin expansion of exp would be

$$1 + x + \frac{x^2}{2!} + \frac{x^3}{3!} + \cdots + \frac{x^n}{n!} + \cdots.$$

This series converges by the ratio test, Corollary 11.9.1, for all complex x, it therefore defines a function, say E. Obviously $E(0) = 0$ and by Theorem 13.25

$$\mathsf{E}'(x) = 1 + 2\frac{x}{2!} + 3\frac{x^2}{3!} + \cdots + n\frac{x^{n-1}}{n!} + \cdots = \mathsf{E}(x).$$

E has the defining property of exp, so we call it exp and we have a very

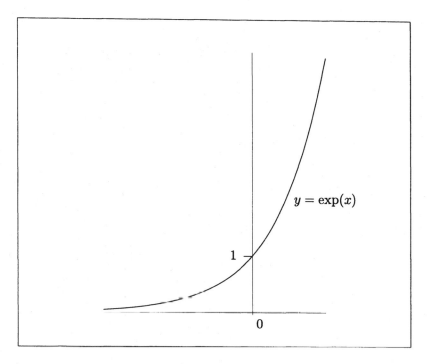

Fig. 14.1   The exponential function

important expansion

$$\exp(x) = 1 + x + \frac{x^2}{2!} + \frac{x^3}{3!} + \cdots + \frac{x^n}{n!} + \cdots . \qquad (14.4)$$

(i) is obvious from the definition of e, that is, by Examples 10.13 and 10.14.

To prove (ii) consider $H : x \mapsto \exp(x + u) \exp(-x)$. We have that $H'(x) = 0$, consequently $H(x) = H(0) = \exp(u)$. Hence we have that $\exp(x+u) \exp(-x) = \exp(u)$. Since x and u are arbitrary complex numbers we can set $z = x + u$ and $-x = y$. Then $u = z + x$ and Equation (14.2) follows.

By (ii) we have $\exp(x) \exp(-x) = \exp(0) = 1$, hence (iii).

$\exp(0) = 1$ by (ii) and by the intermediate value theorem $\exp(x) > 0$ for all $x \in \mathbb{R}$. Since $\exp(x)' = \exp(x) > 0$ the function exp is increasing.

For $x > 0$ we have

$$\exp(x) > 1 + x + \cdots + \frac{x^{n+1}}{(n+1)!} > \frac{x^{n+1}}{(n+1)!}$$

and therefore

$$\frac{\exp(x)}{x^n} > \frac{x}{(n+1)!}.$$

The second limit relation in (14.3) follows from the first since

$$\left|\frac{x^n}{\exp(x)}\right| = \frac{1}{\left|\frac{\exp(x)}{x^n}\right|}.$$

If E is a solution to Equation (14.1) in $\mathbb{C}$ then $[E(x)/\exp(x)]' = 0$, consequently this function is constant and then $E(x) = E(0)\exp(x)$.   □

Equation (14.3) could also been proved using Theorem 13.19.

Example 14.1   For $f(x) = \exp(-1/x^2)$ with $f(0) = 0$, the derivatives $f^{(k)}(0) = 0$ for all $k \in \mathbb{N}$. Firstly

$$f'(0) = \lim_{x \to 0} \frac{\exp\left(-\dfrac{1}{x^2}\right)}{x}. \tag{14.5}$$

It will be convenient to consider

$$\lim_{x \to 0} \frac{\exp\left(-\dfrac{1}{x^2}\right)}{x^k}. \tag{14.6}$$

Since, by Equation (14.4) with x replaced by $1/x^2$, we have

$$\left|x^k \exp(1/x^2)\right| > \frac{|x|^k}{[k!]x^{2k}} = \frac{1}{k![|x|^k]}$$

Consequently the limit in Equation (14.6) is zero. It follows from this equation with $k = 1$ and from Equation (14.6) that $f'(0) = 0$. For $x \neq 0$ the derivative $f'(x) = 2\exp(-1/x^2)/x^3$ is of the form $P(1/x)\exp(-1/x^2)$, where P is a polynomial. Simple induction shows that this is also true for any derivative $f^{(k)}(x)$. By Equation (14.6) then $\lim_{x \to 0} f^{(k)}(x) = 0$. It follows that $f^{(k)}(0) = 0$ for all $k \in \mathbb{N}$ as asserted.

This example is interesting because the Maclaurin series of f converges for all x, it is zero for all x and hence it is *not* equal to $f(x)$ for any $x \neq 0$.

It is customary to write e^x instead of $\exp(x)$ for any $x \in \mathbb{C}$. This is justified by the fact that for $r \in \mathbb{Q}$, $e^r = \exp(r)$. Proof of this is left to Exercise 14.2.2.

Exercises

ⓘ **Exercise 14.2.1**   Prove: $\exp(x)\exp(-x) = 1$.

ⓘ **Exercise 14.2.2**   Prove: $\exp(nx) = (\exp(x))^n$. Then use it repeatedly to show $\exp(r) = e^r$ for $r \in \mathbb{Q}$.

Exercise 14.2.3   For $f(x) = x^{100}\exp(x^2)$ find $f^{(104)}(0)$ and $f^{(1003)}(0)$. [Hint: Use the expansion for $\exp(y)$ with $y = x^2$.]

Exercise 14.2.4   Find local extrema, points of inflection, intervals of convexity and concavity for $f(x) = \exp(-x^9)$. Graph this function.

ⓘ **Exercise 14.2.5**   For $g(x) - (\exp(x) - 1)/x$ show that the discontinuity at $x = 0$ is removable and find the Maclaurin expansion of g.

Exercise 14.2.6   Find the following limits

(1) $\displaystyle\lim_{x \uparrow 0} \exp\left(\frac{1}{x}\right)$;

(2) $\displaystyle\lim_{x \downarrow 0} \frac{\exp(1/x)}{1 + \exp(1/x)}$;

(3) $\displaystyle\lim_{x \to \infty} \frac{x + 10^{-20}\exp(x)}{x^3 + x^2 - 10x + 200}$;

(4) $\displaystyle\lim_{x \to \infty} \exp(ax)$ for $a \in \mathbb{R}$. [Hint: Consider three possibilities for a.]

14.3   The logarithm

We have seen that the restriction of $x \mapsto \exp(x)$ to $\mathbb{R}$ is strictly increasing on $\mathbb{R}$. The inverse function to it exists and is called the natural logarithm or simply the logarithm and is denoted by $\ln$. The graph is shown in Figure 14.2.

We summarise the properties of $\ln$ in the next theorem.

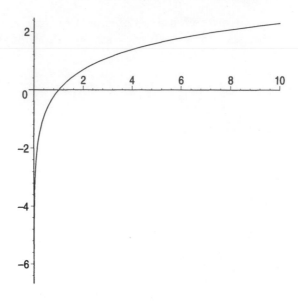

Fig. 14.2   The graph of the logarithm

Theorem 14.3   *The logarithm has the following properties*

(i) *it is defined on* $]0, \infty[$, *is strictly increasing and maps* $]0, \infty[$ *onto* $\mathbb{R}$.

(ii) $(\ln x)' = \dfrac{1}{x}$.

(iii) $\ln 1 = 0$ *and* $\ln x > 0$ *for* $x > 1$; $\ln x < 0$ *for* $0 < x < 1$; $\ln e = 1$.

(iv)

$$\lim_{x \to \infty} \ln x = \infty,$$

$$\lim_{x \to 0} \ln x = -\infty.$$

(v) *For every* $x > 0$ *and* $y > 0$

$$\ln(xy) = \ln x + \ln y, \qquad (14.7)$$

$$\ln \frac{x}{y} = \ln x - \ln y. \qquad (14.8)$$

Proof.   (i) We already know that ln is strictly increasing. By the inverse function theorem $\operatorname{rg} \ln = \operatorname{dom} \exp = \mathbb{R}$.

(ii) By Theorem 13.6 on the derivative of the inverse function ,

$$(\ln x)' = \frac{1}{\exp(\ln(x))} = \frac{1}{x}.$$

(iii) $\exp(0) = 1$ hence $\ln 1 = 0$; the inequalities for $\ln x$ follow from the fact that $\ln$ is increasing; $\exp(1) = e$ hence $\ln e = 1$.

(iv) follows from (i).

(v) Set $x = \exp(u)$ and $y = \exp(v)$. Then

$$\ln xy = \ln\left(\exp(u)\exp(v)\right) = \ln\left(\exp(u+v)\right) = u + v = \ln x + \ln y.$$

The second formula follows by writing x/y instead of x. □

Example 14.2 (Logarithmic differentiation) Sometimes it is easier to evaluate $D\ln(f(x)) = f'(x)/f(x)$ rather than $Df(x)$ and then find $f'(x)$. For instance,

$$f(x) = \sqrt{\frac{1 - x^2}{1 + x^2}},$$

$$\ln\left(f(x)\right) = \frac{1}{2}\ln\left(1 - x^2\right) - \frac{1}{2}\ln\left(1 + x^2\right),$$

$$\frac{f'(x)}{f(x)} = \frac{-x}{(1 - x^2)} - \frac{x}{(1 + x^2)},$$

and after simplification

$$f'(x) = -\frac{2x}{(1 + x^2)\sqrt{1 - x^4}}$$

Example 14.3 (Maclaurin series for the logarithm) We use the derivative of $\ln(1 + x)$ to obtain the expansion

$$(\ln(1 + x))' = \frac{1}{1 + x} = 1 - x + x^2 - \cdots \tag{14.9}$$

The function

$$L(x) = x - \frac{x^2}{2} + \frac{x^3}{3} - \cdots \tag{14.10}$$

has, for $|x| < 1$, the same derivative as $\ln(1 + x)$, and since $L(0) = \ln 1 = 0$ we have that $\ln x = L(x)$ for $|x| < 1$. In other words the right hand side of Equation (14.10) is the Maclaurin expansion of $\ln(1 + x)$ for $|x| < 1$. We now show that the expansion holds also for $x = 1$. Using the fact that in an alternating series with descending terms the remainder does not exceed the first omitted term, we have (for $x > 0$)

$$\left| \ln(1+x) - \sum_{k=1}^{n} (-1)^{k+1} \frac{x^k}{k} \right| \le \frac{x^{n+1}}{n+1} \qquad (14.11)$$

Using continuity of $\ln$ and sending $x \to 1$ first and then $n \to \infty$, we have

$$\ln 2 = 1 - \frac{1}{2} + \frac{1}{3} - \frac{1}{4} + \cdots \qquad (14.12)$$

This interesting series, called the Leibniz series, converges very slowly. To obtain $\ln 2$ with accuracy of five decimal places we would have to add 100000 terms, a forbidding task. A better convergent series is obtained by replacing x with $-x$ in Equation (14.10) and subtracting the resulting series.

$$\ln \frac{1-x}{1+x} = -2 \left(x + \frac{x^3}{3} + \frac{x^5}{5} + \cdots \right). \qquad (14.13)$$

To calculate $\ln 2$ we would set $x = -1/3$. The remainder after the k-th non-zero term r_k can be estimated as follows:

$$|r_k| \le 2 \frac{3^{-2k-1}}{2k+1} \left(1 + \frac{1}{3^2} + \cdots \right) \le \frac{3^{-2k+1}}{8k+4}.$$

To achieve accuracy of five decimal places we need $|r_k| \le 0.5 \cdot 10^{-5}$. A rough estimate suggests $k = 4$ or 5 and a calculator or *Maple* would confirm that $k = 5$ suffices. The series (14.13) was used in the past for calculating logarithmic tables and although there is no need to have logarithmic tables now the series is still useful.

Exercises

Exercise 14.3.1 *Find the Maclaurin expansion of* $\ln(1 + x + x^2)$. [*Hint:* $1 + x + x^2 = (1 - x^3)/(1 - x)$.]

ⓘ **Exercise 14.3.2** *Show that* $\ln$ *is convex everywhere where it is defined.*

Exercise 14.3.3 *Find the following limits without using* Maple.

(1) $\lim_{x \to 1} \left(\ln x + \ln |x - 1| - \ln |x^2 - 1| \right);$

(2) $\lim_{x \to \infty} \dfrac{\ln(x+2)}{\ln(x-100)};$

(3) $\lim_{x \to \infty} \ln(x - \sqrt{x^2 - 1});$

$$(4) \quad \lim_{x \to e} \frac{\ln x - 1}{x - e}.$$

14.4 The general power

If $a > 0$ and $r \in \mathbb{Q}$ then

$$a^r = \exp(r \ln a). \tag{14.14}$$

To prove this we define $f(x) = x^r \exp(-r \ln x)$ for $x > 0$. A simple calculation shows that $f'(x) = 0$ and therefore $f(x) = f(1)$ for $x > 0$. Equation (14.14) follows by setting $x = a$ and using Exercise 14.2.1. We now use Equation (14.14) to *define* the meaning of a^r for $a > 0$ and any $r \in \mathbb{R}$. The symbol a^r represents the exponential with base a and exponent r. With this definition, Theorem 5.8 becomes valid for $r \in \mathbb{R}$ and $s \in \mathbb{R}$. The proofs are easy, as an example we prove part (v) of Theorem 5.8.

$$a^r a^s = a^{r+s},$$

that is

$$a^r a^s = \exp(r \ln a) \exp(s \ln a)$$
$$= \exp\left((r + s) \ln a\right) = a^{r+s}.$$

For $a = e$ we have

$$e^t = \exp(t \ln e) = \exp(t),$$
$$\left(e^t\right)^r = \exp(r \ln(e^t)) = \exp(rt) = e^{rt}.$$

For the ln function the Equation (14.14) leads to

$$\ln a^r = \ln(\exp(r \ln a)) = r \ln a.$$

There are two functions naturally associated with Equation (14.14),

$$p : x \mapsto x^r \quad \text{for} \quad x > 0, r \in \mathbb{R};$$

and

$$E_a : x \mapsto a^x \quad \text{for} \quad a > 0, x \in \mathbb{R}.$$

We are familiar with p for a rational r and have already mentioned that the usual properties expressed in Theorem 5.8 extend to real r. We prove only the 'new' formula $(x^r)' = rx^{r-1}$. Using the chain rule

$$(x^r))' = (\exp(r \ln x))' = \exp(r \ln x)\frac{r}{x}$$
$$= r \exp(r \ln x) \exp(-\ln x) = r \exp((r - 1) \ln x) = rx^{r-1}.$$

A consequence of this formula is that Equation (13.38), namely

$$D^n (1 + x)^m = m(m - 1) \cdots (m - n + 1)(1 + x)^{m-n}, \tag{14.15}$$

now holds for any real m, $n \in \mathbb{N}$ and $x > -1$.

Example 14.4 (Maclaurin expansion of $(1 + x)^m$) Using Equation (14.15) and the definition of binomial coefficients in (13.72) we are led to the following Maclaurin expansion of $(1 + x)^m$

$$M(x) = 1 + \sum_{k=1}^{\infty} \binom{m}{k} x^k, \tag{14.16}$$

and we want to show that $M(x) = (1 + x)^m$. The series on the right-hand side of (14.16) converges for $|x| < 1$ and diverges for $|x| > 1$ by the ratio test. Hence we assume $|x| < 1$ for the rest of this example. Differentiating M gives

$$(1 + x)M'(x) = \sum_{k=1}^{\infty} k \binom{m}{k} x^{k-1} + \sum_{k=1}^{\infty} k \binom{m}{k} x^k$$
$$= m + \sum_{k=1}^{\infty} \left[(k + 1) \binom{m}{k + 1} + k \binom{m}{k} \right] x^k.$$

The coefficient of x^k equals

$$m \left[\binom{m - 1}{k} + \binom{m - 1}{k - 1} \right] = m \binom{m}{k}.$$

Consequently M satisfies $(1 + x)M'(x) = mM(x)$. Using this we have

$$((1 + x)^{-m} M(x))' = (1 + x)^{-m-1} [-mM(x) + (1 + x)M'(x)] = 0.$$

Hence $(1 + x)^{-m} M(x) = M(0) = 1$, and finally

$$(1 + x)^m = 1 + \sum_{k=1}^{\infty} \binom{m}{k} x^k \quad \text{for} \quad |x| < 1.$$

We now turn our attention to the function E_a. The graphs of E_a for $a = 5/9$ and $a = 9/5$ are shown in Figure 14.3.

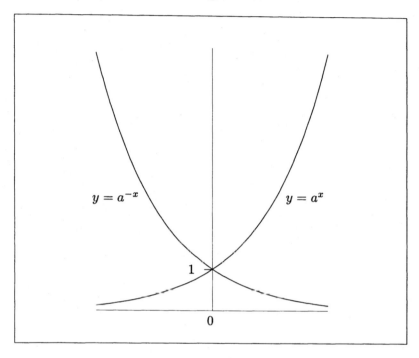

Fig. 14.3   The general exponential function

Theorem 14.4   *The function $x \mapsto a^x$ has the following properties*

  (i) $(a^x)' = a^x \ln a$

 (ii) *It is strictly increasing for $a > 1$ and strictly decreasing for $a < 1$;*

(iii) *it is convex.*

Proof.   Employing the chain rule again

$$(a^x)' = [\exp(x \ln a)]' = \exp(x \ln a) \ln a = a^x \ln a$$

The rest follows from (i) routinely.   □

The function inverse to E_a is called logarithm with base a and denoted by $\log_a$. It exists for $a > 0$, $a \neq 1$. For $a = 10$ it is called the common logarithm and denoted by log. Readers are most likely familiar with this function whose defining property is $x = 10^{\log x}$. To obtain a formula for $\log_a$ we solve the equation $y = a^x$ with respect to x. Clearly $\ln y = x \ln a$ and consequently

$$\log_a : t \mapsto \frac{\ln t}{\ln a}.$$

The properties of the function $\log_a$ are similar to and proved by the properties of ln; see Exercises 14.4.3 and 14.4.4.

Example 14.5   We employ logarithmic differentiation to find the derivative of $X : x \mapsto x^x$. We have

$$\ln X = x \ln x, \tag{14.17}$$

$$\frac{X'}{X} = \ln x + 1, \tag{14.18}$$

$$X' = x^x(\ln x + 1). \tag{14.19}$$

Exercises

ⓘ **Exercise 14.4.1**   *Find the limit of $(1 + x)^{1/x}$ for $x \to 0$. Compare with Exercise 10.13.*

Exercise 14.4.2   *Find the limit of x^x as $x \to 0$ and as $x \to \infty$.*

ⓘ **Exercise 14.4.3**   *If $a > 0$, $b > 0$, $a \neq 1$, $b \neq 1$, $x > 0$ prove*

$$\log_a x = \frac{\log_b x}{\log_b a} = \log_a b \log_b x.$$

ⓘ **Exercise 14.4.4**   *If x, y, a are positive, $a \neq 1$, $r \in \mathbb{R}$, prove*

$$\log_a xy = \log_a x + \log_a y,$$
$$\log_a \frac{x}{y} = \log_a x - \log_a y,$$
$$\log_a x^r = r \log_a x.$$

Exercise 14.4.5   *Find $Dx^{\sqrt{x}}$.*

14.5   Trigonometric functions

There was a good reason for restricting exp to $\mathbb{R}$ when dealing with logarithm. The function exp is not injective on $\mathbb{C}$, indeed, it is periodic. A function f is said to be periodic with period $p \neq 0$ if

(i) $x \in \operatorname{dom} f \Rightarrow x + p \in \operatorname{dom} f$,

(ii) $f(x + p) = f(x)$ for every $x \in \operatorname{dom} f$.

It is easy to see that if f is periodic then $f(x + mp) = f(x)$ for $m \in \mathbb{Z}$. The periodicity of exp is a part of the next theorem. In the rest of this section t denotes a *real* number.

Theorem 14.5   *The exponential function has the following properties*

(i) $|\exp(it)| = 1$ *for* $t \in \mathbb{R}$.

(ii) *There exist a smallest positive number, which shall be called* π, *such that* $\exp(i\pi/2) = i$.

(iii) $z \mapsto \exp(z)$ *is periodic with period* $2\pi i$.

(iv) $\exp(2\pi i z) = 1$ *if and only if z is an integer.*

(v) *For $t \in \mathbb{R}$ the function $t \mapsto \exp(it)$ maps $\mathbb{R}$ onto the unit circle in the complex plane.*

Before we start proving the theorem we define[1] $\cos t = \Re \exp(it)$ and $\sin t = \Im \exp(it)$. It follows from the expansion of exp that

$$\cos t = 1 - \frac{t^2}{2!} + \frac{t^4}{4!} - \frac{t^6}{6!} + \cdots \qquad (14.20)$$

$$\sin t = t - \frac{t^3}{3!} + \frac{t^5}{5!} - \frac{t^7}{7!} + \cdots \qquad (14.21)$$

$$(\sin t)' = \cos t \qquad (14.22)$$

$$(\cos t)' = -\sin t. \qquad (14.23)$$

Proof.   It follows from Equation (14.4) that $\exp(it)$ and $\exp(-it)$ are complex conjugates. Therefore

$$|\exp(it)|^2 = |\exp(it)||\exp(-it)| = |\exp(0)| = 1.$$

(ii) For $t = 2$ the series (14.20) alternates and the terms decrease. Consequently $\cos 2 < 1 - 2 + 16/24 < 0$. There exists a $t_0 \in [0, 2]$ with $\cos t_0 = 0$, by the intermediate value theorem. For $0 < t < 2$ the derivative $(\cos t)' = -\sin t < 0$, hence $\cos t$ strictly decreases. Therefore, t_0 is uniquely determined and it is the smallest positive zero of of the function

---

[1] Readers are most likely used to a different definition. However, later in this section we reconcile our definition with the 'school' definition.

cos. We denote it $\pi/2$. By Equation (14.22), the function sin increases on $[0, \pi/2]$ and by (i) we have $\sin \pi/2 = 1$. So we have $\exp(\imath\pi/2) = \imath$. It follows by successively using Equation (14.2) that $\exp(\pi\imath) = \imath\imath = -1$ and $\exp(2\pi\imath) = (-1)(-1) = 1$.

(iii) now follows from Equation (14.2). (iv) Let $x = \Re z$, $y = \Im z$. Since

$$1 = |\exp(2\pi\imath z)| = |\exp(-2\pi x)| = \exp(-2\pi x)$$

it follows that $x = 0$, since exp is increasing on $\mathbb{R}$. For an integer n we have $\exp(2n\pi\imath) = \exp(4n\pi\imath/2) = \imath^{4n} = 1$. Since $y = \lfloor y \rfloor + t$ with $0 \leq t < 1$, it suffices to show that $\exp(2t\pi\imath) = 1$ implies $t = 0$. If $\exp(2t\pi\imath) = 1$ then $\exp(t\pi\imath) = \exp(-t\pi\imath)$ and since these are complex conjugates it follows that $\Im \exp(t\pi\imath) = \sin(t\pi) = 0$. However $\sin t\pi > 0$ or $0 < t \leq 1/2$, for[2] $t > 1/2$ also $\sin t\pi = \cos(t - \pi/2) > 0$. The only possibility is $t = 0$.

(v) Let $u = a + \imath b$, $|u| = 1$.

(v1) Consider first the case $a \geq 0$, $b \geq 0$. Since cos is continuous and decreasing on $[0, \pi/2]$ there exists t_0 with $\cos t_0 = a$. Then $b \stackrel{\text{def}}{=} \sin t_0 > 0$ and $a + \imath b = \exp(\imath t_0) = u$.

(v2) Let $a < 0$, $b \geq 0$. By (v1) there exists t_1 with

$$\exp(\imath t_1) = b + \imath(-a) = -\imath u = u \exp\left(-\frac{\pi}{2}\imath\right).$$

Consequently $u = \exp(\imath(t_1 + \pi/2))$.

(v3) If $b < 0$ then by (v1) or (v2) there exists t_2 with $-u = \exp(\imath t_2)$ and then $u = \exp(\imath(t_2 + \pi))$.

$\square$

Given $a^2 + b^2 = 1$ we proved, in part (v), the existence of ϕ such that

$$\cos \phi = a$$
$$\sin \phi = b$$
$$-\pi < \phi \leq \pi$$

Part (iv) asserted the uniqueness of ϕ. For a complex number $z \neq 0$ we have $z = |z|(\cos \phi + \imath \sin \phi)$. In ϕ we recognize the argument of z from Section 7.2.4 and in particular from Equation (7.18). Now we have filled the gap from that subsection by proving the existence and uniqueness of $\mathrm{argument}(z)$.

---

[2] $\cos(t - \pi/2) + \imath \sin(t - \pi/2) = \exp((t - \pi/2)\imath) = -\imath \cos t + \sin t$.

We have proved that

$$e^{2\pi i} = 1. \tag{14.24}$$

This is an interesting relation which ties together four of the most important numbers of mathematics, e, π, i and 1.

We know that e^{it} and e^{-it} are complex conjugates, so it follows that

$$\cos t = \frac{e^{it} + e^{-it}}{2}, \tag{14.25}$$

$$\sin t = \frac{e^{it} - e^{-it}}{2i}. \tag{14.26}$$

These are known as the Euler formulae.

All properties of trigonometric functions follow easily from our definitions or from what we have already proved. We have established that sin and cos are periodic with period 2π. We have also proved

$$\sin^2 t + \cos^2 t = 1. \tag{14.27}$$

The formulae for $\sin(x + y)$ and $\cos(x + y)$ with $x, y \in \mathbb{R}$ are obtained by comparing real and imaginary parts in

$$\cos(x + y) + i\sin(x + y) = \exp(i(x + y))$$
$$= \exp(ix)\exp(iy) = (\cos x + i\sin x)(\cos y + i\sin y)$$
$$= \cos x \cos y - \sin x \sin y + i(\sin x \cos y + \cos x \sin y) \tag{14.28}$$

For some other formulae for trigonometric functions see Exercise 14.5.1. The graph of sin and cos (with other trigonometric functions) are shown in Figure 14.4.

Example 14.6   Limit of $(\sin(2x^3 + x^5)/x(1 - \cos x)$ can be easily found by *Maple*:[3]

```
>   restart;
>   limit((sin(2*x^3+x^4))/(x*(1-cos(x))),x=0);
```

$$4$$

Without *Maple*, either Theorem 13.19 or power series can be used. The latter is far more advantageous and we leave it as an exercise to compare both solutions.

[3]It is important to enclose the whole denominator in (), otherwise an insidiously wrong result is obtained.

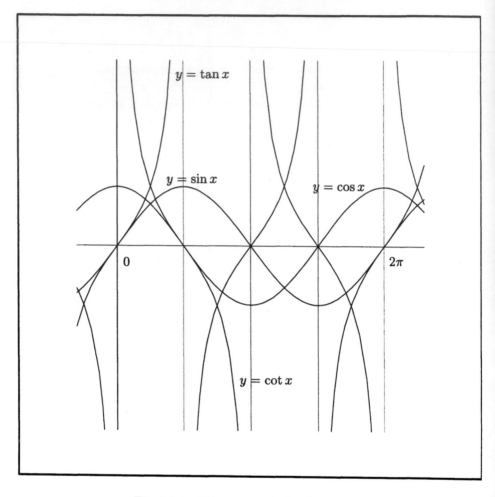

Fig. 14.4   Trigonometric functions

The function tan can now be defined as

$$\tan x = \frac{\sin x}{\cos x},$$

and all its properties developed using sin and cos. The graph is shown in Figure 14.4.

$$\lim_{t \to -\pi/2} \tan t = -\infty \quad \text{and} \quad \lim_{t \to \pi/2} \tan t = \infty$$

Further

$$(\tan x)' = 1 + \tan^2 x = \frac{1}{\cos^2 x}.$$

We define

$$\cot x \stackrel{\text{def}}{=} \frac{\cos x}{\sin x}$$

and have

$$(\cot x)' = -(1 + \cot^2 x) = -1/\sin^2 x.$$

Since $\cot x = -\tan(x - \pi/2)$ the graph of cot is the reflection of the graph of tan on the y-axis and a shift by $\pi/2$ to the right; see Figure 14.4.

The school definition

In the x, y plane we construct the unit circle (see Figure 14.5). Moving from $A = (1,0)$ in the anticlockwise direction we mark a point B on the circle with the length of the circular arc between A and B being α. Then we define $\cos \alpha$ to be the x-coordinate of B and $\sin \alpha$ to be the y-coordinate of B. We call this the school definition.

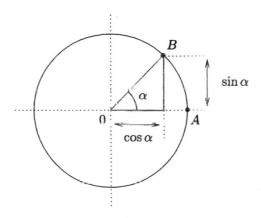

Fig. 14.5   The school definition

Comparing Figure 14.5 with Figure 14.6, where $\exp(it)$ and $\sin t$ and $\cos t$ are depicted according to our definitions, we see that in order to show that both definitions agree, it is sufficient to show that t is the length of the circular arc between 1 and $\exp(ti)$. We divide the interval $[0, t]$ into n

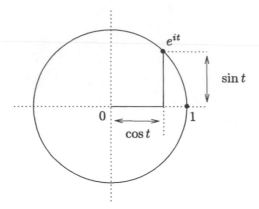

Fig. 14.6   Definition of sin and cos.

parts by dividing points kt/n, $k = 1, 2, \ldots n$. The length of the polygonal line inscribed in the circle between 1 and $\exp(it)$ is

$$l_n = \sum_{k=1}^{n} \left| \exp\left(\frac{kti}{n} \right) - \exp\left(\frac{(k-1)ti}{n} \right) \right|.$$

Since

$$\left| \exp\left(\frac{kti}{n} \right) - \exp\left(\frac{(k-1)ti}{n} \right) \right|$$

$$= \left| \exp\left(\frac{kt}{n} - \frac{t}{2n} \right) i \right| \left| \exp\left(\frac{ti}{2n} \right) - \exp\left(\frac{-ti}{2n} \right) \right| = 2 \left| \sin\left(\frac{t}{2n} \right) \right|$$

Since $0 < t/2n < \pi/2$ for large n, the absolute value on the right hand side of the last equation can be omitted. It is an easy consequence of the definition of sin or of Equation (13.54) that

$$\lim_{x \to 0} \frac{\sin xt}{x} = t$$

and consequently

$$l_n = 2n \sin\left(\frac{t}{2n} \right) \to t \quad \text{as} \quad n \to \infty.$$

As n increases without limit the length of the polygonal line l_n tends to the length of the circular arc between 1 and $\exp(ti)$.

Exercises

ⓘ **Exercise 14.5.1**   *Prove the following formulae for $x, y \in \mathbb{R}$:*

$$\sin 2t = 2\sin t \cos t,$$
$$\cos 2t = \cos^2 t - \sin^2 t,$$
$$\left|\sin\frac{t}{2}\right| = \sqrt{\frac{1 - \cos t}{2}},$$
$$\left|\cos\frac{t}{2}\right| = \sqrt{\frac{1 + \cos t}{2}},$$
$$\sin x - \sin y = 2\sin\frac{x - y}{2}\cos\frac{x + y}{2},$$
$$\cos x - \cos y = -2\sin\frac{x + y}{2}\sin\frac{x - y}{2}.$$

Exercise 14.5.2   *Find* $\displaystyle\lim_{t\to 0}\frac{\cos 5t - \cos t}{1 - \cos t}.$

ⓘ **Exercise 14.5.3**   *The function $f : t \mapsto t^2\sin(1/t)$ for $t \neq 0$ with $f(0) = 0$ has a derivative for all real t. The derivative is discontinuous at 0. Prove it!*

14.6   Inverses to trigonometric functions.

The function sin is strictly increasing and continuous on $[-\pi/2, \pi/2]$. The restriction[4] of it to this interval is also continuous and strictly increasing and maps $[-\pi/2, \pi/2]$ onto $[-1, 1]$. Its inverse is denoted by arcsin. We have $\sin(\arcsin x) = x$ for every x but $\arcsin(\sin x) = x$ only for $x \in [-\pi/2, \pi/2]$. Graphs of sin and arcsin are shown in Figure 14.7.

We now prove

$$D\arcsin x = \frac{1}{\sqrt{1 - x^2}} \quad \text{for} \quad |x| < 1. \tag{14.29}$$

By the theorem on the derivative of the inverse we have

$$\frac{d\arcsin x}{dx} = \frac{1}{\dfrac{d\sin y}{dy}} = \frac{1}{\cos y}.$$

---

[4]It is a common practice to denote this restriction also by sin. When dealing with the inverse it is important to realize that it is an inverse to the *restriction* of sin to $[-\pi/2, \pi/2]$.

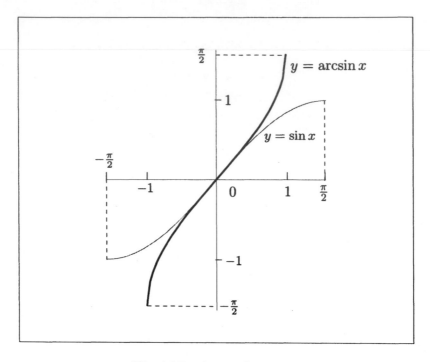

Fig. 14.7   sin x and arcsin x

Now $\sin y = x$, hence $\cos y = \pm\sqrt{1 - x^2}$. Since $-\pi/2 < y < \pi/2$ it follows that $\cos y > 0$, hence $\cos y = \sqrt{1 - x^2}$ and (14.29) is proved.

The function cos restricted to $[0, \pi]$ is strictly decreasing. Its inverse is called arccos, which is strictly decreasing, continuous, and maps $[-1, 1]$ onto $[0, \pi]$. The graphs of arccos and cos are shown in Figure 14.8.

For the derivative we have

$$D \arccos x = -\frac{1}{\sqrt{1 - x^2}}.$$

The proof is similar to establishing (14.29).

The function tan is strictly increasing on $] -\pi/2, \pi/2[$ and continuous. The inverse to tan restricted to this interval is called arctan. It is strictly increasing, continuous, and maps $\mathbb{R}$ onto $] -\pi/2, \pi/2[$. The graphs of tan

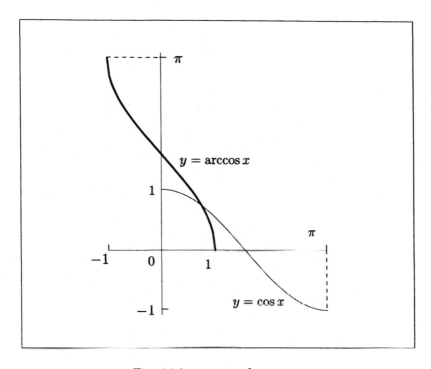

Fig. 14.8 $\cos x$ and $\arccos x$

and arctan are shown in Figure 14.9. For the derivative we have the formula

$$\frac{d \arctan x}{x} = \frac{1}{\dfrac{d \tan y}{dy}} = \cos^2 y = \frac{1}{1 + \tan^2 y} = \frac{1}{1 + x^2}. \qquad (14.30)$$

The function cot restricted to $]0, \pi[$ is strictly decreasing. Its inverse is called arccot, which is strictly decreasing, continuous, and maps $\mathbb{R}$ onto $]0, \pi[$. The graphs are shown in Figure 14.10.

To obtain the Maclaurin expansion of $\arctan x$ is easy. As we have done before we use the expansion for the derivative.

$$D \arctan x = 1 - x^2 + x^4 - x^6 + \cdots. \qquad (14.31)$$

The power series

$$x - \frac{x^3}{3} + \frac{x^5}{5} - \frac{x^7}{7} + \cdots \qquad (14.32)$$

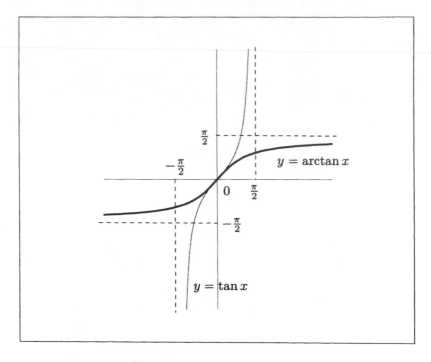

Fig. 14.9 $\tan x$ and $\arctan x$

has, for $|x| < 1$, the same derivative as arctan and both function are equal at $x = 0$, so they must be equal for $|x| < 1$.

$$\arctan x = x - \frac{x^3}{3} + \frac{x^5}{5} - \frac{x^7}{7} + \cdots \qquad (14.33)$$

By the same method we used to prove the Leibniz series for $\log 2$ we can prove that this equation also holds for $x = 1$, obtaining

$$\frac{\pi}{4} = 1 - \frac{1}{3} + \frac{1}{5} - \frac{1}{7} + \cdots . \qquad (14.34)$$

This series converges very slowly. Exercise 14.6.6 contains a guideline for efficient evaluation of π.

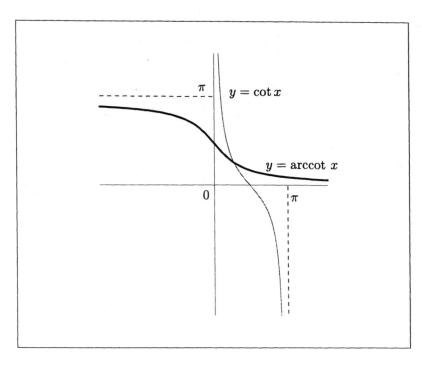

Fig. 14.10   cot x and arccot x

Exercises

Exercise 14.6.1   Use Maple to graph arcsin(sin(x)) on [$3\pi, 3\pi$].

ⓘ **Exercise 14.6.2**   Prove the following identities

$$\arcsin x + \arccos x = \frac{\pi}{2} \quad for \quad -1 \leq x \leq 1;$$

$$\arctan x + \arctan(\frac{1}{x}) = \frac{\pi}{2} \quad for \quad x > 0;$$

$$\arctan x + \arctan(\frac{1}{x}) = -\frac{\pi}{2} \quad for \quad x < 0;$$

$$\arctan x = \arcsin \frac{x}{\sqrt{1 + x^2}} \quad for \quad x \in \mathbb{R}.$$

ⓘ **Exercise 14.6.3**   Prove that: $D\text{arccot } x = -1/(1 + x^2)$.

Exercise 14.6.4   Find f' for

$$f(x) = \frac{1}{2\sqrt{2}} \arctan x\sqrt{2}\sqrt{1 + x^4} - \frac{1}{4\sqrt{2}} \ln \frac{\sqrt{1 + x^4} - x\sqrt{2}}{\sqrt{1 + x^4} + x\sqrt{2}}$$

ⓘ **Exercise 14.6.5**   *Prove* $\lim\limits_{x \to 0} \dfrac{\arctan x}{x} = 1$.

！ⓘ **Exercise 14.6.6**   *By the previous Exercise* $4 \times \arctan 1/5 \approx 4/5 \approx \pi/4$.
Set $\alpha = \arctan 1/5$ *and* $\beta = 4\alpha - \pi/4$. *(It can be expected that* β *and* $\tan \beta$
*would be small, and the series for the arctan would converge quickly.) Show
that* $\tan \beta = 1/239$, $\pi/4 = 4 \arctan 1/5 - \arctan 1/239$. *To evaluate* $\pi/4$ *to
ten decimal places it suffices to take seven terms in the series for* $x = 1/5$
and three terms for $x = 1/239$. *Prove this and and confirm it numerically
with Maple.*

14.7   Hyperbolic functions

The following functions are often used in applications. From a purely mathematical point of view they offer nothing new, interestingly they display
striking similarities with trigonometric functions, although in some formulae the sign + and − are opposite to what they are in corresponding
formulae for trigonometric functions. We have

$$\sinh x \stackrel{\text{def}}{=} \frac{e^x - e^{-x}}{2}, \qquad \cosh x \stackrel{\text{def}}{=} \frac{e^x + e^{-x}}{2},$$

$$\tanh \stackrel{\text{def}}{=} \frac{e^x + e^{-x}}{e^x + e^{-x}}, \qquad \cosh^2 x - \sinh^2 x = 1,$$

$$\cosh^2 x + \sinh^2 x = \cosh 2x, \qquad (\sinh x)' = \cosh x,$$

$$(\cosh x) = \sinh x, \qquad (\tanh x)' = \frac{1}{\cosh^2 x}.$$

Maple knows hyperbolic functions. For instance

```
>  expand(sinh(2*x));
```

$$2 \sinh(x) \cosh(x)$$

Exercises

Exercise 14.7.1   *Show that* sinh *is strictly increasing on* $\mathbb{R}$ *and find an
explicit formula for its inverse.*

Exercise 14.7.2   *Find* $D \left(\dfrac{\cosh x}{\sinh^2 x} + \ln \left(\tanh \dfrac{x}{2} \right) \right)$.

Chapter 15

Integrals

In this chapter we shall present the theory of integration introduced by the contemporary Czech mathematician J. Kurzweil. Sometimes it is referred to as Kurzweil-Henstock[1] theory. The first modern integration theory was developed by the German mathematician B. Riemann. His definition of an integral was similar but more innovative than the one developed previously by A. L. Cauchy. Riemann made a decisive steps forward by considering the totality of functions which can be integrated and building a systematic theory. Initially Riemann's theory was a great success. Later it was found deficient in many aspects. However, unfortunate as it may be, it is this theory which is usually presented to first year classes and dealt with in books on calculus. Our presentation follows, in parts, Lee and Výborný (2000, Chapter 2).[2]

15.1 Intuitive description of the Integral

We considered tagged division of intervals in Section 12.5 of Chapter 12, particularly in Inequalities (12.15). Throughout this chapter T will be a set of dividing points of an interval $[a, b]$

$$a = t_0 < t_1 < \cdots t_n = b$$

and X the set of tags $X_k \in [t_{k-1}, t_k]$. The symbol TX will denote the tagged division with dividing points T and tags X.

---

[1] Henstock is a contemporary British mathematician who made significant contributions.

[2] We recommend this book for further reading on the theory of the integral.

Definition 15.1 (Riemann Sum)   With a function $f : [a, b] \mapsto \mathbb{C}$ and a tagged division TX we associate the sum

$$\mathcal{R}(f, TX) = \sum_{i=1}^{n} f(X_i)(t_i - t_{i-1})$$

This sum is called a Riemann sum, or more explicitly the Riemann sum corresponding to the function f and the tagged division TX.

If no confusion can arise we shall abbreviate $\mathcal{R}(f, TX)$ to $\mathcal{R}(f)$ or $\mathcal{R}(TX)$ or even just $\mathcal{R}$. We shall only do this if the objects left out in the abbreviated notation (like TX in $\mathcal{R}(f)$) are fixed during the discussion.

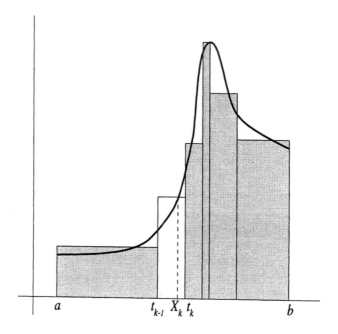

Fig. 15.1   A Riemann sum

The geometric meaning of $\mathcal{R}$ is indicated by Figure 15.1 for a non-negative function f. The sum $\mathcal{R}$ is simply the area of the rectangles which intersect the graph of f. Our intuition leads us to believe that, for a non-negative f, as a tagged division becomes finer and finer the sums $\mathcal{R}$ approach the area of the set $\{(x, y); \ a \leq x \leq b, \ 0 \leq y \leq f(x)\}$.

A word of caution is needed here. Firstly, it is fairly clear what the area of the interior of a simple polygon should be. However, the concept of an area of an arbitrary set in R^2 has to be defined and the theory of area of sets in R^2 (or R^n) should be developed—we cannot merely rely on our intuition.

Just reflect on the following question: what is the area of the set $S = \{(x, y); \ 0 \leq x \leq 1, 0 \leq y \leq 1, y \in \mathbb{Q}\}$. The set S is fairly complicated and some readers may have no idea as to what the area of S should or could be. If we try to graph the set S we would blacken the whole square $Q = \{(x, y); \ 0 \leq x \leq 1, \ 0 \leq y \leq 1\}$ despite the fact that points (x, y) with irrational y do not lie inside S. Since S is so dense in Q, one may conjecture that the area of S is the same as the area of Q, namely 1. Such a conjecture would be completely false.

The question of how and when the sums $\mathcal{R}(f, TX)$ approach a limit as T becomes finer and finer can be studied independently of their geometric meaning. That is the goal of this chapter.

A tentative attempt at definition of the integral

We shall say that the function f is "integrable" if the sums $\mathcal{R}(f, TX)$ approach a number I as TX becomes finer and finer. I is called the integral of f from a to b and is denoted by $\int_a^b f$. One can also say that $\mathcal{R}(f, TX) - I$ becomes arbitrarily small when TX is sufficiently fine.

The notation $\int_a^b f$ is unambiguous and in theoretical discussions preferable. However, in concrete situations when the function f is given by a formula, for instance $f(x) = x^3$, another notation is more common, namely one writes $\int_a^b f(x) \, dx$ instead of $\int_a^b f$, for instance $\int_a^b x^3 \, dx$.

The letter x in the symbol $\int_a^b f(x) \, dx$ is called the variable of integration. It can be replaced by other letters without altering the meaning of the symbol. Hence $\int_a^b f(x) dx = \int_a^b f(u) \, du = \int_a^b f(\alpha) \, d\alpha$.

The reasons for using the notation $\int_a^b f(x) \, dx$ are historical and of convenience. We shall see later that this notation facilitates the use of an important theorem. Also, in $\int_a^b (x^2 + t) \, dx$, the symbol dx indicates that x is the variable of integration and t is a constant, for example, $\int_a^b (x^2 + t) \, dx$ denotes $\int_a^b f$, where $f: x \mapsto x^2 + t$.

Example 15.1 Let f be a constant function $f(x) = c$ on $[a, b]$. For any tagged division TX,

$$\mathcal{R}(f, TX) = \sum_{i=0}^{n-1} c(t_{i+1} - t_i) = c \sum_{i=0}^{n-1} (t_{i+1} - t_i) = c(b - a).$$

All the sums $\mathcal{R}(f, TX)$ are equal to $c(b - a)$, regardless of the division T and the tags X, so clearly $\int_a^b f = c(b - a)$ or $\int_a^b c\, dx = c(b - a)$. For $c = 1$ we have $\int_a^b 1\, dx = \int_a^b dx = b - a$.

Maple has several commands for Riemann sums. They are `leftbox`, `rightbox` and `middlebox` for plotting the function and the rectangles representing the Riemann sums. There are corresponding commands `leftsum`, `rightsum` and `middlesum` for numerical evaluation of Riemann sums. The left, right and middle indicates that the tags are at the left-end, right-end and in the middle of the subintervals. All six commands require the package `student` to be loaded. The basic interval must be given and, optionally, the number of subintervals as well. If the optional number of intervals is not given *Maple* subdivides into four subintervals.

```
>  with(student):
>  leftbox(x^3,x=0..2);rightbox(x^3,x=0..2);
```

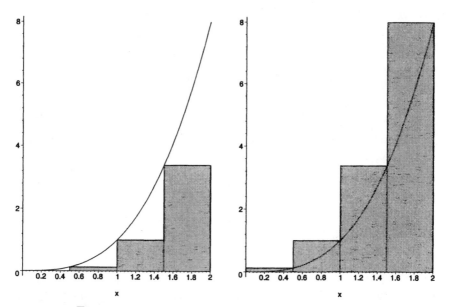

Fig. 15.2   Leftbox and a rightbox for a Riemann sum

For a monotonic function the `middlesum` lies between the `leftsum` and the `rightsum` and gives a better approximation. Increasing the number of intervals from four to ten improves the approximation.

```
>    middlebox(x^3,x=0..2);middlebox(x^3,x=0..2,10);
```

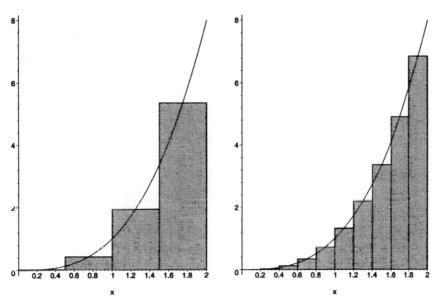

Fig. 15.3   Middleboxes for Riemann sums

The command `middlesum` evaluates the Riemann sum.

```
>    middlesum(x^3,x=0..2,10);
```

$$\frac{1}{5}\left(\sum_{i=0}^{9}(\frac{1}{5}i+\frac{1}{10})^3\right)$$

```
>    evalf(%);
```

$$3.980000000$$

The correct value is 4. Let us try a division with 100 subintervals.

```
>    evalf(middlesum(x^3,x=0..2,100));
```

$$3.999800000$$

In the next example we look at general tagged division with arbitrary tags.

Example 15.2   If $f(x) = x$ on $[a, b]$ then the evaluation of $\mathcal{R}(f, TX)$ is not obvious and $\mathcal{R}$ depends on the tagged division TX. However, for a specially chosen X, say $X = Y$ with

$$Y = \left\{ \frac{t_1 + t_0}{2}, \frac{t_2 + t_1}{2}, \ldots \frac{t_n + t_{n-1}}{2} \right\}$$

it is easy to evaluate $\mathcal{R}(f, TY)$ for any T.

$$\mathcal{R}(f, TY) = \frac{1}{2} \sum_{i=1}^{n} \left(t_i^2 - t_{i-1}^2 \right)$$

$$= \frac{1}{2} \left(t_1^2 - t_0^2 + t_2^2 - t_1^2 + \cdots - t_{n-2}^2 + t_n^2 - t_{n-1}^2 \right)$$

$$= \frac{1}{2} \left(b^2 - a^2 \right),$$

We now show that if T is sufficiently fine then $\mathcal{R}(f, TX)$—for any X— is arbitrarily close to $(b^2 - a^2)/2$. We estimate $\mathcal{R}(f, TY) - \mathcal{R}(f, TX) = \sum_{i=1}^{n} (Y_i - X_i)(t_i - t_{i-1})$. We define $\lambda = \mathrm{Max}((t_i - t_{i-1}); i = 1, 2, \ldots, n)$, then clearly $|Y_i - X_i| \le \lambda$ and

$$|\mathcal{R}(f, TY) - \mathcal{R}(f, TX)| \le \lambda \sum_{1}^{n} (t_i - t_{i-1}) = \lambda(b - a).$$

If λ is small then $\mathcal{R}(f, TX)$ is close to $\mathcal{R}(f, TY) = (b^2 - a^2)/2$. As λ becomes smaller the tagged division becomes finer and $\mathcal{R}(f, TX)$ approaches $\frac{1}{2}(b^2 - a^2)$. Hence

$$\int_a^b f = \int_a^b x \, dx = \frac{1}{2}(b^2 - a^2).$$

We shall try to obtain more insight into the approach of the Riemann sum to a limit by considering one more example.

Example 15.3   Let $f(x) = 1 - \left| \frac{1}{1 + x^2} \right|$, and $[a, b] = [-1, 1]$. Since $f(x)$ is 1 except for $x = 0$, geometric intuition tells us that $\int_{-1}^{1} f = 2$.
    The Riemann sum $\mathcal{R}(f, TX)$ depends on whether or not 0 is a tag. We

distinguish three cases:

> (i) $X_k \neq 0$;   no subinterval is tagged by zero;
> (ii) $X_k = 0 \implies X_i \neq 0$ for $i \neq k$;
>   only the interval $[t_{k-1}, t_k]$ is tagged by zero;
> (iii) $X_k = X_{k+1} = 0$;   zero tags two adjacent intervals.

How can we ensure by one simple condition that

$$|\mathcal{R}(f, TX) - 2| < \varepsilon. \tag{15.1}$$

Define a gauge δ on $[-1, 1]$ by

$$\delta(x) = 2 \quad \text{for} \quad x \neq 0$$
$$\delta(0) = \frac{\varepsilon}{2}$$

If the tagged division TX is δ-fine then the length of the interval which is tagged by zero is less than $\varepsilon/2$ and Equation (15.1) is satisfied. A tagged division is sufficiently fine if it is δ-fine. This is the key for making our tentative definition of the integral precise.

Remark 15.1   In the above example it was convenient to have some intervals smaller than others. If one looks at Figure 15.1 such a requirement is quite natural. When the function to be integrated is unbounded the requirement that some intervals in the tagged division are substantially smaller than others becomes essential. We shall see this later in Example 15.4.

Exercises

(i) **Exercise 15.1.1**   With notation as in Example 15.2 if $\lambda < \dfrac{\varepsilon}{b-a}$ show that $|\mathcal{R}(f, TX) - \frac{1}{2}(b^2 - a^2)| < \varepsilon$.

Exercise 15.1.2   Use Maple's *boxes* for $\sin x$ on the interval $[0, 2\pi]$. Use the `middlesum` with 500 subintervals.

(!) **Exercise 15.1.3**   Given $\varepsilon > 0$, find, for the function $f : x \mapsto \lceil x \rceil - \lfloor x \rfloor$ and the interval $[0, 10]$, a gauge δ such that for every δ-fine tagged division TX the inequality $|\mathcal{R}(f, DX) - 10| < \varepsilon$ holds. [Hint: For an integer n let $\delta(n) = \varepsilon/22$.]

15.2   The definition of the integral

When attempting to define the limit of Riemann sums it is easy to make precise that part of the informal statement concerning the smallness of $\mathcal{R} - I$. We can say: The number I is the integral of f if for every positive ε we have $|\mathcal{R}(f, TX) - I| < \varepsilon$ for all sufficiently fine tagged divisions TX. The difficulty lies in making precise the phrase *sufficiently fine*.

In Example 15.2 it was good enough to interpret the phrase TX *is sufficiently fine* to mean that the length λ of the largest subinterval $[t_{i+1}, t_i]$ was smaller than $\varepsilon/(b-a)$. This can be rephrased as: the tagged division is δ-fine with $\delta = \dfrac{\varepsilon}{2(b-a)}$. (See also Exercise 15.1.1). In Example 15.3 it was convenient to require that some of the subintervals of a tagged division were substantially smaller than others and this was ensured by requiring that the tagged division was δ-fine, with the function δ defined in that example. So we can say that in all our examples, sufficiently fine meant δ-fine with a suitable gauge δ.

We shall abbreviate the saying that TX is δ-fine to $TX \ll \delta$.

Definition 15.2 (Integral)   The number I is said to be the integral of f from a to b if for every positive ε there exits a gauge δ such that

$$|\mathcal{R}(f, TX) - I| < \varepsilon$$

whenever $TX \ll \delta$. If a number I satisfying this condition exists then f is said to be Kurzweil integrable, or just integrable on $[a, b]$. We shall, of course, keep the notation introduced earlier, namely $I = \int_a^b f = \int_a^b f(x)\, dx$.

Remark 15.2   If the gauge in the above definition is a constant positive function then f is said to be Riemann integrable.

Theorem 15.1   *There is at most one number I satisfying the condition from Definition 15.2.*

Proof.   Assume that for every positive ε there exists a gauge δ_1 such that

$$|\mathcal{R}(f, TX) - I| < \varepsilon \tag{15.2}$$

whenever $TX \ll \delta_1$, and also that there exists a gauge δ_2 such that

$$|\mathcal{R}(f, TX) - J| < \varepsilon \tag{15.3}$$

whenever $TX \ll \delta_2$. Let $\delta(x) = \text{Min}(\delta_1(x), \delta_2(x))$ and $TX \ll \delta$. Then both Equations (15.2) and (15.3) hold and consequently

$$|I - J| < 2\varepsilon.$$

By letting $\varepsilon \to 0$ we obtain $|I - J| \leq 0$, that is $I = J$. $\qquad\square$

It follows from considerations in Examples 15.1, 15.2 and 15.3 that

$$\int_a^b c \, dx = c(b - a); \tag{15.4}$$

$$\int_a^b x \, dx = \frac{1}{2}(b^2 - a^2); \tag{15.5}$$

$$\int_{-1}^1 \left(1 - \left|\frac{1}{1+x^2}\right|\right) dx = 2. \tag{15.6}$$

To satisfy Definition 15.2 formally it is sufficient to choose $\delta(x) = 1$ for (15.4) by Example 15.1, $\delta(x) = \dfrac{\varepsilon}{2(b-a)}$ for (15.5) by Example 15.2, and δ as defined in Example 15.3 for (15.6).

Example 15.4   Let $f(x) = 0$ for $x \in [0, 1]$, $x \neq 1/k$, $k \in \mathbb{N}$ and $f(1/k) = k^2$ for $k \in \mathbb{N}$. We show that f is integrable on $[0, 1]$ and $\int_0^1 f = 0$. For every positive ε choose $\delta(1/k) = 2^{-k-2}k^{-2}\varepsilon$ if $x = 1/k$, for $k \in \mathbb{N}$, and $\delta(x) = 1$ otherwise. Let $TX \ll \delta$ then

$$|\mathcal{R}(f, TX) - 0| - S,$$

where S is the sum of finitely many numbers which have the form $f(X_i)(t_i - t_{i-1})$, each is zero or is less than $k^2 2\delta(1/k) \leq 2^{-k-1}\varepsilon$ and there are at most two terms for the same k. Clearly we have that $S < 2\varepsilon \left(\dfrac{1}{2^2} + \cdots + \dfrac{1}{2^N}\right)$ for some positive integer N. Consequently $S < \varepsilon$ and $\int_0^1 f = 0$.

Example 15.5   We prove that $1_\mathbb{Q}$ is integrable, and that $\int_a^b 1_\mathbb{Q} = 0$. Note that $1_\mathbb{Q}$ differs from a function identically equal to 0 only at rational points. Let $r_1, r_2, r_3, \ldots$ be an enumeration of all rational numbers in $[a, b]$. Let ε be a positive number. Define δ as follows

$$\delta(x) = 1 \qquad \text{if } x \text{ is irrational,}$$

$$\delta(r_k) = \frac{\varepsilon}{2^{k+2}} \qquad \text{for } i = 1, 2, \ldots.$$

If $TX \ll \delta$ we have

$$0 \leq \mathcal{R}(TX) = \sum_{i=1}^{n} 1_{\mathbb{Q}}(X_i)(t_i - t_{i-1}). \tag{15.7}$$

The right hand side of this equation is a sum of finitely many numbers which have the form $1_{\mathbb{Q}}(X_i)(t_i - t_{i-1})$, each is zero or smaller than $2\varepsilon/2^{k+2}$ for some k. There are at most two terms for the same k. Similarly to the previous example we have that $S < \varepsilon \left(\dfrac{1}{2} + \cdots + \dfrac{1}{2^N} \right)$ for some positive integer N. Hence $\mathcal{R}(TX) \leq \varepsilon$. This together with Inequality (15.7) proves that $\int_a^b 1_{\mathbb{Q}}$ exists and equals 0.

The function from this example is often quoted in textbooks as an example of a non-integrable function. This happens because authors of those textbooks use the Riemann definition of integral. A great advantage of the Kurzweil integral is that practically every bounded function is integrable.[3]

Example 15.6 (An integral of an unbounded function) We wish to prove that

$$\int_0^A f(x)dx = 2\sqrt{A}, \tag{15.8}$$

where $f(x) = \dfrac{1}{\sqrt{x}}$ for $x \neq 0$ and $f(0) = 0$. Define $\delta(x) = \varepsilon x$ for $x \neq 0$ and $\delta(0) = \varepsilon^2$. Let $\frac{1}{2} > \varepsilon > 0$ and

$$TX \equiv 0 = t_0 \leq X_1 \leq t_1 \leq \cdots \leq X_n \leq t_n = A$$

be a δ-fine partition of $[0, A]$.

Note that this choice of δ implies $X_1 = 0$ (if $X_1 > 0$ then $X_1 \leq t_1 = t_1 - t_0 < 2\varepsilon X_1 < X_1$). Since

$$f(X_i)(t_i - t_{i-1}) \leq 2\sqrt{\frac{t_i}{X_i}}(\sqrt{t_i} - \sqrt{t_{i-1}})$$

and

$$\sqrt{\frac{t_i}{X_i}} \leq \sqrt{1+\varepsilon}$$

---

[3]It can be shown with the axiom of choice that there is a bounded function which is not Kurzweil integrable. However, nobody was able to explicitly define one, so there is no chance of encountering one in applications.

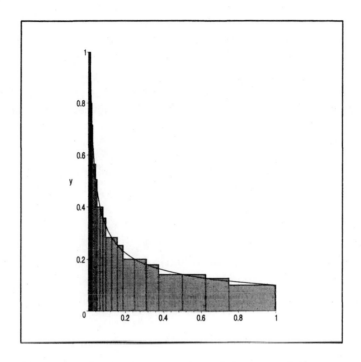

Fig. 15.4   A Riemann sum for a δ-fine tagged division

for $X_i \neq 0$, we have

$$\mathcal{R} \leq 2\sqrt{1+\varepsilon} \sum_{i=1}^{n} (\sqrt{t_i} - \sqrt{t_{i-1}}) = 2\sqrt{A}\sqrt{1+\varepsilon} \tag{15.9}$$

For an estimate from below note that

$$\sqrt{\frac{t_{i-1}}{X_i}} \geq \sqrt{1-\varepsilon},$$

for $i \geq 2$. Hence

$$\begin{aligned}
\mathcal{R} &\geq \sum_{i=2}^{n} f(X_i)(t_i - t_{i-1}) \\
&\geq 2\sqrt{1-\varepsilon} \sum_{i=2}^{n} (\sqrt{t_i} - \sqrt{t_{i-1}}) \\
&= 2\sqrt{1-\varepsilon}(\sqrt{A} - \sqrt{t_1}) \\
&= 2\sqrt{1-\varepsilon}(\sqrt{A} - \varepsilon).
\end{aligned} \tag{15.10}$$

Since ε is arbitrary, Equation (15.8) follows from Inequalities (15.9) and (15.10). We programmed *Maple* to construct a δ-fine partition and the corresponding Riemann sum with $\delta(0) = 0.0001$ and $\delta(x) = 0.3x$. For a better plot we used $f/10$ instead of f. *Maple* divided the interval into 43 subintervals and calculated the Riemann sum $\mathcal{R} = 0.198134$. The graph is shown in Figure 15.4. By contrast `middlesum` and `rightsum` produced with 43 subintervals 0.190776588 and 0.1788903398

Remark 15.3   If $a < c < b$ and δ is a gauge on $[a, b]$ then there exists a δ-fine tagged division having c as one of its dividing points. To see this it is sufficient to find δ-fine tagged divisions of $[a, c]$ and $[c, b]$ and then merge them together.

Example 15.7   Let f be defined by $f(x) = \dfrac{1}{x^2}$ for $x \neq 0$ and $f(0) = 0$. We show that f is not integrable on $[0, 1]$. Assume, for an indirect proof, that it is. Then there exists a number K and a gauge δ_1 such that

$$\mathcal{R}(f, TX) < K \tag{15.11}$$

whenever $TX \ll \delta_1$. Choose α such that $0 < \alpha < 1$, and $1/\alpha > 2K$, and define $\delta(x) = \text{Min}(\delta_1(x), \alpha/4)$ for $x \in [0, 1]$. Let $TX \ll \delta$ be a tagged division of $[0, 1]$ containing α as one of the dividing points, say $\alpha = t_j$ for some j. Then TX is also δ_1 fine and therefore Equation (15.11) holds. On the other hand $X_i \leq X_j \leq \alpha$ for $i \leq j$ and therefore

$$\mathcal{R}(f, TX) \geq \sum_{i=2}^{j} \frac{1}{X_i^2}(t_i - t_{i-1})$$

$$\geq \frac{1}{\alpha^2} \sum_{i=2}^{j}(t_i - t_{i-1})$$

$$= \frac{1}{\alpha^2}(t_j - t_1)$$

and since $t_1 < X_1 + \delta(X_1) < 2\alpha/4$

$$\mathcal{R}(f, TX) \geq \frac{1}{\alpha^2}(\alpha - \frac{\alpha}{2})$$

$$= \frac{1}{2\alpha} > K.$$

This contradicts Equation (15.11) and proves that f is not integrable.

15.2.1 *Integration in* Maple

Maple evaluates integrals efficiently and with ease. The int() function can be used to calculate integrals. The range of integration needs to be specified and then the integral is calculated. If the integral can be calculated exactly, then *Maple* does so. In the next *Maple* session we calculate $\int_0^2 3x^2\,dx$, $\int_0^{2\pi} \sin^2 x\,dx$ and $\int_{-1}^2 x^3 e^{x^2}\,dx$.

```
>  int(3*x^2,x=0..2);
```

$$8$$

```
>  int((sin(x))^2,x=0..2*Pi);
```

$$\pi$$

```
>  int( (x^3)*oxp(x^2),x=-1..2);
```

$$\frac{3}{2}\,e^4$$

Sometimes *Maple* cannot calculate the integral exactly. That is, the integral must be calculated numerically. If *Maple* cannot calculate the integral exactly, then rather than giving an answer, *Maple* displays the expression to be integrated with an integral sign, and does no calculations. We can use the evalf() function to obtain a numerical answer. Consider following example.

We wish to evaluate $\int_0^1 \frac{1}{x+e^x}\,dx$. This integral cannot be evaluated exactly, so *Maple* does no calculations. Instead, the integral is rewritten.

```
>  int(1/(x+exp(x)),x=0..1);
```

$$\int_0^1 \frac{1}{x+e^x}\,dx$$

Use the evalf() function to give a numerical answer.

```
>  evalf(%);
```

$$.5163007634$$

Exercises

Exercise 15.2.1 Let $f(x) = 0$ for all $x \in [a, b]$ except possibly for $x = c$, and $a \le c \le b$. Prove that f is integrable and

$$\int_a^b f = 0.$$

[Hint: This is obvious if $f(c) = 0$, otherwise take $\delta(x) = 1$ for $x \ne c$ and $\delta(c) = \frac{\varepsilon}{2(|f(c)|)}$.]

(i) **Exercise 15.2.2** Prove that the characteristic function of an interval $[\alpha, \beta] \subset [a, b]$ is integrable and its integral from a to b is $\beta - \alpha$. [Hint: Define $\delta(x) = \text{Min}(|x - \alpha|, |x - \beta|)$ for $x \ne \alpha$, $x \ne \beta$; $\delta(\alpha) = \delta(\beta) = \varepsilon/4$.]

(i) **Exercise 15.2.3** Let $f(x) = \dfrac{1}{x^2}$ for $x \ne 0$, and $f(0) = 0$. Show that f is not integrable on $[-1, 1]$. [Hint: Do not work hard, use Example 15.7.]

(i) **Exercise 15.2.4** Prove that $g \colon x \mapsto f(x + A)$ is integrable on $[a - A, b - A]$ if and only if f is integrable on $[a, b]$. [Hint: Use δ for F to define δ for g.]

(i) **Exercise 15.2.5** Let $A > 0$. Prove that $g \colon x \mapsto f(Ax)$ is integrable on $[a/A, b/A]$ if and only if f is integrable on $[a, b]$. State a similar result for $A < 0$. [Hint: Use δ for f to define δ for g.]

Exercise 15.2.6 Prove the following formula using the definition of integral.

$$\int_a^b \cos x \, dx = \sin b - \sin a.$$

[Hint: Let T be a tagged division of $[a, b]$, $T = \{a = t_0 < t_1 < t_2 < \cdots < t_n = b\}$, $\lambda = \text{Max}(t_i - t_{i-1}; i = 1, \ldots, n)$. By the mean value theorem $\sin t_i - \sin t_{i-1} = (t_i - t_{i-1}) \cos y_i$ for some $y_i \in]t_i, t_{i-1}[$. Let $Y = \{y_1, \ldots, y_n\}$, $f(y) = \cos y$. Then $\mathcal{R}(f, DY) = \sin a - \sin b$. $|\mathcal{R}(f, TX) - (\sin b - \sin a)| = \left| \sum_1^n (\cos X_i - \cos y_i)(t_i - t_{i-1}) \right|$. Since $|\cos x_i - \cos y_i| \le |x_i - y_i|$ by the mean value theorem, it follows $|\cos x_i - \cos y_i| \le |x_i - y_i| \le \lambda$. Therefore $|\mathcal{R}(f, TX) - (\sin b - \sin a)| \le \lambda(b - a)$.]

15.3 Basic theorems

The next few theorems are similar to theorems on limits of sequences or functions.

Theorem 15.2   If f is integrable on $[a, b]$ and $c \in \mathbb{R}$, then cf is integrable on $[a, b]$ and

$$\int_a^b cf = c \int_a^b f. \tag{15.12}$$

Proof.   If $c = 0$ the theorem is obvious. Let $c \neq 0$, $\varepsilon > 0$. There exists a positive function δ such that

$$\left| \mathcal{R}(f, TX) - \int_a^b f \right| < \frac{\varepsilon}{|c|}$$

whenever $TX \ll \delta$. It follows that

$$\left| \mathcal{R}(cf, TX) - c \int_a^b f \right| = |c| \left| \mathcal{R}(f, TX) - \int_a^b f \right| < \varepsilon$$

$\square$

Remark 15.4   If cf is integrable and $c \neq 0$ then f is also integrable and Equation (15.12) holds.

Theorem 15.3   If f and g are integrable on $[a, b]$ then so is $f + g$ and

$$\int_a^b (f + g) = \int_a^b f + \int_a^b g. \tag{15.13}$$

Proof.   For every positive ε there exist positive functions δ_1 and δ_2 such that

$$\left| \mathcal{R}(f, TX)f - \int_a^b f \right| < \frac{\varepsilon}{2} \tag{15.14}$$

for $TX \ll \delta_1$, and

$$\left| \mathcal{R}(f, TX)g - \int_a^b g \right| < \frac{\varepsilon}{2} \tag{15.15}$$

for $TX \ll \delta_2$. Define $\delta(x) = \text{Min}(\delta_1(x), \delta_2(x))$. If $TX \ll \delta$ then Inequalities (15.14) and (15.15) hold and since $\mathcal{R}(f + g, TX) = \mathcal{R}(f, TX) + \mathcal{R}(g, TX)$, we have

$$\left| \mathcal{R}(f + g, TX) - \int_a^b f - \int_a^b g \right| < \frac{\varepsilon}{2} + \frac{\varepsilon}{2} = \varepsilon.$$

$\square$

Remark 15.5   It is an easy exercise to extend Formula (15.13) to a sum of n functions.

Theorem 15.4   *If S is countable and $f(x) = 0$ for $x \in [a, b] \setminus S$ then f is integrable and $\int_a^b f = 0$.*

The proof combines ideas from Examples 15.4 and 15.5.

Proof.   Obviously we can assume that S is enumerable, (otherwise we enlarge it by joining it with an arbitrary enumerable part of $[a, b]$). Let $n \mapsto s_n$ be an enumeration of S. Define

$$\delta(s_n) = \frac{\varepsilon}{2^{n+2}(|f(s_n)| + 1)}, \quad \text{and} \quad \delta(x) = 1 \quad \text{for} \quad x \notin S.$$

In a Riemann sum denote by $\sum' f(X_i)(t_i - t_{i-1})$ the sum of terms for which $X_i \in S$, and $\sum'' f(X_i)(t_i - t_{i-1})$ the sum of terms for which $X_i \notin S$; clearly $\sum'' f(X_i)(t_i - t_{i-1}) = 0$ and therefore

$$\mathcal{R}(f, TX) = \sum' f(X_i)(t_i - t_{i-1}).$$

If $TX \ll \delta$ then (s_i can tag two adjacent intervals)

$$|\mathcal{R}(f, TX)| \leq 2 \sum_{i=1}^{\infty} |f(s_i)| \frac{\varepsilon}{2^{i+1}(|f(s_i)| + 1)} \leq \varepsilon \sum_{i=1}^{\infty} \frac{1}{2^i} = \varepsilon.$$

This proves f is integrable and $\int_a^b f = 0$. $\square$

Theorem 15.5   *Let f be an integrable function on $[a, b]$ and let g be a function which differs from f only at points of a countable set $S \subset [a, b]$. Then g is integrable and*

$$\int_a^b g = \int_a^b f.$$

Proof.   Since $g = f + (g - f)$, this theorem is an immediate consequence of Theorems 15.3 and 15.4. $\square$

Remark 15.6 The last theorem can also be expressed as follows: A change in the definition of a function[4] f at countably many points affects neither the existence nor the value of $\int_a^b f$. This is used very often; it also allows us to assign a meaning to the integral of a function which is not defined for all x in the basic interval.

For example, $\displaystyle\int_{-1}^2 \frac{x}{|x|}\, dx = \int_{-1}^2 f(x)\, dx$, where $f(x) = \dfrac{x}{|x|}$ for $x \neq 0$ and $f(0) = 0$ (or something else). Therefore, $\displaystyle\int_{-1}^2 \frac{x}{|x|}\, dx = 1$.

Theorem 15.6 *If f and g are integrable on $[a, b]$ and $f \leq g$ then*

$$\int_a^b f \leq \int_a^b g.$$

Proof. For every positive ε there exist gauges δ_1 and δ_2 such that

$$\int_a^b f - \varepsilon < \mathcal{R}(f, TX) < \int_a^b f + \varepsilon, \qquad (15.16)$$

whenever TX is δ_1-fine, and

$$\int_a^b g - \varepsilon < \mathcal{R}(g, TX) < \int_a^b g + \varepsilon \qquad (15.17)$$

whenever TX is δ_2-fine. Let $\delta(x) = \mathrm{Min}(\delta_1(x), \delta_2(x))$. If TX is δ-fine then (15.16) and (15.17) hold simultaneously. Since $\mathcal{R}(f, TX) \leq \mathcal{R}(g, TX)$, it follows from (15.16) and (15.17) that

$$\int_a^b f - \varepsilon < \int_a^b g + \varepsilon.$$

Letting $\varepsilon \to 0$ completes the proof. $\qquad\square$

Corollary 15.6.1 *If f and g are integrable and $f(x) \leq g(x)$ for all $x \in [a, b]$ except on a countable set then*

$$\int_a^b f \leq \int_a^b g.$$

---

[4]Strictly speaking, "change in the definition of the function" defines a new function, but such points are usually ignored.

Corollary 15.6.2   *If f is integrable and $f \leq M$ on $[a, b]$ for some $M \in R$, then*

$$\int_a^b f \leq M(b - a).$$

Similarly,

$$m(b - a) \leq \int_a^b f,$$

for an integrable f satisfying $f \geq m$ on $[a, b]$.

Corollary 15.6.3   *If both the functions f and $|f|$ are integrable then*

$$\left| \int_a^b f \right| \leq \int_a^b |f|.$$

Proof.   We have

$$-|f| \leq f \leq |f|$$

and therefore

$$\int_a^b -|f| = - \int_a^b |f| \leq \int_a^b f \leq \int_a^b |f|$$

as required   □

Warning: Even if f is integrable, the function $|f|$ need not be. An example of such a function is given later in Example 15.17

Remark 15.7   If $\delta(x) \leq x - a$ for $x > a$ then the first tag X_1 of any δ-fine tagged division must be a. Indeed, assuming that $a = t_0 < X_1 \leq t_1$, we have $X_1 - a \leq t_1 - t_0 < \delta(X_1) \leq X_1 - a$, a clear contradiction. Similarly, if $\delta(x) < b - x$ for $x < b$, then for a δ-fine tagged division of $[a, b]$ the last tag X_n is equal to b. Further, if $\delta(x) \leq |x - c|$ for $x \neq c$ and $a < c < b$, then for any δ-fine P-tagged division of $[a, b]$ there exists an integer j such that $X_j = c$. We shall refer to choosing δ in such a way that c becomes a tag of any δ-fine tagged division as *anchoring* the tagged division on c.

It is now easy to see that if S is a finite set in $[a, b]$ then there exists a positive function δ_s such that every δ-fine tagged division TX is anchored on S—by that we mean that all points of S become tags.

Theorem 15.7   *If f is integrable over $[a, c]$ and $[c, b]$ then f is integrable over $[a, b]$ and*

$$\int_a^c f + \int_c^b f = \int_a^b f.$$

Proof.   For every positive ε there exist positive functions $\underline{\delta}$ and $\overline{\delta}$ such that if $\underline{TX}$ is a $\underline{\delta}$-fine tagged division of $[a, c]$ and $\overline{TX}$ is a $\overline{\delta}$-fine tagged division of $[c, b]$ then

$$\left| \mathcal{R}(f, \underline{TX}) - \int_a^c f \right| < \frac{\varepsilon}{2} \tag{15.18}$$

and

$$\left| \mathcal{R}(f, \overline{TX}) - \int_c^b f \right| < \frac{\varepsilon}{2}. \tag{15.19}$$

Let

$$\delta = \text{Min}(\underline{\delta}(x), (c - x)), \quad \text{for} \quad a \leq x < c;$$
$$\delta = \text{Min}(\overline{\delta}(x), (x - c)), \quad \text{for} \quad c < x \leq b;$$
$$\delta(c) = \text{Min}(\overline{\delta}(c), \underline{\delta}(c)).$$

If $TX \ll \delta$ then $c = X_k$ for some $k \in \mathbb{N}$ and

$$\mathcal{R}(f, TX) = \sum_{i=1}^n f(X_i)(t_i - t_{i-1})$$

$$= \sum_{i=1}^{k-1} f(X_i)(t_i - t_{i-1}) + f(X_k)(X_k - t_{k-1}) + f(X_k)(t_k - X_k)$$

$$+ \sum_{i=k+1}^n f(X_i)(t_i - t_{i-1}).$$

Clearly,

$$\sum_{i=1}^{k-1} f(X_i)(t_i - t_{i-1}) + f(x_k)(X_k - t_k) = \mathcal{R}(f, \underline{TX})$$

and

$$f(X_k)(t_k - x_{k-1}) + \sum_{i=k+1}^{n} f(X_i)(t_i - t_{i-1}) = \mathcal{R}(f, \overline{TX})$$

Hence by (15.18) and (15.19)

$$\left| \mathcal{R}(f, TX) - \int_a^c f - \int_c^a f \right| < \frac{\varepsilon}{2} + \frac{\varepsilon}{2} = \varepsilon$$

$\square$

In the proof of Theorem 15.7 we split the Riemann sum into two, one for the interval $[a, c]$ and another for $[c, b]$. This was possible because we anchored the tagged division on c. This trick is very useful and we will use it without further explanation in future. If f is integrable on $[a, b]$ then it is integrable on any subinterval of $[a, b]$. A simple proof uses the Bolzano–Cauchy principle which we consider in the next section.

A function f is called a *step function* on $[a, b]$ if there exists a division $T \equiv \{a = t_0 < t_1 < \cdots < t_n = b\}$ such that f is constant on $]t_k, t_{k+1}[$ for $k = 0, 1, \ldots, n - 1$. Every step function is a linear combination of characteristic functions of intervals and one-point sets. The next theorem is a direct consequence of Theorems 15.2, 15.3, 15.4, 15.7 and integrability of constant functions.

Theorem 15.8   *Every step function is integrable.*

Exercises

Exercise 15.3.1   *Show that the functions $\lfloor x \rfloor$ and $g(x) = (x - \lfloor x \rfloor)^2$ are integrable on $[1, 3]$.*

Exercise 15.3.2   *If f is integrable on $[a_1, a_2], [a_2, a_3], \ldots, [a_{n-1}, a_n]$ then it is integrable on $[a_1, a_n]$. Prove this and use it to show that $f(x) = \lfloor x \rfloor$ is integrable over any interval $[a, b]$.*

15.4   Bolzano–Cauchy principle

We encountered the Bolzano–Cauchy principle for limits of sequences and functions. For the integral it reads

---

Theorem 15.9   *A function $f : [a, b] \mapsto \mathbb{C}$ is integrable if and only if for every positive ε there exists a gauge δ such that for any two tagged divisions $TX \ll \delta$ and $SY \ll \delta$*

$$|\mathcal{R}(f, TX) - \mathcal{R}(f, SY)| < \varepsilon.$$

The proof of the necessity of the condition is very similar to the proof of Theorem 10.16 and we therefore omit it.

Proof.   Let δ_n be the gauge associated with $\varepsilon = 1/n$ by the condition of the theorem. We can assume that $\delta_n \geq \delta_{n+1}$, otherwise we just replace $\delta_{n+1}(x)$ by $\text{Min}(\delta_1(x), \ldots \delta_n(x))$. For each n let us choose $T_n X_n \ll \delta_n$. For $n > N$ we have

$$|\mathcal{R}(f, T_N X_N) - \mathcal{R}(f, T_n X_n)| < \frac{1}{N} \tag{15.20}$$

This implies that the sequence with terms equal to $\mathcal{R}(f, T_n X_n)$ is Cauchy, so it has a limit, say I. Letting $n \to \infty$ gives

$$|\mathcal{R}(f, T_N X_N) - I| \leq \frac{1}{N}$$

For a given positive ε let $N > \dfrac{2}{\varepsilon}$ and $TX \ll \delta_N$. Then

$$|\mathcal{R}(f, TX) - I|$$
$$\leq |\mathcal{R}(f, TX) - \mathcal{R}(f, T_N X_N)| + |\mathcal{R}(f, T_N X_N) - I| < \frac{1}{N} + \frac{1}{N} < \varepsilon. \quad \square$$

With the Bolzano–Cauchy principle it is easy to prove integrability on subintervals.

Theorem 15.10   *If f is integrable on $[a, b]$ and $[\alpha, \beta] \subset [a, b]$ then f is integrable on $[\alpha, \beta]$.*

Proof.   We prove the theorem for $a < \alpha = c$ and $\beta = b$. A proof for $\alpha = a$ and $\beta < b$ is similar and the general case follows by a combination of these special cases. For every positive ε there is a gauge δ with the following property: if TX and SY are δ-fine tagged divisions of $[a, b]$ then

(i) TX and SY are anchored on c,

(ii)

$$|\mathcal{R}(TX) - \mathcal{R}(SY)| < \varepsilon \qquad (15.21)$$

Let

$$a = t_0 \leq X_1 \leq t_1 \leq \ldots \leq X_m \leq t_m = c$$
$$\leq X_{m+1} \leq \ldots \leq X_n \leq t_n = b$$

Denote by $\underline{TX}$ the tagged division

$$t_0 \leq X_1 \leq t_1 \leq \ldots t_m$$

and let UW and VZ be δ-fine tagged divisions of $[c, b]$.

$$\mathcal{R}(\underline{TX}) + \mathcal{R}(UW)$$

is a Riemann sum for a tagged division of $[a, b]$, which we may denote $\mathcal{R}(TX)$. Similarly $\mathcal{R}(\underline{TX}) + \mathcal{R}(VZ) = \mathcal{R}(SY)$. For both tagged divisions, $\mathcal{R}(TX) \ll \delta$ and $\mathcal{R}(SY) \ll \delta$. Clearly

$$\mathcal{R}(UW) - \mathcal{R}(VZ) = \mathcal{R}(TX) - \mathcal{R}(SY)$$

It follows from Inequality (15.21) and by the definition of δ that

$$|\mathcal{R}(UW) - \mathcal{R}(VZ)| < \varepsilon. \qquad \square$$

The Bolzano–Cauchy principle is the necessary and sufficient condition for integrability. Theorem 15.11 below gives only a sufficient condition but it is easier to apply.

Theorem 15.11 *If for every positive ε there exist integrable functions h, H such that*

$$h(x) \ \leq \ f(x) \ \leq \ H(x), \qquad (15.22)$$

for all $x \in [a, b]$ except possibly a countable set and

$$\int_a^b H - \int_a^b h \leq \varepsilon, \qquad (15.23)$$

then f is integrable on $[a, b]$.

Proof. Since change of a function on a countable set does not affect either the existence or the value of the integral, we can and shall assume that Inequality (15.22) holds everywhere on $[a, b]$. For $\varepsilon > 0$ there exists a gauge δ such that all inequalities

$$\mathcal{R}(H, DX) < \int_a^b H + \varepsilon,$$

$$\mathcal{R}(H, SY) < \int_a^b H + \varepsilon,$$

$$\mathcal{R}(h, DX) > \int_a^b h - \varepsilon,$$

$$\mathcal{R}(h, SY) > \int_a^b h - \varepsilon.$$

hold whenever $DX \ll \delta$ and $SY \ll \delta$. The Riemann sums for f are smaller than Riemann sums for H and they are bigger than Riemann sums for h. Hence

$$\mathcal{R}(f, DX) - \mathcal{R}(f(SY) \leq \mathcal{R}(H, DX) - \mathcal{R}(h, SY) \leq \int_a^b H - \int_a^b h + 2\varepsilon < 3\varepsilon.$$

Similarly

$$\mathcal{R}(f, DX) - \mathcal{R}(f, SY) > -3\varepsilon.$$

$\square$

The last theorem is often applied to step functions h, H.

Theorem 15.12   *If f is monotonic on $[a, b]$ then it is integrable on $[a, b]$.*

Proof.   Without loss of generality, let f be increasing. Let T be a division of $[a, b]$ into n equal parts with dividing points t_k. Define $H(t) = f(t_k)$ and $h(t) = f(t_{k-1})$ for $t_{k-1} \leq t < t_k$. Then

$$\int_a^b H = \sum_{k=1}^n f(t_k)(t_k - t_{k-1}),$$

$$\int_a^b h = \sum_{k=1}^n f(t_{k-1})(t_k - t_{k-1}),$$

$$\int_a^b H - \int_a^b h \le \sum_{k=1}^n [f(t_k) - f(t_{k-1})](t_k - t_{k-1}),$$

$$\le \frac{b-a}{n} \sum_{k=1}^n (f(t_k) - f(t_{k-1})) \le \frac{1}{n}(b-a)(f(b) - f(a))$$

Clearly, for $n > (b-a)(f(b) - f(a))/\varepsilon$ Inequality (15.23) is satisfied. $\square$

Remark 15.8   It is important in the above theorem that f is monotonic on a finite and *closed* interval $[a, b]$. The function f, defined by $f(x) = 1/x^2$ for $x \in]0, 1]$ and $f(0) = 0$ is decreasing on $]0, 1]$ but it is not integrable on $[0, 1]$ (by Example 15.7). If, however, f is monotonic and bounded on $]a, b[$, then it is integrable because f has finite one sided limits at the end points and therefore differs at at most two points from its monotonic extension to $[a, b]$.

Remark 15.9   A function f is called *piecewise monotonic* on $[a, b]$ if there exists a division T of $[a, b]$ such that f restricted to $]t_i, t_{i+1}[$ is monotonic for $i = 0, 1, \ldots, n - 1$. If f is piecewise monotonic on $[a, b]$ and bounded then it is integrable.

Integrability of continuous functions is proved by the same method as Theorem 15.12.

Theorem 15.13   *If f is continuous on $[a, b]$ then it is integrable on $[a, b]$.*

Proof.   For $\varepsilon > 0$ let δ be a gauge with the property that

$$|X - t| < \delta(X) \Rightarrow |f(X) - f(t)| < \frac{\varepsilon}{2(b-a)} = \eta.$$

For a δ-fine tagged division TX let

$$H(t) = f(X_k) + \eta \quad \text{for} \quad t \in [t_{k-1}, t_k[,$$
$$h(t) = f(X_k) - \eta \quad \text{for} \quad t \in [t_{k-1}, t_k[.$$

Both Inequalities (15.22) and (15.23) are satisfied and f is integrable by Theorem 15.11. $\square$

Exercises

Exercise 15.4.1 *Show that the function* $x \mapsto x - \lfloor x \rfloor$ *is integrable on any interval* $[a, b]$.

(i) **Exercise 15.4.2** *Let S be a countable set. Prove that a function bounded and continuous on $[a, b] \setminus S$ is integrable on $[a, b]$. Give an example of a non-integrable f which fails to be continuous only at one point.*

15.5 Antiderivates and areas

In this section we shall assume that we know intuitively what the area of a planar set is. We denote by $I°$ the set of all interior points of an interval I; for example $[0, \infty[° =]0, \infty[$, $[0, 1]° =]0, 1[$.

The function F is said to be an *antiderivative* of f on I if F is continuous on I and $F'(x) = f(x)$ for every point $x \in I°$. The word "primitive" is used interchangeably for the word antiderivative. If F and G are antiderivatives of f on I then there is a constant c such that

$$F(x) = G(x) + c.$$

To prove this consider $H = F - G$, then $H' = 0$ on $I°$ and consequently H is constant on $I°$ and therefore on I.

Let us now consider a function f which is continuous and non-negative on $[a, b]$ and let F be an antiderivative of f on $[a, b]$. Let us denote by $A(v)$ the area of

$$\{(x, y); \ a \leq x \leq v, \ 0 \leq y \leq f(x)\}.$$

For h positive, $A(v + h) - A(v)$ is the area of the set $\{(x, y); \ v \leq x \leq v + h, \ 0 \leq y \leq f(x)\}$. Clearly

$$h \cdot \mathrm{Min}\{f(x); \ v \leq x \leq v + h\} \leq A(v + h) - A(v)$$
$$\leq h \cdot \mathrm{Max}\{f(x); \ v \leq x \leq v + h\}. \quad (15.24)$$

Since f is continuous at v, for every $\varepsilon > 0$ there exists $\delta > 0$ such that

$$f(v) - \varepsilon < f(x) < f(v) + \varepsilon$$

for $|v - x| < \delta$. Therefore if $0 < h < \delta$ then

$$f(v) - \varepsilon < \frac{A(v + h) - A(v)}{h} < f(v) + \varepsilon.$$

This proves $A'_+(v) = f(v)$ for $a \le v < b$; similarly, one can show that $A'_-(v) = f(v)$ for $a < v \le b$. It follows that $A'(v) = f(v)$ and A is an antiderivative of f. Consequently $A(v) = F(v) + c$. By setting $x = a$ it follows that $c = -F(a)$ and hence $A(x) = F(x) - F(a)$. In particular, for $x = b$ we have

$$A(b) = F(b) - F(a). \tag{15.25}$$

We have discovered a way to compute the area of the set

$$\{(x,y); \ a \le x \le b, \ 0 \le y \le f(x)\},$$

by (15.25), the difference of the values of an antiderivative to f at b and a.

Example 15.8   Consider the area of the set

$$S = \{(x, y); \ 0 \le x \le 1, \ x^2 \le y \le x\}.$$

By the previous discussion the areas A_1 and A_2 of the sets $\{(x, y); \ 0 \le x \le 1, \ 0 \le y \le x\}$ and $\{(x, y); \ 0 \le x \le 1, 0 \le y \le x^2\}$ are respectively

$$A_1(v) = \frac{v^2}{2}, \quad A_1(1) = \frac{1}{2},$$

$$A_2(v) = \frac{v^3}{3}, \quad A_2(1) = \frac{1}{3}.$$

Obviously the area of S is $\dfrac{1}{6}$.

A more detailed analysis of our discussion would show that we have actually proved the following theorem.

Theorem 15.14   *If it is at all possible to assign to every function f which is continuous and non-negative on $[a, b]$, and which has an antiderivative F, and to every interval $[u, v] \subset [a, b]$ a number $A_u^v(f)$ in such a way that*

(i) $A_u^v \ge 0$;
(ii) $A_u^v + A_v^w(f) = A_u^w(f)$ for all $a \le u \le v \le w \le b$;
(iii) $A_u^v = c(v - u)$ if $f(x) = c$ for $u \le x \le v$;
(iv) $A_u^v \le A_u^v(g)$ if $f(x) \le g(x)$ for $u \le x \le v$;

then $A_u^v(f) = F(v) - F(u)$. In particular, $A_a^b(f) = F(b) - F(a)$.

For any reasonable theory of area of planar sets, the area $A_u^v(f)$ of the set $\{(x, y); \ u \le x \le v, \ 0 \le y \le f(x)\}$ must satisfy the requirements (i)-(iv).

Our theorem asserts that $A_v^u(f)$ is uniquely determined and $A_u^v(f) = F(v) - F(u)$, where F is any primitive of f. In other words: No matter how the area of a planar set is defined, as long as requirements (i)-(iv) are satisfied we always have $A_u^v(f) = F(v) - F(u)$.

The theory which deals with the concepts of length, area and volume is measure theory. We shall not attempt to expound it and refer our readers to Lee and Výborný (2000). Theorem 15.14 is a kind of justification[5] for evaluating areas by antiderivatives.

15.6 Introduction to the fundamental theorem of calculus

In Examples 15.1 and 15.1 and in Exercise 15.1.1 we found that $\int_a^b f$ was always the difference between the values of the antiderivative F of f at b and a. This is by no means so by chance. In Theorem 15.14 we found that the area of the set $\{(x,y);\ a \le x \le b,\ 0 \le y \le f(x)\}$ was the same difference. On the other hand, our intuition led us to believe that this area is $\int_a^b f$. All this indicates that

$$\int_a^b f = F(a) - F(b), \tag{15.26}$$

where F is an antiderivative of f. Equation (15.26) is often referred to as the *Newton-Leibniz* formula. The theorem which states that Formula (15.26) is valid for every function f possessing an antiderivative F is called the Fundamental Theorem of Calculus. We shall prove it in the next section. The difference $F(b) - F(a)$ is often denoted by $F(x)|_a^b$ or $F|_a^b$, or by $\left[F(x)\right]_a^b$. Hence (15.26) can be rewritten in the form $\int_a^b f = F(x)|_a^b$.

Example 15.9 By the Newton-Leibniz formula

$$\int_a^b x^n\, dx = \frac{1}{n+1}(b^{n+1} - a^{n+1}), \qquad \text{for} \quad n \in \mathbb{N}.$$

This result can also be established by using *Maple*.

Example 15.10 By the Fundamental Theorem of Calculus we have

$$\int_a^b \frac{d x}{x^2} = \frac{1}{a} - \frac{1}{b}, \tag{15.27}$$

---

[5]Provisional until readers become familiar with measure theory

if either $a < b < 0$ or $0 < a < b$. We know from Example 15.7 that the function f, defined by $f(x) = 1/x^2$ for $x \neq 0$ and $f(0) = 0$ is not integrable on any interval containing zero.

15.7   The fundamental theorem of calculus

Theorem 15.15 (The Fundamental Theorem of Calculus.)
Let F be an a antiderivative of f on $[a, b]$. Then f is (Kurzweil) integrable and

$$\int_a^b f = F(b) - F(a). \tag{15.28}$$

We commence with a plan of the proof. For any tagged division T we have

$$F(b) - F(a) = \sum_{i=1}^{n} \left(F(t_i) - F(t_{i-1}) \right). \tag{15.29}$$

We wish to show that $\mathcal{R}(f, TX)$ is close to $F(b) - F(a)$. We are going to do this by showing that $F(t_i) - F(t_{i-1})$ is very close to $f(X_i)(t_i - t_{i-1})$. For that we employ the relation $F'(X_i) = f(X_i)$. However this need not hold for $i = 1$ or $i = n$. (X_i can be either a or b and F need not have a derivative at a or b). Therefore we split the sum (15.29) and $\mathcal{R}(f, TX)$ into three summands and estimate the differences between them separately. We write

$$\mathcal{R}(f, TX) = f(X_1)(t_1 - a) + \sum_{i=2}^{n-1} f(X_i)(t_i - t_{i-1}) + f(X_n)(b - t_{n-1})$$

and

$$F(b) - F(a) = F(t_1) - F(a) + \sum_{i=2}^{n-1} [F(t_i) - F(t_{i-1})] + F(b) - F(t_{n-1}).$$

During the proof we shall consider only tagged divisions anchored at a and b, that is, we shall have $X_1 = a$ and $X_n = b$. We can always ensure this by demanding that the tagged division is δ-fine for some gauge.

We denote

$$A_1 = |F(t_1) - F(a)|,$$
$$A_2 = |F(b) - F(t_{n-1})|,$$
$$A_3 = |f(X_1)(t_1 - a)| = |f(a)(t_1 - a)|,$$
$$A_4 = |f(X_n)(b - t_{n-1})| = |f(b)(b - t_{n-1})|,$$
$$A_5 = \left| \sum_{i=2}^{n-1} (F(t_i) - F(t_{i-1}) - f(X_i)(t_i - t_{i-1})) \right|.$$

Proof. For any tagged division

$$|\mathcal{R}(f, TX) - (F(b) - F(a))| \leq \sum_{i=1}^{5} A_i. \tag{15.30}$$

Let $\varepsilon > 0$. By continuity of F at a and b there exists a positive γ such that $a \leq t < a + \gamma$ implies that

$$|F(t) - F(a)| < \varepsilon \tag{15.31}$$

and $b - \gamma < t \leq b$ implies that

$$|F(b) - F(t)| < \varepsilon. \tag{15.32}$$

Since $\lim_{t \downarrow a} f(a)(t - a) = 0$ and $\lim_{t \uparrow b} f(b)(b - t) = 0$ there exists a positive ω such that

$$|f(a)(t - a)| < \varepsilon$$

whenever $a \leq t < a + \omega$, and

$$|f(b)(b - t)| < \varepsilon$$

whenever $b - \omega < t \leq b$. Since $F'(x) = f(x)$ for $x \in (a, b)$ we can find a positive $\eta = \eta(x)$ such that

$$|F(w) - F(X) - f(X)(w - X)| < \varepsilon(w - X), \tag{15.33}$$

and

$$|F(u) - F(X) - f(X)(u - X)| < \varepsilon(X - u), \tag{15.34}$$

whenever $X - \eta(X) < u \leq X \leq w < X + \eta(X)$. Combining Inequalities (15.34) and (15.33)

$$|F(w) - F(u) - f(X)(w - u)| < \varepsilon(w - u), \tag{15.35}$$

Now define δ by

$$\delta(a) = \delta(b) = \text{Min}(\gamma, \omega)$$
$$\delta(x) = \text{Min}\left(\eta(x), (x - a), (b - x)\right), \quad \text{for} \quad x \in (a, b)$$

For the rest of the proof, let TX by δ-fine. (Recall from Remark 15.7 that $(x - a)$ and $(b - x)$ appear in the definition of δ to anchor X_1 at a and X_n at b). Since $t_1 - a < \delta(X_1) = \delta(a) \leq \gamma$ we can use (15.31) with $t = t_1$ and we have

$$|F(t_1) - F(a)| = A_1 < \varepsilon. \tag{15.36}$$

Similarly

$$|F(b) - F(t_{n-1})| = A_2 < \varepsilon. \tag{15.37}$$

Since $t_1 - a < \delta(X_1) \leq \omega$ we can use (15.7) with $t = t_1$ and we have

$$|f(a)(t_1 - a)| = A_3 < \varepsilon. \tag{15.38}$$

Similarly

$$|f(b)(b - t_{n-1})| = A_4 < \varepsilon. \tag{15.39}$$

We can use (15.35) with $X = X_i$, $w = t_i$, and $u = t_{i-1}$ and we obtain

$$|F(t_i) - F(t_{i-1}) - f(X_i)(t_i - t_{i-1})| \leq \varepsilon(t_i - t_{i-1})$$

Summing these inequalities for $i = 2, \ldots, n - 1$, we have

$$A_5 < \varepsilon(b - a). \tag{15.40}$$

Combining (15.30) with (15.36), (15.37), (15.38), (15.39) and (15.40) gives

$$|\mathcal{R}(f, TX) - (F(b) - F(a))| < \varepsilon(4 + b - a)$$

which completes the proof since ε was arbitrary. $\square$

Remark 15.10   If F is continuous on $[a, b]$ and $F'(x) = f(x)$ for all $x \in$ $]a, b[$ except possibly a finite set then f is integrable and $\int_a^b f = F(b) - F(a)$. This follows immediately by dividing $[a, b]$ into a finite number of intervals, where the Fundamental Theorem is applicable. It can be shown that (15.28) holds if there is a countable set S such that $F'(x) = f(x)$ for every $x \in [a, b] - S$ and F is continuous on $[a, b]$. (See Exercise 15.7.6.)

Direct integration

For evaluation of an integral of f we need a primitive of f. It is denoted by $\int f(x)\,dx$. A primitive to a given function is not uniquely determined. However, two primitives to the same function differ at most by a constant. Some authors write, for instance, $\int x\,dx = x^2/2 + C$, emphasizing in this way that the primitive to id is $x \mapsto x^2$ or a function differing from it by a constant function. We prefer to regard equations involving primitives as equivalences: two functions are equivalent if they differ by a constant. Then we do not have to write C in the equations, but this requires some caution. The symbol $\int f(x)\,dx$ denotes a primitive of f but not necessarily the same function at each occurrence. For instance, we have both

$$\int (x+1)dx = \frac{(x+1)^2}{2},$$

$$\int (x+1)dx = \frac{x^2}{2} + x.$$

At many occasions the primitive can be easily obtained by reading a formula for differentiation backwards. Table 15.7 summarises such formulae, which we established earlier by differentiation. The formulae for $\int \dfrac{dx}{x}$ (for the interval $]-\infty, 0[$) and $\int \dfrac{1}{\sqrt{x^2 \pm 1}}$ are partly new. These are easily verified by differentiation. Extensive tables of primitives have been compiled (see, for example, Gradshtein and Ryzhik (1996)). Now, however, *Maple* will find the primitive with the command int(f(x),x). Of course, there is no point in finding a primitive with *Maple* for evaluating an integral by the Fundamental Theorem. *Maple* will find the integral instantly. However if you want to have a primitive, *Maple* will find it for you on most occasions.[6] *Maple* also omits the constant C as we do. Sometimes the primitive can be found by using Table 15.7 easily, so easily that it is not worth opening the computer. Here are some examples:

Example 15.11

$$\int \frac{(x+2)(x^2-1)}{\sqrt[5]{x^2}}dx = \int \left(x^{13/5} + 2x^{8/5} - x^{3/5} - 2x^{-2/5} \right) dx$$

$$= \frac{5}{18}x^{18/5} + \frac{10}{13}x^{13}5 - \frac{5}{8}x^{8/5} - \frac{10}{3}x^{3/5}.$$

---

[6]If it does not the primitive cannot be expressed by a simple formula anyhow.

Table 15.1   A short table of primitives

$\displaystyle\int f(x)\,dx = F(x)$	Range of validity; comments
$\displaystyle\int x^n\,dx = \dfrac{x^{n+1}}{n+1}$	$n \in \mathbb{Z}$, $n \neq -1$; x unrestricted for $n \geq 0$; $x \neq 0$ for $n < 0$
$\displaystyle\int x^\alpha\,dx = \dfrac{x^{\alpha+1}}{\alpha+1}$	$x \in \mathbb{P}$, $\alpha \in \mathbb{R}$, $\alpha \neq -1$
$\displaystyle\int \dfrac{dx}{x} = \ln|x|$	either $x \in\,]-\infty, 0[$ or $x \in\,]0, \infty[$
$\displaystyle\int e^x\,dx = e^x$	no restriction
$\displaystyle\int a^x\,dx = \dfrac{a^x}{\ln a}$	$a \in \mathbb{P}$, $a \neq 1$
$\displaystyle\int \sin x\,dx = -\cos x$	no restriction
$\displaystyle\int \cos x\,dx = \sin x$	no restriction
$\displaystyle\int \dfrac{1}{\cos^2 x}\,dx = \tan x$	any interval not containing $(2k+1)\pi/2$, $k \in \mathbb{Z}$
$\displaystyle\int \dfrac{1}{\sin^2 x}\,dx = \cot x$	any interval not containing $k\pi$, $k \in \mathbb{Z}$
$\displaystyle\int \dfrac{1}{\sqrt{1-x^2}}\,dx = \arcsin x$	$x \in\,]-1, 1[$
$\displaystyle\int \dfrac{1}{1+x^2}\,dx = \arctan x$	no restriction
$\displaystyle\int \dfrac{1}{\sqrt{x^2 \pm 1}}\,dx = \ln\left|x + \sqrt{x^2 \pm 1}\right|$	no restriction for $+$ sign; $]-\infty, -1[$ or $]1, \infty[$ for $-$ sign
$\displaystyle\int c\,f(x)\,dx = c\int f(x)\,dx$	
$\displaystyle\int (f(x) + g(x))\,dx = \int f(x)\,dx + \int g(x)\,dx$	

Example 15.12

$$\int \frac{dx}{\sin^2 x \cos^2 x} = \int \frac{\sin^2 x + \cos^2 x}{\sin^2 x \cos^2 x} dx = \int \frac{dx}{\cos^2 x} + \int \frac{dx}{\sin^2 x} = \tan x - \cot x.$$

If F is the primitive of f on an interval $[a, b]$ then $x \mapsto \dfrac{1}{A} F(Ax + B)$ is the primitive of $x \mapsto f(Ax + B)$ on the interval with endpoints $Aa + B$ and $Ab + B$.[7] Indeed, by the chain rule

$$\left(\frac{1}{A} F(Ax + B) \right)' = \frac{1}{A} F'(Ax + B)A = f(Ax + B).$$

Example 15.13

$$\int \sin^2 x \, dx = \int \frac{1 - \cos 2x}{2} dx = \frac{x}{2} - \frac{1}{4} \sin 2x.$$

There is another simple formula very useful in direct integration, namely

$$\int \frac{f'(x)}{f(x)} dx = \ln |f(x)|$$

valid on any interval on which $f(x) \neq 0$. If for instance $f(x) < 0$ then $|f(x)| = -f(x)$ and by the chain rule

$$[\ln(-f(x))]' = \frac{1}{-f(x)}(-f'(x)) = \frac{f'(x)}{f(x)}.$$

Example 15.14

$$\int \tan x \, dx = -\int \frac{-\sin x}{\cos x} dx = \ln |\cos x|,$$

$$\int \frac{1}{\sin 2x} dx = \int \frac{\sin^2 x + \cos^2 x}{2 \sin x \cos x} dx = \int \frac{\sin x}{2 \cos x} dx + \int \frac{\cos x}{2 \sin x} dx$$

$$= -\frac{1}{2} \ln |\cos x| + \frac{1}{2} \ln |\sin x|,$$

$$\int \frac{x}{x^2 - 4x + 5} dx = \frac{1}{2} \int \frac{2x - 4}{x^2 - 4x + 5} dx + 7 \int \frac{1}{1 + (x - 2)^2} dx$$

$$= \frac{1}{2} \ln(x^2 - 4x + 5) + 7 \arctan(x - 2).$$

We now illustrate the use of the Fundamental Theorem by several examples

---

[7]This is a simple special case of change of variables, but it is convenient to consider it here, before Theorem 15.17

Example 15.15   We have

$$\int_0^a \frac{dx}{\sqrt{x}} = 2\sqrt{a}.$$

Just apply the Newton-Leibniz formula. Note that $\sqrt{x}$ is not differentiable at zero. In this example we can see how convenient it was to assume in the theorem the existence of the derivative only in the *open* interval $]a, b[$. However, for the validity of the theorem it is essential that F is *continuous on the closed interval* $[a, b]$.

Example 15.16   Using the Fundamental Theorem gives an instant solution to Example 15.7. If the integral $\int_0^1 (1/x^2)dx$ existed then (with $0 < \varepsilon < 1$)

$$\int_0^1 \frac{1}{x^2}dx \geq \int_\varepsilon^1 \frac{1}{x^2}dx = \frac{1}{\varepsilon} - 1.$$

For $\varepsilon \to 0$ the right hand side diverges to infinity – a contradiction.

Example 15.17   Let F be defined by $F(x) = x^2 \cos(\pi/x^2)$ for $x \neq 0$ and $F(0) = 0$. By the Fundamental Theorem of Calculus F' is integrable on $[0, 1]$ and $\int_0^1 F' = -1$. However $|F'|$ is not integrable. For an indirect proof assume it is and define f_i to be $|F'|$ on $]1/\sqrt{i+1}, 1/\sqrt{i}[$ (we denote this interval $\equiv]a_i, b_i[$), and $f_i = 0$ otherwise. Then

$$\int_0^1 f_i = \int_{a_i}^{b_i} f_i = \int_{a_i}^{b_i} |F'| \geq \left| \int_{a_i}^{b_i} F' \right| = |F(b_i) - F(a_i)| = \frac{1}{i+1} + \frac{1}{i} \geq \frac{2}{i+1}.$$

For every $n \in \mathbb{N}$ and every $x \in [0, 1]$ we have

$$|F'(x)| \geq \sum_{i=1}^n f_i(x).$$

Consequently

$$\int_0^1 |F'| \geq \sum_{i=1}^n \frac{2}{i+1}.$$

Since the right hand side of this inequality diverges as $n \to \infty$ we have a contradiction.

Example 15.18   Let $f(x) = \sum_{n=0}^\infty c_n x^n$ with radius of convergence r. Then the function F defined by $F(x) = \sum_{n=0}^\infty \frac{c_n x^{n+1}}{n+1}$ is a primitive of f by the theorem on differentiation of power series (that is, Theorem 13.25). Hence by the Fundamental Theorem of Calculus

$$\int_\alpha^\beta \sum_{n=0}^\infty c_n x^n \, dx = \sum_{n=0}^\infty c_n \frac{1}{n+1} \left(\beta^{n+1} - \alpha^{n+1} \right) = \sum_{n=0}^\infty c_n \int_\alpha^\beta x^n \, dx$$

provided that $-r < \alpha < \beta < r$.

Example 15.19   Using the previous example with $f(x) = e^{-x^2}$ we obtain

$$\int_0^x e^{-t^2} \, dt = \sum_{n=0}^\infty \int_0^x \frac{(-1)^n t^{2n}}{n!} \, dt = \sum_{n=0}^\infty (-1)^n \frac{x^{2n+1}}{n!\,(2n+1)} \qquad (15.41)$$

for every $x > 0$.

The function $\Phi(x) = \int_0^x e^{-t^2} \, dt$ plays an important role in probability and areas of applied mathematics such as heat conduction. Equation (15.41) gives a meaningful expression for Φ and a method for calculating it, at least for small x. Extensive tables for the function Φ have been produced. It is interesting to see how *Maple* handles this integral.

```
>   int(exp(-t^2),t=0..x);
```

$$\frac{1}{2} \sqrt{\pi} \,\mathrm{erf}(x)$$

If you look in Help for the meaning of $\mathrm{erf}(x)$ *Maple* will tell you that

---

$$\mathrm{erf}(x) = \frac{2}{\sqrt{\pi}} \int_0^x \exp(-t^2) dt.$$

The answer, at the first glance, looks silly: it says that the integral in question equals to itself. However, the use of erf makes sense: erf is a higher transcendental function which is encountered often. It cannot be expressed in terms of more familiar functions[8] but other integrals or quantities of interest can be expressed in terms of erf. For instance

---

```
>   int(x^10*exp(-x^2),x=0..2);
```

$$-\frac{16201}{16} e^{(-4)} + \frac{945}{64} \sqrt{\pi}\, \mathrm{erf}(2)$$

---

[8]Like rational, exponential, logarithmic or trigonometric functions.

Of course, if you need a numerical value of the last expression you can obtain it with the evalf command.

Example 15.20 Noting from Example 14.3 that

$$-\frac{\ln(1-x)}{x} = \sum_{n=1}^{\infty} \frac{x^{n-1}}{n},$$

we can apply Example 15.18 to $f(x) = -\dfrac{\ln(1-x)}{x}$, and then we have

$$-\int_0^t \frac{\ln(1-x)}{x}\, dx = \sum_{n=1}^{\infty} \frac{t^n}{n^2}.$$

We cannot set $t = 1$ directly in this equation because the radius of convergence of the power series is 1 and Example15.18 is not applicable with $[\alpha, \beta] = [0, 1]$. However, if we set $F(t) = \displaystyle\sum_{n=0}^{\infty} \frac{t^n}{n^2}$ and show that F is continuous at 1 from the left, then F will be an antiderivative of f on $[0, 1]$ and by applying the Fundamental Theorem of Calculus we obtain

$$-\int_0^1 \frac{\ln(1-x)}{x}\, dx = F(1) - F(0) = \sum_{n=1}^{\infty} \frac{1}{n^2}.$$

To prove that F is continuous from the left at 1 we write

$$F(t) - F(1) = \sum_{n=1}^{N} \frac{t^n - 1}{n^2} + \sum_{n=N+1}^{\infty} \frac{t^n - 1}{n^2}.$$

Since $\displaystyle\sum_{n=1}^{\infty} \frac{1}{n^2}$ converges, we can find N such that $\displaystyle\sum_{n=N+1}^{\infty} \frac{1}{n^2} < \frac{\varepsilon}{4}$. Then

$$\left| \sum_{n=N+1}^{\infty} \frac{t^n - 1}{n^2} \right| \leq 2 \sum_{n=N+1}^{\infty} \frac{1}{n^2} < \frac{\varepsilon}{2}.$$

With N now fixed,

$$\lim_{t \uparrow 1} \sum_{n=1}^{N} \frac{t^n - 1}{n^2} = 0.$$

Consequently there exists a δ such that for $1 - \delta < t \le 1$ we have

$$\left| \sum_{n=1}^{N} \frac{t^n - 1}{n^2} \right| < \frac{\varepsilon}{2}.$$

Combining these results we have

$$|F(1) - F(t)| \le \left| \sum_{n=1}^{N} \frac{t^n - 1}{n^2} \right| + \left| \sum_{n=N+1}^{\infty} \frac{t^n - 1}{n^2} \right| < \frac{\varepsilon}{2} + \frac{\varepsilon}{2} = \varepsilon$$

which completes the proof. The continuity of F from the right follows immediately from the Abel theorem. This theorem asserts that a power series with radius of convergence r is continuous from the left at r if the power series converges for $x = r$. For Abel's theorem we refer to Lee and Výborný (2000, Theorem A.5.2). The *Maple* solution is

---

```
>   int(-(ln(1-x))/x,x=0..1);
```

$$\frac{1}{6} \pi^2$$

This answer only looks different from ours: the sum of the series is $\pi^2/6$, as can be confirmed by *Maple*.

```
>   sum(1/n^2,n=1..infinity);
```

$$\frac{1}{6} \pi^2$$

An elementary *proof* of the equation $\sum_{1}^{\infty} \frac{1}{n^2} = \frac{\pi^2}{6}$ can be found in Matsuoka (1961).

Exercises

Exercise 15.7.1   *Find*

1. $\int (a + bx^3)^4 \, dx;$ 2. $\int a^x b^{2x} \, dx;$ 3. $\int \tan^2 x \, dx;$

4. $\int \cot^2 x \, dx;$ 5. $\int \dfrac{x^2 + 7x - 5}{x + 2} \, dx;$ 6. $\int \dfrac{2x + 3}{\sqrt[3]{x + 1}} \, dx;$

7. $\int \dfrac{dx}{1 - \sin x};$ 8. $\int \cos^2 3x \, dx;$ 9. $\int \dfrac{dx}{4x^2 + 4x + 5};$

10. $\int \dfrac{dx}{\sqrt{x^2 + 16}};$ 11. $\int \dfrac{dx}{\sqrt{9 + 4x - x^2}};$ 12. $\int \dfrac{dx}{\sqrt{x^2 + 4x + 29}}.$

Exercise 15.7.2   Let $f(x) = 2x \sin \frac{1}{x^n} - nx^{1-n} \cos \frac{1}{x^n}$ for $x \neq 0$, and $f(0) = 0$. Find $\int_0^{1/\pi} f(x) \, dx$ and $\int_{-2/\pi}^{2/\pi} f(x) \, dx$. [Hint: It is not difficult to guess the primitive.]

Exercise 15.7.3   *Show that for any $x > 0$*

$$\int_0^x \frac{\sin t}{t} \, dt = \sum_{n=1}^{\infty} (-1)^{n+1} \frac{x^{2n}}{(2n - 1)! \, 2n}.$$

Exercise 15.7.4   *Show that*

$$\int_0^r \sum_{n=0}^{\infty} c_n x^n \, dx = \sum_{n=0}^{\infty} \frac{c_n}{n + 1} r^{n+1}$$

if $r > 0$ and the series on the right hand side converges absolutely. [Hint: Use the same method as in Example 15.20.]

ⓘ **Exercise 15.7.5**   *Using the Fundamental Theorem of Calculus prove that $f(x) = 1/x$ is not integrable on $[0, 1]$. Also give a simpler proof of Example 15.7.* [Hint: $\int_0^1 f \geq \int_\varepsilon^1 \geq -\ln \varepsilon$.]

ⓘ **Exercise 15.7.6**   *Prove that if*

(i) $F'(x) = f(x)$ for all $x \in [a, b] - S$ with countable S, and
(ii) F is continuous on $[a, b]$

then $\displaystyle\int_a^b f(x) \, dx = F(b) - F(a)$. [Hint. Find $\eta(x)$ as in (15.35) for $x \in [a, b] - S$ and let $\eta(x) = 1$ otherwise. Let $n \mapsto x_n$ be an enumeration of S, define $\delta(x_n) > 0$ to be such that for $|x - x_n| < \delta(x_n)$ we have $|F(x) - F(x_n)| < \varepsilon/2^n$ and $|f(x_n)| < \varepsilon/2^n$ and let $\delta(x) = 1$ for $x \notin S$. Define

$\delta(x) = \text{Min}(\eta(x), \delta(x))$ and proceed as in the proof of the Fundamental Theorem of Calculus.]

15.8 Consequences of the fundamental theorem

The rule for differentiation of a product and the chain rule lead to theorems on integration by parts and substitution.

Theorem 15.16 (Integration by parts) *If F and G are continuous on $[a,b]$, $F' = f$ and $G' = g$ on $[a,b]$ except on a finite set[9] then*

$$\int_a^b (Fg + Gf) = F(b)G(b) - F(a)G(a). \tag{15.42}$$

Proof. We have $(FG)' = Fg + Gf$ except on a finite set, the function FG is continuous on $[a, b]$ and by the Fundamental Theorem, Equation (15.42) follows. $\square$

Provided that one of the integrals in Equation (15.43) below exists we obtain useful formula

$$\int_a^b Fg = F(b)G(b) - F(a)G(a) - \int_a^b fG. \tag{15.43}$$

The above formula is often rewritten in the easily remembered form

$$\int_a^b u'v = uv\Big|_a^b - \int_a^b uv' \tag{15.44}$$

Example 15.21 The assumptions of the theorem alone do not guarantee the existence of either integral in Equation (15.43) as the following example shows. For $F(x) = x^2 \sin x^{-4}, G(x) = x^2 \cos x^{-4}$ for $x \neq 0$ and $F(0) = G(0) = 0$ neither $\int_0^1 Fg$ nor $\int_0^1 Gf$ exists because then both would exist by Equation (15.43) and so would their difference. However $F(x)g(x) - G(x)f(x) = 4x^{-1}$ is not integrable. For F and G chosen as above, Equation (15.42) holds but Formula (15.43) does not make sense. Moreover this example also shows that the product of an integrable function f and a continuous function G need not be integrable.

Remark 15.11 We illustrate the use of integration by parts in three cases.

(i) integrating a product in which one factor has a simple derivative and the other is not too difficult to integrate,

[9]If one is prepared to use Exercise 15.7.6 this set can be countable.

(ii) by inventing a factor which is easy to integrate (this is actually a subcase of (i)),

(iii) by integrating by parts we obtain an equation from which the integral can be found (sometimes to obtain such an equation one must integrate by parts twice).

Example 15.22 We illustrate (i) with the integral $\int_1^t x^\alpha \ln x\, dx$, $\alpha \neq -1$, $\alpha \in \mathbb{R}$. In applying Equation (15.44) we set $u' = x^\alpha$, $v = \ln x$ and have

$$\int_1^t x^\alpha \ln x\, dx = \frac{t^{\alpha+1}}{\alpha+1} \ln t - \int_1^t \frac{x^\alpha}{\alpha+1} dx = \frac{t^{\alpha+1}}{\alpha+1} \ln t - \frac{t^{\alpha+1}-1}{(\alpha+1)^2}.$$

As a teaching tool, *Maple* has a command `intparts` in the student package. Besides the integral and the limits of integration, *Maple* also requires the factor to be differentiated, in other words the function v in Formula (15.44).

```
>    student[intparts](Int(x^(alpha)*ln(x),x=1..t),ln(x));
```

$$\frac{\ln(t)\, t^{(\alpha+1)}}{\alpha+1} - \int_1^t \frac{x^{(\alpha+1)}}{x\,(\alpha+1)}\, dx$$

```
>    simplify(%);
```

$$\frac{\ln(t)\, t^{(\alpha+1)} - \displaystyle\int_1^t x^\alpha\, dx}{\alpha+1}$$

The command `intparts` is useful in seeing how integration by parts works or as a practical introduction to integration by parts but a better insight into the method is obtained by working with pen and paper. If we want just the integral *Maple* would compute it instantly with the `int()` command.

Example 15.23 An example illustrating (ii) is

$$\int_0^1 \arctan x\, dx = \frac{\pi}{4} - \int_0^1 \frac{x\, dx}{1+x^2} = \frac{\pi}{4} - \frac{1}{2} \ln 2.$$

Example 15.24 We deal with the integral from Example 15.22 for $\alpha = -1$ to illustrate (iii)

$$J = \int_1^t \frac{\ln x}{x}\, dx = \ln^2 t - \int_0^t \frac{\ln x}{x}\, dx = \ln^2 t - J,$$

consequently $J = \dfrac{1}{2} \ln^2 t$.

Example 15.25 (Wallis' formula) By integration by parts we diminish k in the integral $S_k = \int_0^{\frac{\pi}{2}} \sin^k x\, dx$. Such formulae are called reduction formulae. In Equation (15.44) we set $u' = \sin x$, $v = \sin^{k-1} x$ and have

$$S_k = -\sin^{k-1} x \cos x \Big|_0^{\pi/2} + (k-1) \int_0^{\pi/2} \sin^{k-2} x \cos^2 x\, dx$$

$$= (k-1)S_{k-2} - (k-1)S_k,$$

$$S_k = \frac{k-1}{k} S_{k-2}.$$

This formula leads to an interesting limit for π, established by the British mathematician Wallis. For an even $k = 2n \in \mathbb{N}$ we have

$$S_{2n} = \frac{2n-1}{2n} S_{2n-2} = \frac{2n-1}{2n} \frac{2n-3}{2n-2} \cdots \frac{1}{2} \frac{\pi}{2}$$

since $S_0 = \pi/2$. Similarly

$$S_{2n+1} = \frac{2n}{2n+1} \frac{2n-2}{2n-1} \cdots \frac{2}{3},$$

since $S_1 = 1$. For $x \in [0, \pi/2]$ we have $0 \le \sin x \le 1$ and it follows that $\sin^{2n+2} x \le \sin^{2n+1} x \le \sin^{2n} x$, and consequently $S_{2n+2} \le S_{2n+1} \le S_{2n}$. This implies

$$\frac{2n+1}{2n+2} \frac{\pi}{2} \le \frac{2 \cdot 2 \cdot 4 \cdot 4 \cdots 2n \cdot 2n}{1 \cdot 3 \cdot 3 \cdot 5 \cdots (2n-1)(2n-1)(2n+1)} \le \frac{\pi}{2},$$

and by the Squeeze principle

$$\lim_{n \to \infty} \frac{2 \cdot 2 \cdot 4 \cdot 4 \cdots 2n \cdot 2n}{1 \cdot 3 \cdot 3 \cdot 5 \cdots (2n-1)(2n-1)(2n+1)} = \frac{\pi}{2}. \qquad (15.45)$$

Since $\dfrac{2n+2}{2n+1} \to 1$ we also have

$$\lim_{n \to \infty} \frac{2 \cdot 2 \cdot 4 \cdot 4 \cdots 2n \cdot 2n(2n+2)}{1 \cdot 3 \cdot 3 \cdot 5 \cdots (2n-1)(2n-1)(2n+1)(2n+1)} = \frac{\pi}{2}. \qquad (15.46)$$

Equations (15.45) and (15.46) are the Wallis formulae. It follows from these formulae that $\lim\limits_{n \to \infty} \sqrt{n} S_n = \sqrt{\pi/2}$.

Application of the chain rule leads to

Theorem 15.17 (Change of variables.)   *If*

  (i) *the function $\phi : [a,\, b] \mapsto [A,\, B]$,*
  (ii) *ϕ is continuous on $[a,\, b]$,*
  (iii) *ϕ has a derivative on $]a,\, b[$;*
  (iv) *there is a function F, continuous on $[A,\, B]$ with $F'(x) = f(x)$ for $x \in (A,\, B)$,*

then

$$\int_{\phi(a)}^{\phi(b)} f(x)dx = \int_{a}^{b} f(\phi(t))\phi'(t)dt \qquad (15.47)$$

The theorem is often referred to as integration by substitution.

Remark 15.12   The theorem as stated suggests that $\phi(a) < \phi(b)$. However, this is not guaranteed by the assumption of the theorem. Therefore we extend the definition of the integral first.

Definition 15.3   If $a = b$ we define

$$\int_{a}^{a} f = 0 \qquad (15.48)$$

and if $b < a$ we define

$$\int_{a}^{b} f = -\int_{b}^{a} f \qquad (15.49)$$

Most of what we learned about the integral extends easily to the case of $b < a$. For instance, integration by parts (15.43) is valid without any change, if $b < a$. On the other hand the formula

$$\left| \int_{a}^{b} f \right| \leq \int_{a}^{b} |f|$$

must be modified to

$$\left| \int_{a}^{b} f \right| \leq \left| \int_{a}^{b} |f| \right|$$

in order to be valid for any pair $a,\, b$.

Proof.   Noting that $[F(\phi(t))]' = f(\phi(t))\phi'(t)$ for every $t \in (a,\, b)$ and applying the Fundamental Theorem to the integral on the right-hand side of (15.47) we have

$$\int_a^b f(\phi(t))\phi'(t)dt = F(\phi(b)) - F(\phi(a)).$$

If $\phi(t) = \phi(b)$ then both sides of Equation (15.47) are zero. If $\phi(a) < \phi(b)$ then by the Fundamental Theorem the left-hand side of (15.47) is $F(\phi(b)) - F(\phi(a))$ and both sides of Equation (15.47) are equal. If $\phi(a) > \phi(b)$ then

$$\int_{\phi(a)}^{\phi(b)} f \stackrel{\text{def}}{=} -\int_{\phi(b)}^{\phi(a)} f = -\left(F(\phi(a)) - F(\phi(b))\right),$$

the last equation holding by the Fundamental Theorem. □

Equation (15.47) can be used in two different ways: to evaluate the integral on the left hand side by finding the value of the integral on the right hand side; or by going in the opposite direction. In either case, when making the substitution $x = \phi(t)$ one has to substitute not only $\phi(t)$ for x but also $dx = \phi'(t)dt$. This equation is easy to remember: it is obtained from $\dfrac{dx}{dt} = \phi'(t)$, as if by multiplication by dt. When using Equation (15.47) from right to left an experienced mathematician recognizes the factor $\phi'(t)dt$ for substitution as dx; often an expert creates that factor by some manipulation of the integrand and achieves considerable simplification. Very important for the use of the theorem is to change not only the variables but also the limits of integration. If we wish to change variables $x = \phi(t)$ in the integral $\int_\alpha^\beta f$ we need to find a and b such that $\phi(a) = \alpha$ and $\phi(\beta) = b$. This is easy if ϕ has an inverse, since then, $a = \phi_{-1}(\alpha)$ and $b = \phi(\beta)$.

The strength of this theorem lies in the fact that no assumption is made concerning the monotonicity of ϕ or existence of ϕ_{-1}. This is illustrated in Example 15.27.

Example 15.26   The integral $\displaystyle\int_0^r \sqrt{r^2 - x^2}\,dx$ represents the area of a quarter circle $\left\{(x, y); x^2 + y^2 \leq r^2\right\}$. To apply the theorem we set $x = \phi(t) = r\sin t$ with $t \in [-\pi/2, \pi/2]$ and $dx = r\cos t\,dt$. When $x = 0$ or $x = r$ then $t = 0$ or $t = \pi/2$, and we have

$$\int_0^r \sqrt{r^2 - x^2}\,dx = r^2 \int_0^{\pi/2} \cos^2 t\,dt$$

$$= \frac{r^2}{2} \int_0^{\pi/2} (1 + \cos 2t)\,dt = \frac{r^2}{2}\left[t + \frac{1}{2}\sin 2t\right]_0^{\pi/2} = \frac{\pi r^2}{4}.$$

We have recaptured the geometric significance of π, for the second time. We originally defined $\pi/2$ as the smallest positive root of the equation $\cos x = 0$.

Maple has a command **changevar** for changing variables. It resides in the student package. The structure of the command is as follows:

$$\text{changevar(equation defining the substitution,}$$
$$(\text{Int}(f(x), x = a..b), \text{ new variable});$$

We give two examples of changing variables in

$$\int_0^2 \frac{dx}{\sqrt{1+x^2}} \quad \text{and} \quad \int_0^1 \sqrt{\frac{1-x}{1+x}}\, dx.$$

```
>   student[changevar](x+sqrt(1+x^2)=t,
>   Int(1/sqrt(1+x^2),x=0..2),t);
```

$$\int_1^{\sqrt{5}+2} \frac{1}{t}\, dt$$

```
>   student[changevar](x=sin(t),
>   Int(sqrt((1-x)/(1+x)),x=0..1),t);
```

$$\int_0^{1/2\,\pi} \sqrt{\frac{1-\sin(t)}{\sin(t)+1}}\, \cos(t)\, dt$$

Maple will not easily simplify the integrand in the last example, but it is clear that the integrand is $1 - \sin t$ and so the integral can be easily evaluated. Part of the skill of using change of variables is anticipation of the form of the new integral and it is our belief that such an anticipation is acquired by working examples on change of variables with pen and paper. So we give two more examples working without *Maple*.

Example 15.27 In order to calculate $\int_{-2}^3 \frac{t\, dt}{\sqrt{t^2 + 16}}$ we use Equation (15.47) from left to right. When making the substitution $t^2 + 16 = \phi(t) = x$ we recognize in $t\, dt$ half of dx and we have

$$\int_{-2}^3 \frac{t\, dt}{\sqrt{t^2 + 16}} = \frac{1}{2} \int_{20}^{25} \frac{dx}{\sqrt{x}} = \sqrt{x}\Big|_{20}^{25} = 5 - 2\sqrt{5}.$$

Note that ϕ is not monotonic on $[-2, 3]$.

Example 15.28 Using the substitution $y = \tan x$ and $dy = \dfrac{dx}{\cos^2 x}$ in the integral $I = \displaystyle\int_0^\pi \dfrac{dx}{1 + 8\cos^2 x}$ leads to

$$I = \int_0^\pi \frac{1}{1/\cos^2 x + 8\cos^2 x}\,\frac{dx}{\cos^2 x} = \int_0^0 \frac{dy}{9 + y^2} = 0.$$

The result is obviously wrong: an integral of a continuous positive function must be positive. We must look for an explanation. The function tan is discontinuous at $\pi/2$, violating assumption (ii) in Theorem 15.47, so the theorem cannot be applied. The correct evaluation is as follows:

$$\int_0^\pi \frac{dx}{1 + 8\cos^2 x} = \int_0^{\pi/2} \frac{dx}{1 + 8\cos^2 x} + \int_{\pi/2}^\pi \frac{dx}{1 + 8\cos^2 x}.$$

For the second integral we apply the substitution $x = \pi - t$, $dx = -dt$ and have

$$\int_{\pi/2}^\pi \frac{dx}{1 + 8\cos^2 x} = -\int_{\pi/2}^0 \frac{dt}{1 + 8\cos^2 t} = \int_0^{\pi/2} \frac{dx}{1 + 8\cos^2 x}.$$

Consequently

$$\int_0^\pi \frac{dx}{1 + 8\cos^2 x} = 2 \int_0^{\pi/2} \cdot \frac{dx}{1 + 8\cos^2 x}$$

First we show that[10]

$$\lim_{b \to \pi/2} \int_0^b \frac{dx}{1 + 8\cos^2 x} = \int_0^{\pi/2} \frac{dx}{1 + 8\cos^2 x}.$$

Indeed

$$\left| \int_b^{\pi/2} \frac{dx}{1 + 8\cos^2 x} \right| \le \pi/2 - b.$$

Now we evaluate (with $b < \pi/2$)

$$\int_0^b \frac{dx}{1 + 8\cos^2 x} = \int_0^{\tan b} \frac{dy}{y^2 + 9} = \frac{1}{3} \int_0^{\tan b} \frac{\frac{1}{3}dy}{(y/3)^2 + 1} = \frac{1}{3} \arctan\left(\frac{\tan b}{3} \right).$$

---

[10]If f is integrable on $[a, b]$ then $\lim\limits_{t \uparrow b} \displaystyle\int_a^t f = \int_a^b f$. We shall prove that later in Section 15.10

The limit of the right hand side for $b \uparrow \pi/2$ is $\pi/6$. Finally we have the result

$$\int_0^\pi \frac{d\,x}{1 + 8\cos^2 x} = 2\frac{\pi}{6} = \frac{\pi}{3}.$$

It is interesting to compare our work in the last three examples with the ease with which *Maple* evaluates these integrals.

```
>   restart;
>   int(sqrt(r^2-x^2),x=0..r) assuming r>0;
```

$$\frac{1}{4}r^2\,\pi$$

```
>   int(x/sqrt(x^2+16),x=-2..3);
```

$$5 - 2\sqrt{5}$$

```
>   int(1/(1+8*(cos(x))^2),x=0..Pi);
```

$$\frac{1}{3}\,\pi$$

The last example on *Maple* use is particularly pleasing. Although mechanical use of change of variables could easily lead to a wrong result, *Maple* was sophisticated enough to avoid any mistake.

Assumption (iv) in Theorem 15.17 is a little inconvenient. On some occasions we might like to change the variables because we do not know F but then it might be difficult to justify the use of (15.47).[11] The next theorem does not have this weakness. However there is an additional assumption that ϕ is monotonic.

Theorem 15.18 (Monotonic change of variables) *If we assume (i)–(iii) of Theorem 15.17 and if ϕ is strictly monotonic then Formula (15.47) holds in the following sense: If one of the integrals appearing in (15.47) exists then so does the other and (15.47) is valid.*

For the proof we refer to Lee and Výborný (2000, Theorem 2.7.8).

Exercises

Exercise 15.8.1 *Use integration by parts to evaluate the following integrals, with and without Maple.*

---

[11]If f is continuous on $[a, b]$ then (iv) holds because of Corollary 15.21.1.

(1) $\int_0^t \dfrac{x}{e^x} dx;$

(2) $\int_0^x t3^{-t} dt;$

(3) $\int_0^x (t^2 - 3t + 1)e^{-3t} dt;$

(4) $\int_0^x t\sin t\cos t\, dt;$

(5) $\int_a^x \dfrac{t}{\cos^2 t} dt,$ provided $-\frac{\pi}{2} < a < x < \frac{\pi}{2}.$

Exercise 15.8.2 *Change the variables in the following integrals*

(1) $\int_{-1}^u (1 + x)^{100}\, dx,\ t = 1 + x.$

(2) $\int_0^u \dfrac{1 + x}{1 + \sqrt{x}}\, dx,\ 1 + \sqrt{x} = t;$

(3) $\int_0^u \dfrac{\sin x}{\sqrt{\cos x}}\, dx,\ t = \cos x$ assuming $-\pi/2 < x < \pi/2;$

(4) $\int_0^u \dfrac{1}{\sqrt{1 + x^2}}\, dx,\ t = x + \sqrt{1 + x^2};$

(5) $\int_0^u \dfrac{1}{\sqrt{1 + x^2}}\, dx,\ x = \tan t;$

15.9 Remainder in the Taylor formula

The Peano form of the remainder in the Taylor formula says that the remainder is smaller than the last term by an order of magnitude. Sometimes a more precise estimate is needed. It is provided by the next theorem.

Theorem 15.19 (Integral remainder) *If $f, f', \ldots, f^{(n)}$ are continuous on an interval I with end-points a, b and $f^{(n+1)}$ exists on I except possibly a finite subset of I, then the function $t \mapsto f^{(n+1)}(t)(b - t)^n/n!$ is integrable on I and*

$$f(b) = f(a) + f'(a)(b - a) + \cdots + f^n(a)\frac{(b - a)^n}{n!} + R_{n+1}, \quad (15.50)$$

where

$$R_{n+1} = \int_a^b f^{(n+1)}(t)\frac{(b - t)^n}{n!}\, dt. \quad (15.51)$$

Remark 15.13   If one is prepared to use Exercise 15.7.6 then the exceptional set in the above theorem can be countable.

Proof.   By induction. For $n = 0$ the result is just the Fundamental Theorem.[12] For $n = 1$ we have

$$f(b) - f(a) = \int_a^b f'(t)dt,$$

$$f'(t) = f'(a) + \int_a^t f''(s)ds,$$

$$f(b) - f(a) = f'(a)(b - a) + \int_a^b \left(\int_a^t f''(s)ds \right).$$

Employing integration by parts (15.44) with $u = t - b$ and $v' = f''$ we have

$$\int_a^b \left(\int_a^t f(s)ds \right) = \int_a^b (b - t) f''(t)dt.$$

Turning to the induction hypothesis, by Equation (15.44) with $u = f^{(n+1)}$ and $v(t) = (b - t)^{(n+1)}/(n + 1)!$ we have (the existence of the first integral is guaranteed by the induction hypothesis)

$$\int_a^b f^{(n+1)}(t) \frac{(b - t)^n}{n!} dt = f^{(n+1)}(a) \frac{(b - a)^{n+1}}{(n + 1)!}$$

$$+ \int_a^b f^{(n+2)}(t) \frac{(b - t)^{n+1}}{(n + 1)!} dt.$$

This means

$$R_{n+1} = f^{(n+1)}(a) \frac{(b - a)^{n+1}}{(n + 1)!} + R_{n+2}.$$

$\square$

Equation (15.50) is often given a different form

$$f(x) = f(a) + f'(a)(x - a) + \cdots + f^n(a) \frac{(x - a)^n}{n!} + R_{n+1},$$

$$f(a + h) = f(a) + f'(a)h + \cdots + f^n(a) \frac{h^n}{n!} + R_{n+1}.$$

These ways of writing the formula have the advantage that they do not suggest that h is positive or that $x > a$.

---

[12]The case $n = 1$ is considered as motivation for the rest of the proof, from a strictly logical view this step is not needed.

A particularly useful estimate of R_{n+1} follows from from the remainder formula (15.51) if f^{n+1} is bounded on the interval $[a, b]$. If $\left|f^{n+1}(x)\right| \leq M$ for $x \in [a, b]$ then

$$|R_{n+1}| \leq \left|\int_a^b \frac{(b-t)^n}{n!} M\right| \leq \frac{M|b-a|^{n+1}}{(n+1)!} \tag{15.52}$$

Example 15.29 We wish to approximate $\tan x$ just by x. Using the *Maple* command `taylor` for obtaining the Taylor approximation gives

---

```
>  taylor(tan(x),x=0,3);
```
$$x + O(x^3)$$
.

This tells us that $|\tan x - x| \leq K|x|^3$, for some constant K. This can be .001 or 1000. In concrete application a precise estimate of K can be important. We use Inequality (15.52).

```
>  t3:=diff(tan(x),x,x,x);
```
$$t3 := 2(1 + \tan(x)^2)^2 + 4\tan(x)^2(1 + \tan(x)^2)$$
```
>  simplify(%);
```
$$2 + 8\tan(x)^2 + 6\tan(x)^4$$

We need some preliminary estimate of tan.

$$|\tan x| \leq \frac{|\sin x|}{|\cos x|} \leq \frac{|x|}{1 - \frac{x^2}{2}}.$$

```
>  a:=0.01/1-(0.01^2)/2;
```
$$a := .0099500000$$
```
>  M:=2+8*a^2+6*a^4;
```
$$M := 2.000792079$$

Finally

$$|\tan x - x| \leq \frac{M}{3!}|x|^3 \leq .34|x|^3,$$

for $|x| \leq .01$. In the next example we wish to approximate $\exp(\tan x))$ by a polynomial of fifth degree.

> `taylor(exp(tan(x)),x=0,6);`

$$1 + x + \frac{1}{2} x^2 + \frac{1}{2} x^3 + \frac{3}{8} x^4 + \frac{37}{120} x^5 + O(x^6)$$

> `t5:=diff(exp(tan(x)),x,x,x,x,x,x);`

$$
\begin{aligned}
t6 := {} & 272 \tan(x) \left(1 + \tan(x)^2\right)^3 e^{\tan(x)} + 136 \left(1 + \tan(x)^2\right)^4 e^{\tan(x)} \\
& + 416 \tan(x)^3 \left(1 + \tan(x)^2\right)^2 e^{\tan(x)} + 1168 \tan(x)^2 \left(1 + \tan(x)^2\right)^3 e^{\tan(x)} \\
& + 480 \tan(x) \left(1 + \tan(x)^2\right)^4 e^{\tan(x)} + 40 \left(1 + \tan(x)^2\right)^5 e^{\tan(x)} \\
& + 32 \tan(x)^5 \left(1 + \tan(x)^2\right) e^{\tan(x)} + 496 \tan(x)^4 \left(1 + \tan(x)^2\right)^2 e^{\tan(x)} \\
& + 720 \tan(x)^3 \left(1 + \tan(x)^2\right)^3 e^{\tan(x)} + 260 \tan(x)^2 \left(1 + \tan(x)^2\right)^4 e^{\tan(x)} \\
& + 30 \tan(x) \left(1 + \tan(x)^2\right)^5 e^{\tan(x)} + \left(1 + \tan(x)^2\right)^6 e^{\tan(x)}
\end{aligned}
$$

> `subs(tan(x)=a,t6);`

$$185.0005512 \, e^{.0099}$$

> `M:=evalf(%);`

$$M := 186.8504949$$

The error is smaller than $10^{-12} M/(6!) < .026 \cdot 10^{-12}$ for $|x| \leq 0.01$. Such high accuracy is rarely needed but it was easy to achieve with *Maple*.

Remark 15.14 (The Lagrange remainder)   If $f^{n+1}(x)$ exists for all $x \in [a,\, b]$ then there exists a $c \in]a,\, b[$ such that

$$R_{n+1} = f^{(n+1)}(c) \frac{(b-a)^n}{n!}. \tag{15.53}$$

For $n = 0$ this formula reduces to the Lagrange mean value theorem and it is called the Lagrange remainder. This form is easy to remember: the remainder looks like the next term in the expansion but the derivative is evaluated at a shifted point c. The Lagrange remainder is valid if $f^{(n+1)}$ exists on $[a, b]$. In proving it we assume, for sake of simplicity, that $f^{(n+1)}$ is bounded. Let m, M be the greatest lower bound or the least upper bound of $f^{(n+1)}$ on $[a, b]$. Then

$$m \frac{(b-a)^{n+1}}{(n+1)!} \leq R_{n+1} \leq M \frac{(b-a)^{n+1}}{(n+1)!}.$$

By the intermediate value property of the derivative there exists a c with

$$f^{(n+1)}(c) = R_{n+1}\frac{(n+1)!}{(b-a)^{n+1}}.$$

Exercises

Exercise 15.9.1   *Find the accuracy of the following formulae on the given intervals:*

(1) $\sin x \approx x$ $[-\pi/6,\ \pi/6]$;

(2) $\dfrac{1}{1+x+x^3} \approx 1 - x$ $[0,\ 0.01]$;

(3) $\cos x \approx 1$ $[-10^{-3}, 10^{-3}]$.

(i) **Exercise 15.9.2**   *Give an example showing that the Lagrange remainder is not valid if $f^{(n+1)}$ fails to exist at one point of $]a, b[$. [Hint: Go back to the mean value theorem.]*

15.10   The indefinite integral

The function

$$F(x) = \int_a^x f \tag{15.54}$$

is called the indefinite integral of f, or simply the indefinite integral. We shall study its properties in this section. Whatever we say or prove about F applies with little or no change to $\underline{F}(x) = \int_x^b f$. If TX is a tagged division

$$TX \equiv a = t_0 \leq X_1 \leq t_1 \leq \cdots \leq c$$
$$= t_{k-1} \leq X_k \leq \ldots X_m \leq t_m = d \leq \cdots \leq t_n = b$$

then by $TX\big|_c^d$ we denote the tagged division

$$c = t_{k-1} \leq X_k \leq \cdots \leq t_{m-1} \leq X_m \leq t_m = d.$$

Lemma 15.1   *If f is integrable on $[a, b]$, $a \leq c < d \leq b$ and δ a gauge with the property that*

$$DX \ll \delta \Rightarrow \left| \mathcal{R}(f, DX) - \int_a^b f \right| < \varepsilon \tag{15.55}$$

then, if[13] $DX\big|_c^d \ll \delta$

$$\left| \mathcal{R}(f, DX\big|_c^d) - \int_c^d f \right| \leq \varepsilon. \tag{15.56}$$

Proof. For $\eta > 0$ let $\delta_1 \leq \delta$ and $\delta_2 \leq \delta$ be gauges such that

$$\left| \mathcal{R}(f, SY) - \int_a^c f \right| < \eta \quad \text{and} \quad \left| \mathcal{R}(f, UV) - \int_d^b f \right| < \eta, \tag{15.57}$$

tagged divisions $SY \ll \delta_1$ and $UV \ll \delta_2$ of $[a, c]$ and $[d, b]$, respectively.[14] The sum

$$\mathcal{R}(f, SY) + \mathcal{R}(f, DX\big|_c^d) + \mathcal{R}(f, UV)$$

is a Riemann sum for a δ-fine tagged division of f on $[a, b]$. Using (15.55) and (15.57) leads to

$$\left| \mathcal{R}(f, SY) - \int_a^c f + \mathcal{R}(f, DX\big|_c^d) - \int_c^d f + \mathcal{R}(f, UV) - \int_d^b f \right| < \varepsilon,$$

$$\left| \mathcal{R}(f, DX\big|_c^d) - \int_c^d f \right| - 2\eta \leq \varepsilon.$$

Sending $\eta \to 0$ gives Equation (15.56). $\qquad\square$

Theorem 15.20 *If f is integrable on $[a, b]$ then its indefinite integral is continuous on $[a, b]$.*

Proof. We prove continuity of F from the right at c, $a \leq c < b$. Continuity form the left at a point which is not the left end-point of $[a, b]$ is proved similarly. For $\varepsilon > 0$ find a gauge δ such that

$$DX \ll \delta \Rightarrow \left| \mathcal{R}(f, DX) - \int_a^b f \right| < \frac{\varepsilon}{2}. \tag{15.58}$$

Obviously $\lim_{t \downarrow c} f(c)(t - c) = 0$ and there exists $\omega > 0$ such that $f(c)(t - c) < \varepsilon/2$ for $c < t < c = \omega$. Let $c < t < \text{Min}(\omega, \delta(c))$ then the lemma can be

[13]We should have said: if $DX\big|_c^d$ is fine with the restriction of δ to $[c, d]$. We shall not make a distinction between a gauge and its restriction.
[14]If $a = c$ or $d = b$ terms involving $\mathcal{R}(SY)$ or $\mathcal{R}(UV)$, respectively, should be omitted.

applied with $DX\big|_c^t \equiv c = t_{k-1} = X_k < t = t_k$ and we have

$$|f(c)(t - c) - (F(t) - F(c))| < \frac{\varepsilon}{2}.$$

Consequently

$$|F(t) - F(c)| < \frac{\varepsilon}{2} + |f(c)(t - c)| \leq \varepsilon.$$

$\square$

The next natural question to ask, after continuity, is differentiability. Generally speaking, the indefinite integral need not be differentiable on all of $]a,\, b[$. For instance, if $f : x \mapsto x - \lfloor x \rfloor$ then

$$F(x) = \begin{cases} \dfrac{x^2}{2} & \text{for } 0 \leq x \leq 1 \\ 1 + \dfrac{x^2}{2} - x & \text{for } 1 < x \leq 2 \end{cases}$$

Hence $F'_-(1) = 1$, $F'_+(1) = 0$ and F is not differentiable at 1. The next question is: if F is differentiable, is $F'(x) = f(x)$? The answer is again no. If $f = 1_{\mathbb{Q}}$, then $F(x) = F'(x) = 0 \neq 1_{\mathbb{Q}}(x) = 1$ if x is rational. However the next theorem shows that F is differentiable at points of continuity of f and then $F'(x) = f(x)$.

Theorem 15.21   *If f is integrable on $[a,\, b]$ then*

(i) $F'_-(c) = f(c)$ *if f is continuous from the left at c, $a < c \leq b$;*
(ii) $F'_+(c) = f(c)$ *if f is continuous from the right at c, $a \leq c < b$.*
(iii) $F'(c) = f(c)$ *if f is continuous at c, $a < c < b$.*

Corollary 15.21.1   *If f is continuous on $[a,\, b]$ then it has a primitive, more precisely there exists a function G such that*

$$G'(x) = f(x) \tag{15.59}$$

for $a < x < b$ and $G'_-(a) = f(a)$, $G'_+(b) = f(b)$.

Proof.   We prove only (i) since (ii) is entirely similar and (iii) follows from (i) and (ii). By continuity of f from the left at c for $\varepsilon > 0$ there exists $\eta > 0$ such that

$$f(c) - \varepsilon < f(t) < f(c) + \varepsilon \quad \text{for} \quad c - \eta < t \leq c.$$

It follows by integrating this inequality that

$$(f(c) - \varepsilon)(c - x) \le F(c) - F(x) \le (f(c) + \varepsilon)(c - x)$$

and $F'_-(c) = f(c)$. $\square$

Theorem 15.20 says, for an integrable f, that

$$\lim_{t \downarrow a} \int_t^b f = F(b) - F(a) = \int_a^b f.$$

If we knew without assuming the integrability of f that this equation holds, we can use it profitably for evaluation of integrals. We established in Example 15.22 that

$$\int_x^1 \ln t \, dt = -1 - x \ln x + x.$$

The right hand side has limit equal to -1 as $x \to 0$. If we knew that the limit of the left hand side is $\int_0^1 \ln t \, dt$ then we have evaluated the integral. The next theorem gives an affirmative answer.

Theorem 15.22 (Hake's theorem)   *If f is integrable on $[c, b]$ for every c with $a < c < b$ and*

$$\lim_{c \downarrow a} \int_c^b f \text{ exists and equals } A \tag{15.60}$$

then f is integrable on $[a, b]$ and $\int_a^b f = A$.

Similarly

Remark 15.15   If f is integrable on $[a, c]$ for every c with $a < c < b$ and

$$\lim_{c \uparrow b} \int_a^c f \text{ exists and equals } A \tag{15.61}$$

then f is integrable on $[a, b]$ and $\int_a^b f = A$.

Proof.   We can and shall assume without loss of generality that $f(a) = 0$. Take a strictly decreasing sequence $\{c_n\}$ with $c_n \downarrow a$ and $c_0 = b$. For every positive ε there exists δ_n such that if $TX(n)$ is a δ_n-fine tagged division of $[c_n, c_{n-1}]$ then

$$\left| \mathcal{R}(TX(n)) - \int_{c_n}^{c_{n-1}} f \right| < \frac{\varepsilon}{2^{n+1}}. \tag{15.62}$$

According to the hypothesis there exists r such that

$$\left| \int_c^b f - A \right| < \frac{\varepsilon}{2} \qquad (15.63)$$

whenever $a < c < r$. Now define a gauge δ which has, for all $n \in \mathbb{N}$, the following properties.[15]

$$\delta(x) \le \delta_n(x) \text{ for } x \in [c_n, c_{n-1}],$$

$$[x - \delta(x), x + \delta(x)] \subset]c_n, c_{n-1}[\text{ for } x \in]c_n, c_{n-1}[, \qquad (15.64)$$

$$[c_n - \delta(c_n), c_n + \delta(c_n)] \subset]c_{n+1}, c_{n-1}[. \qquad (15.65)$$

In addition we also require that

$$\delta(a) = r - a, \qquad (15.66)$$

$$\delta(b) = \frac{1}{2}(b - c_1). \qquad (15.67)$$

Let now TX be a δ-fine partition of $[a, b]$,

$$TX \equiv a = t_0 \le X_1 \le t_1 \le \cdots X_n \le t_n = b.$$

Relations (15.64) and (15.65) imply that $X_1 = a$ and hence $f(X_1) = 0$. The Riemann sum $\mathcal{R}(TX)$ starts with the term $f(X_2)(t_2 - t_1)$. Let N be the first integer for which $c_{N+1} \le t_1$. By Inequality (15.66)

$$t_1 < t_0 + \delta(X_1) = a + \delta(a) < r$$

and consequently by (15.63)

$$\left| \int_{t_1}^b f - A \right| < \frac{\varepsilon}{2}.$$

Since

$$|\mathcal{R}(TX) - A| \le \left| \mathcal{R}(TX) - \int_{t_1}^b \right| + \left| \int_{t_1}^b f - A \right|$$

it suffices to show that

$$\left| \mathcal{R}(TX) - \int_{t_1}^b f \right| < \frac{\varepsilon}{2}. \qquad (15.68)$$

---

[15]The first inequality implies that $\delta(c_n) \le \text{Min}(\delta_n(c_n), \delta_{n+1}(c_n))$.

Claim: The δ-fine partition TX anchors on $c_0, c_1, \ldots c_N$.

Accepting the claim for a moment we have

$$\mathcal{R}(TX) = \mathcal{R}\left(TX|_{t_1}^{c_N}\right) + \sum_{n=1}^{N} \mathcal{R}(TX(n))$$

and consequently

$$\left|\mathcal{R}(TX) - \int_{t_1}^{b} f\right| \le \left|\mathcal{R}\left(TX|_{t_1}^{c_N}\right) - \int_{t_1}^{c_N} f\right| + \sum_{i=1}^{N}\left|\mathcal{R}(TX(i)) - \int_{c_i}^{c_{i-1}} f\right|.$$

The sum on the right hand side is smaller then $\varepsilon/2$ by Inequalities 15.62. Applying Lemma 15.1 with the roles of α and β played by t_1 and c_N we have

$$\left|\mathcal{R}\left(TX|_{t_1}^{c_N}\right) - \int_{t_1}^{c_N} f\right| < \frac{\varepsilon}{2^N}.$$

Thus we have Equation (15.68).

It remains to prove the claim. For $k \in \{0, 1, \ldots, N\}$ each c_k belongs to $[t_{i-1}, t_i]$ for some $i \in \{2, 3, \ldots, n\}$. Then

$$X_i - \delta(X_i) < t_{i-1} \le c_k. \tag{15.69}$$

If $c_k < X_i$ then $X_i \in]c_j, c_{j+1}]$ for some $j \le k$. Then by (15.64) or by (15.65)[16]

$$X_i - \delta(X_i) > c_{j+1} \ge c_k$$

contradicting Inequality (15.69). The possibility $X_i < c_k$ can be ruled out similarly. For $k = 0, 1, \ldots c_N$ the term c_k tags the subinterval in which it lies. The claim has been proved. $\qquad \square$

Theorem 15.22 yields the next theorem

Theorem 15.23 (Integrability test) *If, for every c, $a < c < b$, f is integrable on $[c, b]$ and there are functions G, g integrable on $[a, b]$ and such that*

$$g(x) \le f(x) \le G(x)$$

for all $x \in [a, b]$ except a finite set then f is integrable on $[a, b]$.

---

[16]If $X_i = c_{j+1}$.

Proof. The limits

$$\lim_{x \downarrow a} \int_x^b g \quad \text{and} \quad \lim_{x \downarrow a} \int_x^b G$$

exist by Theorem 15.20. By the Bolzano–Cauchy convergence principle for limits of functions, for every positive ε there is a d such that

$$\left| \int_{a'}^{a''} g \right| < \varepsilon \quad \text{and} \quad \left| \int_{a'}^{a''} G \right| < \varepsilon$$

whenever $a < a'$, $a'' < d$. Consequently

$$\left| \int_{a'}^{a''} f \right| < \varepsilon$$

and $\lim_{x \downarrow a} \int_x^b f$ exists by the Bolzano–Cauchy principle. $\square$

Example 15.30   The integral

$$\int_0^t \frac{dx}{\sqrt{1 - x^4}}$$

is called elliptic. It cannot be easily evaluated by the Fundamental Theorem, as the primitive is not an elementary function. We show that $1/\sqrt{1 - x^4}$ is integrable on $[0, 1]$. Factorizing, $1 - x^4 = (1 - x)(1 + x)(1 + x^2)$. Since $(1 + x)(1 + x^2) > 1$ for $0 < x < 1$ we have

$$0 \le \frac{1}{\sqrt{1 - x^4}} \le \frac{1}{\sqrt{1 - x}}$$

and Theorem 15.23 can be applied with $g(x) = 0$ and $G(x) = 1/\sqrt{1 - x}$ to show integrability. *Maple* can evaluate the integral

---

```
>  int(1/sqrt(1-x^4),x=0..1);
```

$$\frac{1}{4} B(\frac{1}{4}, \frac{1}{2})$$

```
>  evalf(%);
```

$$1.311028777$$

In the intermediate result *Maple* employed the function Beta, also called Euler's integral of the first kind. It is defined for $p > 0$ and $q > 0$ by the equation

$$B(p,q) = \int_0^1 x^{p-1}(1-x)^{q-1}\,d\,x.$$

Except for some special values of p and q the Beta function cannot be expressed in terms of elementary functions of p and q. However, as in the above example, many other integrals occurring in applications can be expressed in terms of the Beta function.

Exercises

ⓘ **Exercise 15.10.1**   *Prove: If f is integrable on $[c, 1]$ for $0 < c < 1$ and $\lim_{c\downarrow 0} x^\alpha f(x) = L \neq 0$ then f is integrable on $[0, 1]$ if and only if $\alpha < 1$.* [Hint: Use Theorem 15.23.]

Exercise 15.10.2   *Use the previous exercise to show the integrability of $1/\sqrt{\sin x}$ and nonintegrabilty of $1/\sin x$ on $[0, 1]$.*

15.11   Integrals over unbounded intervals

In applications one encounters integrals over unbounded intervals. For instance

$$\int_0^\infty \frac{\sin x}{x}\,dx \quad \text{and} \quad \int_{-\infty}^\infty e^{-x^2}$$

appear in physics and probability, respectively. So far we have not defined integrals over unbounded intervals. We remedy that now. We denote by $\boldsymbol{I}$ any of the following intervals

$$[a,\,b] \subset \mathbb{R}, \quad]-\infty, \infty[, \quad [a, \infty[, \quad]-\infty,\, b] \tag{15.70}$$

Definition 15.4   A number I is the integral of f over $\boldsymbol{I}$ if for every positive ε there are a gauge δ and a positive number K such that

$$|\mathcal{R}(f, TX) - I| < \varepsilon \tag{15.71}$$

whenever TX is a δ-fine tagged division of a bounded interval $[A,\, B]$ with $\boldsymbol{I} \supset [A,\, B] \supset \boldsymbol{I} \cap [-K,\, K]$. Integral of f over $\boldsymbol{I}$ is denoted by

$$\int_{\boldsymbol{I}} f \quad \text{or} \quad \int_{\boldsymbol{I}} f(x)\,d\,x.$$

If $I = [a, b]$ then this definition has the same meaning as Definition 15.2. The geometric meaning behind the definition above is that for a non-negative function f on an unbounded interval I the area under the graph of f and above the x-axis is approximated by Riemann sums for δ-fine tagged divisions of sufficiently large bounded intervals. There is at most one number I satisfying the requirements of the definition: in other words, the number I is well-defined. The proof is rather similar to the proof of Theorem 15.1 and is omitted. If I is one of the intervals in (15.70) then

$$\int_I f$$

is denoted by

$$\int_a^b f, \quad \int_{-\infty}^\infty f, \quad \int_a^\infty f, \quad \int_{-\infty}^b f,$$

respectively, or by similar symbols where f is replaced by $f(x)\,dx$ or by $f(u)\,du$, etc.

Example 15.31 We wish to show that $\int_1^\infty x^{-2}dx = 1$. We choose $\delta(x) = \varepsilon x/3(1 + \varepsilon)$ and $K > 3/\varepsilon$. Let TX be a δ-fine tagged division of $[1, t_n]$ with $t_n > K$. Motivated by the Fundamental Theorem and by the fact that $-\dfrac{d}{dx}x^{-1} = x^{-2}$ we shall approximate

$$\frac{1}{X_i^2}(t_i - t_{i-1}) \quad \text{by} \quad \frac{1}{t_{i-1}} - \frac{1}{t_i}.$$

Firstly

$$1 \ge \sum_1^n \left(\frac{1}{t_{i-1}} - \frac{1}{t_i} \right) = 1 - \frac{1}{t_n} > 1 - \frac{\varepsilon}{3}, \tag{15.72}$$

secondly

$$1 \le \frac{t_i}{X_i} < \frac{X_i + \delta(X_i)}{X_i} < 1 + \frac{\varepsilon}{3},$$

$$1 \le \frac{X_i}{t_{i-1}} < \frac{X_i}{X_i - \delta(X_i)} = \frac{3(1 + \varepsilon)}{3 + 2\varepsilon} < 1 + \frac{\varepsilon}{3}.$$

This leads to

$$\left| \frac{1}{X_i^2}(t_i - t_{i-1}) - \left(\frac{1}{t_{i-1}} - \frac{1}{t_i} \right) \right| = \left| \frac{t_{i-1}t_i}{X_i^2} - 1 \right| \left(\frac{1}{t_{i-1}} - \frac{1}{t_i} \right)$$

$$\leq \left| \frac{t_{i-1}}{X_i} \right| \left| \frac{t_i}{X_i} - \frac{X_i}{t_{i-1}} \right| \left(\frac{1}{t_{i-1}} - \frac{1}{t_i} \right) \leq \frac{\varepsilon}{3} \left(\frac{1}{t_{i-1}} - \frac{1}{t_i} \right).$$

Using these estimates on the Riemann sum yields

$$\sum_1^n \left[\frac{1}{X_i^2}(t_i - t_{i-1}) - \left(\frac{1}{t_{i-1}} - \frac{1}{t_i} \right) \right] \leq \frac{\varepsilon}{3} \sum_1^n \left(\frac{1}{t_{i-1}} - \frac{1}{t_i} \right).$$

Combining this inequality with (15.72) gives

$$\left| \sum_1^n \frac{1}{X_i^2}(t_i - t_{i-1}) - 1 \right| \leq \frac{\varepsilon}{3} + \frac{\varepsilon}{3} < \varepsilon.$$

A combined version of Theorems 15.20 and 15.22 for an infinite interval reads

---

Theorem 15.24 (Hake)   *f is KH-integrable on $[a, \infty]$ if and only if*

$$\lim_{b \to \infty} \int_a^b f$$

exists and then

$$\lim_{b \to \infty} \int_a^b f = \int_a^\infty f. \qquad (15.73)$$

Remark 15.16   Similar theorems hold for $\int_{-\infty}^b f$ and for $\int_{-\infty}^\infty f = \int_{-\infty}^0 f + \int_0^\infty f$. The integral

$$\int_{-\infty}^\infty f = \lim_{t \to \infty} \int_{-t}^t f$$

if it exists but the limit on the right hand side might exist and the integral might not. If, for instance, $f(x) = \sin x$ or x then the limit is zero but the integral does not exist.

Example 15.32   We give a one line solution to Example 15.31. By the Fundamental Theorem

$$\lim_{b \to \infty} \int_1^b \frac{1}{x^2} dx = \lim_{b \to \infty} \left(1 - \frac{1}{b} \right) = 1.$$

Example 15.33   We prove integrability of $(\sin x)/x$ on $[1, \infty]$. Integrating by parts gives

$$\int_a^b \frac{\sin x}{x} dx = \frac{\cos a}{a} - \frac{\cos b}{b} - \int_a^b \frac{\cos x}{x^2} dx.$$

Consequently, for $a > K$ and $b > a$

$$\left| \int_a^b \frac{\sin x}{x} dx \right| \le \frac{4}{K}.$$

By the Bolzano–Cauchy Theorem the limit $\displaystyle\lim_{b \to \infty} \int_1^\infty \frac{\sin x}{x} dx$ exists.

Theorem 15.25 (Integrability test)   *If, for every c, $a < c < \infty$, f is integrable on $[a, c]$ and there are integrable functions G, g such that*

$$g(x) \le f(x) \le G(x)$$

for all $x \in [a, \infty[$ then f is integrable on $[a, \infty[$.

The proof is very similar to the proof of the integrability test in Section 15.10, Theorem 15.23. For an application of this integrability test see Exercise 15.11.1. An analogous test holds for $c \to -\infty$ if f is integrable on $[c, b]$ for every $c < b$.

Example 15.34   Since $\exp(-x^2) \le 1/x^2$ for $x \ne 0$ the integrability test can be applied with $g(x) = 0$ and $G(x) = 1/x^2$ on both intervals $[1, \infty[$ and $]-\infty, -1]$ to conclude integrability of $\exp(-x^2)$ on $]-\infty, \infty[$.

In this chapter we encountered the special functions erf and the Euler integral $(p, q) \mapsto B(p, q)$. One of the most important higher transcendental functions is the Γ function, also called Euler's integral of the second kind. It is defined as

$$\Gamma(s) = \int_0^\infty \exp(-x)x^{s-1} dx, \qquad (15.74)$$

for $s > 0$. For a positive integer $s = n$ it becomes

$$\Gamma(n) = 1 \cdot 2 \cdots (n - 1) = (n - 1)! \qquad (15.75)$$

Many integrals can be expressed in terms of the Gamma function as in our next example, where we evaluate $\int_0^\infty \dfrac{\cos 2x}{x^{1/5}}\,dx$

```
>  int( (cos(2*x))/x^(1/5),x=0..infinity);
```

$$\frac{1}{2}\,\frac{\sqrt{\pi}\,\csc(\frac{2}{5}\,\pi)\,\sin(\frac{1}{10}\,\pi)\,\Gamma(\frac{9}{10})}{\Gamma(\frac{3}{5})}$$

```
>  evalf(%);
```

$$.2066317466$$

Exercises

(i) **Exercise 15.11.1**   *Let $f : [a, \infty[\mapsto \mathbb{R}$ be integrable on $[a, c]$ for every $c > a$ and*

$$\lim_{x \to \infty} x^\alpha f(x) = L \neq 0.$$

Show that if $\alpha > 1$ then $\int_a^\infty f$ exists and if $\alpha \leq 1$ then the integral does not exist.

Exercise 15.11.2   *Use Maple to evaluate $\int_{-\infty}^\infty \exp(-x^2)\,dx$.*

15.12   Interchange of limit and integration

In pure and applied mathematics, interchange of limit and integration, according to the equation

$$\lim_{n \to \infty} \int_a^b f_n = \int_a^b \left(\lim_{n \to \infty} f_n(x) \right) dx, \tag{15.76}$$

is often needed. Since the sum of a series is nothing but the limit of partial sums, Equation (15.76) translates into

$$\sum_{n=1}^\infty \int_a^b u_n = \int_a^b \sum_{n=1}^\infty u_n.$$

It would be wrong to assume that Equation (15.76) always holds. If $f_n = n1_{]0,1/n[}$ then $\int_0^1 f_n = 1$ for every $n \in \mathbb{N}$, hence $\lim_{n\to\infty} \int_0^1 f_n = 1$. On the other hand $\lim_{n\to\infty} n1_{]0,1[}(x) = 0$ for every $x \in [0,1]$ and consequently

$$\int_0^1 \lim_{n\to\infty} f_n = 0.$$

Let f_n and f be functions defined on an interval I. Define

$$r_n = \sup\{|f(x) - f_n(x)|; \; x \in I\}.$$

Definition 15.5   The sequence of functions $n \mapsto f_n$ is said to be uniformly convergent on I to f if $r_n \to 0$.

For instance, the sequence $n \to (\sin nx)/n$ is uniformly convergent to the zero function on $\mathbb{R}$, since, in this case, $r_n = 1/n$.

Under uniform convergence Equation (15.76) is valid. More precisely, we have

Theorem 15.26   *If the functions f_n are integrable on an interval $[a, b]$ and the convergence $f_n \to f$ is uniform on $[a, b]$ then f is integrable and Equation (15.76) holds.*

Proof.   It follows from the definition of uniform convergence that for every $\varepsilon > 0$ there exists a number N such that

$$f_n(x) - \varepsilon < f(x) < f_n(x) + \varepsilon \tag{15.77}$$

for all $x \in [a, b]$ and all $n > N$. The integrability of f follows from these inequalities and Theorem 15.11. By Theorem 15.6

$$\int_a^b f_n - \varepsilon(b-a) \le \int_a^b f \le \int_a^b f_n + \varepsilon(b-a),$$

for all $n > N$.   □

This theorem has simple wording and a very easy proof. It is a useful theorem, however, the condition of uniform convergence is sometimes too restrictive. If $f_n(x) = x^n$ then, as is easily checked by direct evaluation, Equation (15.76) is valid with $[a, b] = [0, 1]$. The theorem is not applicable because the convergence is not uniform. The theorem is no longer valid if the interval $[a, b]$ is replaced by an unbounded interval. For instance, if $f_n = 1_{n,2n}/n$ then f_n converge uniformly to zero but

$$\lim_{n\to\infty} \int_0^\infty f_n = 1 \neq \int_0^\infty \lim_{n\to\infty} f_n(x)dx = 0.$$

The concept of uniform convergence is, nevertheless, important, as the next theorem shows.

> **Theorem 15.27** *If f_n are functions continuous at $c \in]a, b[$ and converging uniformly on $[a, b]$ to f then f is continuous at c.*

Proof. The following inequality is obvious

$$|f(x) - f(c)| \leq$$
$$|f(x) - f_n(x)| + |f_n(x) - f_n(c)| + |f_n(c) - f(c)| \leq$$
$$2r_n + |f_n(x) - f_n(c)|.$$

First we find n such that $r_n < \varepsilon/3$ and then with this n fixed we find $\delta > 0$ such that

$$|f_n(x) - f_n(c)| < \frac{\varepsilon}{3} \quad \text{for} \quad |x - c| < \delta.$$

The continuity of f at c follows. $\qquad \square$

We now return to the interchange of limit and integration.

Theorem 15.28 (Monotone Convergence Theorem) *Assume that*

   (i) *the functions f_n are integrable on an interval I for every $n \in \mathbb{N}$,*
   (ii) $\lim\limits_{n\to\infty} f_n(x) = f(x)$ *for every $x \in I$,*
   (iii) $f_n(x) \leq f_{n+1}(x)$ *for $n \in \mathbb{N}$ and all $x \in I$, or $f_n(x) \geq f_{n+1}(x)$ for $n \in \mathbb{N}$ and all $x \in I$.*

Then

$$\lim_{n\to\infty} \int_I f_n$$

exists. If it is finite then f is integrable and

$$\lim_{n\to\infty} \int_I f_n = \int_I f,$$

if it is infinite then f is not integrable.

For the proof we refer to Lee and Výborný (2000, Theorem 3.5.2). Here we illustrate the theorem by two examples.

Example 15.35 Denote by $|I|$ the length of the interval I. Let

$$A = \cup_1^\infty I_n,$$

where the intervals are pairwise disjoint. Then

$$\int_{-\infty}^\infty 1_A = \sum_1^\infty |I_n|, \tag{15.78}$$

provided the series converges. Let $A_n = \cup_{k=1}^n I_k$, then 1_{A_n} converge increasingly to 1_A and

$$\int_{-\infty}^\infty 1_{A_n} = \sum_{k=1}^n I_k.$$

Taking limits as $n \to \infty$ on both sides of this equation and applying the Monotone Convergence Theorem proves Equation (15.78)

Example 15.36 (The Laplace integral) Integral $\int_0^\infty \exp(-x^2)dx$ is called the Laplace integral. *Maple* already told us that it equals to $\sqrt{\pi}/2$. Now we show how this can be proved.[17] From the power series expansion of

---

[17] There are more elegant proofs, however they require knowledge which is not yet available to us. For a short proof see Lee and Výborný (2000, Example 6.6.9).

$\ln(1 + x)$ it follows that

$$\lim_{x \to 0} \frac{\ln(1 + ax)}{x} = a$$

and consequently

$$\lim_{n \to \infty} \left(1 - \frac{x^2}{n}\right)^n = \exp(-x^2).$$

Using the arithmetic-geometric mean inequality, it can be proved similarly as in Inequalities (10.23) and (10.24) that the sequence, with terms $f_n(x) = (1 - x^2/n)^n$ for $0 \le x \le \sqrt{n}$ and $f_n(x) = 0$ otherwise, is increasing for every x. It follows from the Monotone Convergence Theorem that

$$\int_0^\infty \exp(-x^2) dx = \lim_{n \to \infty} \int_0^{\sqrt{n}} \left(1 - \frac{x^2}{n}\right)^n. \tag{15.79}$$

We evaluate the integral I_n on the right hand side by changing variables $x = \sqrt{n} \cos t$, $dx = -\sin t\, dt$ and have

$$I_n = \sqrt{n} \int_0^{\frac{\pi}{2}} \sin^{2n+1} t\, dt = \sqrt{n} S_{2n+1}. \tag{15.80}$$

We evaluated the integral S_n in Example 15.25 where we also proved that $\sqrt{n} S_n \to \sqrt{\pi/2}$. Hence

$$I_n = \sqrt{\frac{n}{2n+1}} \sqrt{2n+1} S_n \to \frac{1}{\sqrt{2}} \sqrt{\frac{\pi}{2}}. \tag{15.81}$$

Altogether, we have by (15.79), (15.80) and (15.81)

$$\int_0^\infty \exp(-x^2) dx = \frac{\sqrt{\pi}}{2}.$$

Another important theorem on interchange of limit and integration is the next theorem.

Theorem 15.29 (Dominated Convergence Theorem) *Assume that*

   (i) *the functions f_n are integrable on an interval I for every $n \in \mathbb{N}$,*

   (ii) $\lim\limits_{n \to \infty} f_n(x) = f(x)$ *for every $x \in I$,*

   (iii) *there exists integrable functions g and G such that*

$$g(x) \le f_n(x) \le G(x)$$

   for all $x \in I$ and all $n \in \mathbb{N}$.

Then f is integrable on I and

$$\lim_{n \to \infty} \int_I f_n = \int_I f,$$

Again, we only illustrate this theorem by an example.

Example 15.37 The Γ function was defined in Equation (15.74). We wish to prove the formula

$$\Gamma(t) = \lim_{n \to \infty} \frac{n!\, n^t}{t(t+1)\cdots(t+n)},$$

for $t > 0$. It is not difficult to show that

$$\int_0^1 (1-y)^n y^{t-1}\, dy = \frac{n!}{t(t+1)\ldots(t+n)},$$

$$\int_0^n \left(1 - \frac{u}{n}\right)^n u^{t-1}\, du = n^t \int_0^1 (1-y)^n y^{t-1}\, dy.$$

The first formula can be proved by induction, the second by substitution, with $u = ny$ and $du = n\, dy$. Hence we want to show that

$$\Gamma(t) = \lim_{n \to \infty} \int_0^n \left(1 - \frac{u}{n}\right)^n u^{t-1}\, du.$$

Since

$$\lim_{n \to \infty} \left(1 - \frac{x}{n}\right)^n = e^{-x},$$

we seek an integrable non-negative function G such that $0 \le f_n \le G$, where

$$f_n(x) = \begin{cases} 0 & \text{if } x > n \\ \left(1 - \dfrac{x}{n}\right)^n x^{t-1} & \text{if } 0 \le x \le n. \end{cases}$$

The inequality $\ln(1 - x/n) \le -x/n$ follows from the power series expansion of $\ln(1 + x)$ and it yields $(1 - x/n)^n \le e^{-x}$ for $0 \le x \le n$. We can take $G(x) = e^{-x}x^{t-1}$, this function is integrable on $[0, \infty[$ because it is continuous and

$$G(x) \le x^{t-1} \quad \text{for} \quad 0 < x \le 1, \tag{15.82}$$

$$G(x) \le \frac{k!}{x^{k+1-t}} \quad for \quad x > 1. \tag{15.83}$$

The first inequality guarantees integrability of G on $[0, 1]$, the second with $k > t + 1$ on $[1, \infty[$.

15.13 Comments and supplements

The basic ideas for integration can be found in antiquity, in the exhaustion method and in the work of Archimedes. There is a long list of mathematicians contributing to the subject before Newton and Leibniz, including Fermat, Cavalieri, Kepler; see Eves (1981). Surprisingly, the Fundamental Theorem of Calculus was discovered by Barrow, a predecessor of Newton at Cambridge. Newton and Leibniz developed systematic methods for calculating derivatives and integrals and open ways to apply calculus to problems in other sciences. The logical foundations of Calculus had some important critics. The mathematicians of the seventeenth and eighteenth century were aware of the shortcomings of the logical foundations but had different priorities. D'Alembert even said: "Let's move ahead, the confidence will arrive later." The first definition of integral which can stand up to modern scrutiny was Riemann's. Shortly before him, Cauchy proved that the right-sums (and the leftsums) converge to a limit when the maximal length of the subintervals tended to zero. Riemann not only allowed the tag to float anywhere in the subinterval but made a decisive step forward by considering the set of (Riemann) integrable functions, namely those functions for which the Riemann sums had a limit as the maximal length of subintervals tended to zero. Riemann's definition was originally a success, but as time went by, it was realized that the definition was far from ideal. One of the deficiencies of Riemann theory is that none of the monotone convergence theorem, the dominated convergence theorem or the Hake theorems are valid for the Riemann integral. Lebesgue set himself a goal to create an integration theory such that:

(i) To every function bounded on the interval $[a, b]$ there is a number

I associated with it such that the following axioms are satisfied. If
$I = \int_a^b f = \int_a^b f(x)\,dx$ then

(ii)

$$\int_a^b (f + g) = \int_a^b f + \int_a^b g;$$

(iii) for all a, b, c

$$\int_a^b f + \int_b^c f + \int_c^a f = 0;$$

(iv) If $f \geq 0$ and $a < b$ then

$$\int_a^b f \geq 0;$$

(v) For all h, a, b

$$\int_a^b f - \int_{a+h}^{b+h} f(x - h)\,dx;$$

(vi)

$$\int_0^1 1\,dx = 1;$$

(vii) If $f_n(x) \leq f_{n+1}$ and $f_n(x) \to f(x)$ for every x in $[a, b]$ then

$$\lim_{n \to \infty} \int_a^b f_n = \int_a^b f.$$

In Lebesgue's requirement (vii) we recognize the monotone convergence theorem. Lebesgue created the integration satisfying his axioms. His theory had an immediate and beneficial impact on mathematical analysis and is still the integration theory predominantly used by professional mathematicians. Lebesgue, however, did not define the integral for *every* bounded function. Instead, he created a class of summable, or Lebesgue integrable functions. These include all Riemann integrable functions and also many unbounded functions. However, in Lebesgue theory, Hake's theorems are still not valid and neither is the Fundamental Theorem, at least not in such a simple and powerful form as in Theorem 15.15. Lots of effort was devoted to making Lebesgue theory accessible to non-specialist physicists and engineers. Some of the users of integration can and do master Lebesgue theory. Experience shows, however, that Lebesgue theory can be too subtle and too difficult for general consumption. In contrast, the Kurzweil integration theory is relatively simple and has all the advantages of Lebesgue theory. This is the reason why we made perhaps an unusual step in adopting it for an elementary exposition. Kurzweil's definition is equivalent to two definitions proposed during the time after Lebesgue and before Kurzweil: one by Perron and another by Denjoy. Theories based on these definitions are even more demanding that the Lebesgue theory. In recent years, several publications presented Kurzweil theory in detail: we recommend Lee and Výborný (2000) and Bartle (2001), where further references can be found.

Appendix A

Maple Programming

A.1 Some *Maple* programs

A.1.1 *Introduction*

It is not our aim to expound *Maple* programming. However, some examples can be well illuminated by exposing the entire structure of the computation by a *Maple* program. So we give here the mere rudiments of *Maple* programming, just what we need in these examples.

One occasion when readers might find it profitable to write their own program is when they need to carry out the same operation[1] many times. A *Maple* procedure starts with its name, followed by the assignment operator :=, the word **proc** and parameters to be operated on, enclosed in parenthesis. Hence it looks like

```
name := proc(parameters)
```

Wide choice is allowed for the parameters. They can be integers, real or complex numbers, polynomials or even sets. You must end the procedure either with **end proc:** or **end proc;**.[2] The semicolon is better: *Maple* comes back and prints the procedure or gives an error message. The program itself consists of commands operating on the parameters, each individual command ending with a ; character or a : character.

The symmetric difference of two sets is the set which contain all elements which lie in one of the sets but not in both. Our first procedure computes symmetric difference of two sets.

```
>   sdiff:=proc(A,B)
```

---

[1] Not available in *Maple* itself.

[2] In older versions of *Maple*, use just **end;**

501

```
>   (A union B) minus (A intersect B);
>   end proc;
```

$$sdiff := \mathbf{proc}(A,\ B)\,(A\,\text{union}\,B)\,\text{minus}\,(A\,\text{intersect}\,B)\,\mathbf{end\,proc}$$

Let us test our program

```
>   A:={1,2,3}; B:={2,3,4};
```

$$A := \{1,\ 2,\ 3\}$$

$$B := \{2,\ 3,\ 4\}$$

```
>   sdiff(A,B);
```

$$\{1,\ 4\}$$

A.1.2 *The conditional statement*

When programming, it is important that the computation be able to branch according to some condition which can only be tested midway during the computation itself. This is achieved in *Maple* by the statement if. The most basic form of this statement has the structure

> if *condition(s)* then *command(s)* else *commands* end if;

It is important that you end each if statement with **end if**;.[3] The use of if is best explained by our next example in which the procedure picks up from two sets the one which has the smaller number of elements. In the example we use the command nops(A) which counts the number of elements of A.

```
>   smaller:=proc(A,B)
>   if nops(A)<nops(B) then A;
>   else B; end if;
>   end proc;
```

---

[3] In older versions of *Maple* you have to use fi instead of end if.

smaller := **proc**(A, B) **if** nops(A) < nops(B) **then** A **else** B **end if end proc**

Testing the procedure, we have

```
>   smaller(A,B);
```

$$\{2, 3, 4\}$$

Note that our program picked up B as the smaller set; this is as expected. If the number of elements of A is not smaller than the number of elements of B then the **else** alternative is activated. Let us say that we overlooked this and that we want both sets printed when the number of elements in the sets is the same. The next procedure does that, and provides an example of nested **if** statements. Note the indentation of the nested statements, which improves readability. If you need to use nested **if** statements more often, then you might like to learn the **elif** statement from **Help**.

```
>   Smaller:=proc(A,B)
>   if nops(A)<nops(B) then A;
>   else
>               if nops(A)=nops(B) then {A,B};
>               else B;
>               end if
>   end if;
>   end proc;
```

$Smaller$:= **proc**(A, B)
    if nops(A) < nops(B) **then** A
    else if nops(A) = nops(B) **then** $\{B, A\}$
    else B **end if**
    end if
  end proc

Testing again gives

```
> Smaller(A,B);
```

$$\{\{1, 2, 3\}, \{3, 4, 5\}\}$$

A.1.3   *The* while *statement*

The while statement allows one or more commands to be executed multiple times (as a loop). Commands are executed while a stated condition is satisfied, with execution terminating when the condition fails. The while statement has the structure

while condition do command(s)   end do;

Do not forget to end the statement with end do;[4] The following procedure halves a number a until the quotient becomes smaller than 10^{-5}. In the procedure we assign q to be a, in order to prevent this assignment from taking effect outside the procedure, we use the command local q;. Actually, *Maple* automatically makes any variable introduced in a procedure, local to that procedure.

```
> pulka:=proc(a)
> local q;
> q:=a;
> while 10^5*q>1 do
>       q:=q/2;
> end do;
> q;
> end proc:
```
Let us test our program for $a = 3$.

```
> pulka(3);
```

$$\frac{3}{524288}$$

---

[4]Older versions of *Maple* use od instead of end do.

In the old days when computations were limited to pen and paper the following formula was used for efficiently computing the square root of a.

$$x_{n+1} = \frac{1}{2}(x_n + \frac{a}{x_n}).$$

Some algebraic manipulation leads to the estimate

$$0 \le x_n - \sqrt{a} = \frac{2x_n(x_n - x_{n+1})}{x_n + \sqrt{a}} \le 2(x_n - x_{n+1}).$$

Hence in order to obtain $\sqrt{a}$ within k decimal places we let the computation run while $2(x_n - x_{n+1}) \ge \frac{1}{2}10^{-k}$.

```
>   rt2:=proc(a,k::nonnegint)
>   x:=a:   oldx:=x+10^(-k) :
>   while 2*(oldx-x)>(1/2)*10^(-k) do
>        oldx:=x:   x:=0.5*(x+a/x)
>   end do:
>   end:
```

Warning, 'x' is implicitly declared local

Warning, 'oldx' is implicitly declared local

We test our procedure and compare it with the square root procedure provided by Maple:

```
>   rt2(2,9);sqrt(2.0); rt2(7225,8);
>   sqrt(7225.0);
```

1.414213563

1.414213562

85.00000001

85.00000000

The discrepancies at the last decimal place are not caused by some imperfection of our program but by *Maple* using only ten significant digits (as a default) for the computation. Increasing the number of significant digits to, say twelve, will remove the blemish. Of course, we wrote our program as an illustration on the use of `while` and not for calculating the square root.

A.2 Examples

In our first example we automate the calculation of the greatest common divisor of two positive[5] integers which we did on Page 59 by repeated use of the command `irem`.

Example A.1 (Greatest common divisor, the Euclid algorithm)
The idea is to run `irem` by using `while` as long as the result is not zero.

```
>   mygcd:=proc(m::posint,n::posint)
>   local a,b,r;
>   a :=m; b := n;
>   while b > 0 do
>           r := irem(a,b);
>           a := b;
>           b := r;
>   end do;
>   a;
>   end;
```

$mygcd := \mathbf{proc}(m{::}posint, n{::}posint)$
$\mathbf{local}\, a, b, r;$
$\quad a := m\,;\, b := n\,;\, \mathbf{while}\, 0 < b\, \mathbf{do}\, r := \mathrm{irem}(a, b)\,;\, a := b\,;\, b := r\, \mathbf{end\ do}\,;\, a$
$\mathbf{end\ proc}$

Let us test our procedure and compare the results with *Maple* igcd

```
>   mygcd(987654321123456789,123456789987654321);
>   igcd(987654321123456789,123456789987654321);
```

$$1222222221$$

$$1222222221$$

---

[5]The case when one of the numbers is zero need no computing.

The following example provides a program for calculating the *exact* value of the square root of a complex number.

Example A.2   The equation

$$\beta = \frac{1}{\sqrt{2}}\sqrt{a + \sqrt{a^2 + b^2}} + \imath \frac{b}{|b|\sqrt{2}}\sqrt{\sqrt{a^2 + b^2} - a} \qquad (A.1)$$

from Theorem 7.2 can be modified that it covers any complex number $a + \imath b$, not only the case $b \neq 0$. This is done by replacing $b/|b|$ by $S(b)$ where S is defined as follows

$$S(x) = \begin{cases} -1, & \text{for } x < 0 \\ 1, & \text{for } x \geq 0 \end{cases} \qquad (A.2)$$

The easy verification is left to readers. The function S differs from the *Maple* function `signum` only at zero. However the command

`_Envsignum0= a;`

redefines signum at zero to be a. We can now write a procedure for computing the exact value of the square root of a complex number, we call it `compsqrt`

```
>  compsqrt:=proc(A)
>  local a, b, t:
>  a:=Re(A); b:=Im(A);
>  _Envsignum0:=1;
>  t:=sqrt(a^2+b^2);
>  sqrt((t+a)/2)+I*signum(b)*sqrt((t-a)/2)
>  end proc:
```

Here are a few numerical examples

```
>  compsqrt(-I);
```

$$\frac{1}{2}\sqrt{2} - \frac{1}{2}I\sqrt{2}$$

```
>  compsqrt(-1);
```

$$I$$

```
>  compsqrt(3-4*I);
```

$$2 - I$$

Now we save the procedure compsqrt in a file with the extension .m. We call the file `Ourprocs.m`. Note the quotation marks around the file name.

```
>   save compsqrt, "Ourprocs.m";
```

In a future session you can recall the file and the function by the command **read**. The following *Maple* worksheet illustrates this. There is a similarity to calling a package via the command **with(package);**. In each case, one has to recall the file first, and then the procedure can be used.[6]

```
>   restart;
```

First we call the file with the read command then we use the procedure for finding the exact value of the square root

```
>   read "Ourprocs.m";
>
>   compsqrt(-I);
```

$$\frac{1}{2}\sqrt{2} - \frac{1}{2}I\sqrt{2}$$

Other *Maple* objects can be saved the same way we saved **compsqrt**. Several functions, expressions or programs can be saved in a file. The structure of the command is

```
save function, expression, ..., "myfile.m"
```

Note the comma after the words function and expression.

The next Example illustrates the practical use of the idea in the proof of Theorem 12.15.

Example A.3 (Bisection method for finding roots)   We halve the interval (A, B) to locate a zero of a given f within ten significant digits.

---

[6]Function, expression.

```
> bisect:=proc(f,A,B)
> local a,b,g;
> if 0>evalf(f(A))
> then g:=f
> else g:=-f
> end if;
> #
> #now we make sure that f takes values
> #of opposite signs at the ends of the interval
> #
> if evalf(g(B))<0 then
> Error('f(A)*f(B) must be negative')
> else a:=A; b:=B;
> while evalf(10^10*(b-a))>1 do
>    if 0>evalf(g(a/2+b/2)) then a:=(1/2)*a +(1/2)*b;
>    b:=b;
>    else b:=(1/2)*a+(1/2)*b; a:=a;
>    end if;
> end do;
> #
> #it is easier to read it in decimal form,
> #
> (a+b)/2, evalf((a+b)/2);
> end if;
> end proc;
```

$$bisect := \mathbf{proc}(f,\, A,\, B)$$

$$\mathbf{local}\, a,\, b,\, g;$$

$$\mathbf{if}\, \mathrm{evalf}(f(A)) < 0\ \mathbf{then}\, g := f\, \mathbf{else}\, g := -f\, \mathbf{end\ if};$$

$$\mathbf{if}\, \mathrm{evalf}(g(B)) < 0\ \mathbf{then}\, \mathrm{Error}('f(A) * f(B)\ \mathit{must\ be\ negative}')$$

$$\mathbf{else}$$

$$\quad a := A;$$

$$\quad b := B;$$

$$\quad \mathbf{while}\, 1 < \mathrm{evalf}(10000000000 * b - 10000000000 * a)\, \mathbf{do}$$

$$\qquad \mathbf{if}\, \mathrm{evalf}(g(1/2 * a + 1/2 * b)) < 0$$

$$\qquad\quad \mathbf{then}\, a := 1/2 * a + 1/2 * b;\, b := b$$

$$\qquad \mathbf{else}\, b := 1/2 * a + 1/2 * b;\, a := a$$

$$\qquad \mathbf{end\ if}$$

$$\quad \mathbf{end\ do};$$

$$\quad 1/2 * a + 1/2 * b,\, \mathrm{evalf}(1/2 * a + 1/2 * b)$$

$$\mathbf{end\ if}$$

$$\mathbf{end\ proc}$$

An advantage of this method is that it always finds a zero. Other methods might become unstable if more than one zero of f is present in $[A, B]$. We now illustrate the use of the procedure on several examples.

```
>   f:=x->x^3-3*x+1;
```

$$f := x \to x^3 - 3x + 1$$

The polynomial has zeros in $[-2, -1], [0, 1]$ and $[1, 2]$. Let us see what happens for $[-3, 7]$ and $[-2, 5]$.

```
>   bisect(f,-3,7); bisect(f,-2,5);
```

$$\frac{-258300740771}{137438953472}, \; -1.879385242$$

$$\frac{421137386305}{274877906944}, \; 1.532088886$$

The use of bisect is not restricted to polynomials.

```
>   f:=x->(1/2)*x-sin(x);
```

$$f := x \to \frac{1}{2}x - \sin(x)$$

```
>   bisect(f,Pi/2,Pi);
```

$$\frac{41462209955}{68719476736}\pi, \; 1.895494267$$

```
>   save bisect, "Bisect.m";
```

References

Arrow, K. (1963). *Social choice and individual values*. New Haven: Yale University Press.

Bartle, R. G. (2001). *A modern theory of integration*. Providence, Rhode Island: American Mathematical Society.

Boas, R. P. (1972). *A primer of real functions*. The Mathematical Association of America.

Eves, H. (1981). *Great moments in mathematics (after 1650)*. Mathematical Association of America.

Gradshtein, I., & Ryzhik, I. (1996). *Table of integrals, series and products (computer file)*. Boston: Academic Press.

Hadlock, C. R. (1978). *Field theory and its classical problems*. Mathematical Association of America.

Halmos, P. R. (1974). *Naive set theory*. Springer.

Kelly, J. (1978). *Arrow impossibility theorems*. New York: Academic Press.

Knopp, K. (1956). *Infinite sequences and series*. New York: Dover.

Kurosh, A. (1980). *Higher algebra*. Mir.

Landau, E. (1951). *Foundations of analysis*. Chelsea.

Lee, P. Y., & Výborný, R. (2000). *The integral: An easy approach after Kurzweil and Henstock*. Cambridge, UK: Cambridge University Press.

Matsuoka, Y. (1961). An elementary proof of the formula $\sum_{k=1}^{\infty} 1/k^2 = \pi^2/6$. *Amer. Math. Monthly*, *68*, 485–487. (Reprinted in *Selected papers on calculus*, The Mathematical Association of America)

McCarthy, J. (1953). An everywhere continuous nowhere differentiable function. *Amer. Math. Monthly*, *60*, 709.

Spivak, M. (1967). *Calculus*. New York, Amsterdam: W.A. Benjamin, Inc.

Stromberg, K. R. (1981). *Foundations of analysis*. Belmont, California: Wadsworth.

Vyborny, R. (1987). Differentiation of power series. *Amer. Math. Monthly*, *94*, 369–370.

Výborný, R. (2001). Bolzano's anniversary. *The Australian Mathematical Society Gazette*, *28*, 177–183.

Vyborny, R., & Nester, R. (1989). L'Hôpital rule, a counterexample. *Elemente der Mathematik*, *44*, 116–121.

Youse, B. (1972). *Introduction to real analysis*. Boston: Allyn and Bacon.

Index of *Maple* commands used in this book

The following table provides a brief description of the *Maple* commands used in this book, together with the pages on which they are described, and some of the pages on which they are used.

Most of the commands in the following table are set in the type used for *Maple* commands throughout this book, such as `collect`. Any which are set in normal type, such as Exit, are not, strictly, *Maple* commands, but are provided here to assist in running a *Maple* session.

Table A.1: List of *Maple* commands used in this book

Command	Description	Pages
`abs(x)`	The absolute value of the number x	75
`annuity`	Calculates quantities relating to annuities: requires the `finance` package to be loaded	31
`Apollonius`	Calculates and graphs the eight circles which touch three specified circles: requires the `geometry` package to be loaded	30
`assume`	Make an assumption about a variable: holds for the remainder of the worksheet (or until `restart`)	112
`assuming`	Make an assumption about a variable: holds only for the command it follows	111
`binomial`	Evaluates the binomial coefficient	151
`ceil(x)`	The ceiling of x (the smallest integer not less than x)	75, 118
`changevar`	Carry out a change of variables in an integral	474
`collect`	Collect similar terms together	24, 26
`combinat`	A collection of commands for solving problems in combinatorial theory: loaded by using the command `with(combinat)`	37
`combine`	Combine expressions into a single expression	25, 26
`conjugate`	The complex conjugate $\bar{z}$ of a complex number z	199

Continued on next page

List of *Maple* commands (continued)

Command	Description	Pages
`convert`	Convert from one form to another	17, 17, 173, 184
`cos(x)`	The trigonometric function $\cos x$	75
`diff`	Differentiate a function	361
`D(f)(a)`	Evaluates the derivative of the function f at $x = a$	359
`Digits`	Set the number of digits used in calculations	15, 16
`divisors`	Find all the divisors of an integer: requires the `numtheory` package to be loaded	30, 215
`erf(x)`	The error function: $$\text{erf}(x) = \frac{2}{\sqrt{\pi}} \int_0^x e^{-x^2}\, dx$$	465
`evalc`	Changes a complex valued expression into the form $a + bi$	199, 203
`evalf`	Evaluate an expression in numerical terms	16, 17, 141
`Exit`	Exit from *Maple* session	8
`exp(1)`	The base of natural logarithms: $e = 2.7182818284\ldots$	12
`exp(x)`	The exponential function e^x	75
`expand`	Expand an expression into individual terms	21, 25, 26
`factor`	Factorise a polynomial	22, 26, 187
`Factor`	The same as `factor` except that calculations are carried out in a field modulo a prime p	189
`finance`	A collection of commands for financial calculations: loaded by using the command `with(finance)`	31
`floor(x)`	The floor of x (the largest integer not greater than x)	75, 118
`fsolve`	Find numerical solution(s) of an equation	94, 209, 210
`geometry`	A collection of commands for solving problems in geometry: loaded by using the command `with(geometry)`	30

Continued on next page

List of *Maple* commands (continued)

Command	Description	Pages
int	Integrate a function	443
intersect	Intersection of two sets	38
intparts	Carry out an integration by parts	470
iquo	The quotient when one integer is divided by another	29
irem	The remainder when one integer is divided by another	29
is	Test the value of a Boolean function	80
isolve	Find integer solutions of equations	27, 28
ithprime	Finds the ith prime number	188
ln(x)	The logarithmic function $\ln x$	75
log(x)	The logarithmic function $\log x$ (the same as $\ln x$)	75
max(x,y,z)	The maximum of two or more real numbers	75
map(f,S)	Map elements of the set S into another set by the function f	377
min(x,y,z)	The minimum of two or more real numbers	75
minus	Difference of two sets	38
nops	Counts the number of items in a list or set	40
normal	Collects a sum of fractions over a common denominator	23, 26
numtheory	A collection of commands for solving problems in number theory: loaded by using the command with(numtheory)	30
op	Select one or more items from a list or set	41, 81
Pi	The value of π: 3.1415926535 ...	12
plot	Plot a function	81
powerset	Finds the sets comprising the power set of a given set: requires the combinat package to be loaded	37
product	The product of a number of terms	143
quo	Quotient when one polynomial is divided by another	29, 176

Continued on next page

List of *Maple* commands (continued)

Command	Description	Pages
Quo	The same as quo except that calculations are carried out in a field modulo a prime p	177
rationalize	Rationalise a result with a complicated denominator	27
rem	Remainder when one polynomial is divided by another	29, 176
Rem	The same as rem except that calculations are carried out in a field modulo a prime p	177
remove	Remove one or more items from a set or a list	81
restart	Reset the *Maple* environment to its starting setting with no variables defined	19
roots	Finds rational roots of a polynomial	216
select	Selects some elements from a list	80
seq	Used to generate an expression sequence	41
simplify	Simplify a complicated expression	22, 26, 142, 143
sin(x)	The trigonometric function $\sin x$	75
solve	Find a solution of an equation	108, 209
sort	Sort a sequence of terms	24, 26
spline	Approximate a function by straight lines or curves	352
sqrt(x)	The square root function	11, 12, 13, 75,
subs	Substitutes a value, or an expresseion, into another expression	26
sum	The sum of a series	141, 143, 289
surd	Finds the real value of an odd root of a real number	256
tan(x)	The trigonometric function $\tan x$	75
tau	The number of factors of an integer	30
taylor	Used to find the product of two polynomials, or a Taylor series	173, 183

Continued on next page

List of *Maple* commands (continued)

Index